APPLYING ALGEBRAIC THINKING TO DATA

Phil DeMarois

William Rainey Harper College

Mercedes McGowen

William Rainey Harper College

Darlene Whitkanack

Indiana University

CONCEPTS AND PROCESSES FOR THE INTERMEDIATE ALGEBRA STUDENT

An imprint of Addison Wesley Longman, Inc.

Reading, Massachusetts • Menlo Park, California • New York • Harlow, England
Don Mills, Ontario • Sydney • Mexico City • Madrid • Amsterdam

Publisher: Jason Jordan
Project Editor: Kari Heen
Managing Editor: Ron Hampton
Production Supervisor: Kathleen Manley
Production Assistant: Jane Estrella
Design Supervisor: Susan Carsten
Text Designer: Catherine Hawkes
Cover Designer: Judy Arisman
Prepress Services Buyer: Caroline Fell
Composition: TKM Productions
Technical Art Consultant: Joseph Vetere
Marketing Managers: Andy Fisher and Liz O'Neil
Manufacturing Supervisor: Ralph Mattivello

Partial support for this work was provided by the National Science Foundation's Course and Curriculum Development Program through grant DUE #9354471. The opinions, findings, and conclusions or recommendations expressed in this publication are those of the authors and do not necessarily reflect the views of the National Science Foundation.

DeMarois, Phil.
Applying algebraic thinking to data : concepts and processes for the intermediate algebra student / Phil DeMarois, Mercedes McGowen, Darlene Whitkanack.
p. cm.
Includes index.
ISBN 0-321-01047-7
1. Algebra. I. McGowen, Mercedes A. II. Whitkanack, Darlene.
III. Title.
QA154.2.D445 1998
512--dc21 97-40865
CIP

1 2 3 4 5 6 7 8 9 10—CRK—0100999897

Contents

Preface

A Special Word about the First Edition

After six years of writing and testing materials for introductory and intermediate algebra, rewriting and more testing, we look back with great respect and appreciation of the colleagues who have supported our ideas, attended our workshops, and implemented the materials in their classrooms. For us it has been a time of tremendous professional and mathematical growth as we have wrestled with translating our concerns about well-known student misconceptions into workable solutions for the classroom. The opportunity to debate our ideas and our approach with reflective researchers and practitioners has resulted in the incorporation of many ideas about teaching and learning currently held by several in the mathematics education community. The success of the materials is directly proportional to our ability to incorporate those ideas into these materials. The cooperation of a large number of students enrolled in the program who willingly discussed their experiences and participated in interviews that probed their understandings has resulted in materials that reflect student input as well.

In many ways our original vision has been modified to reflect current theory and practice as we have attended conferences, read papers generously shared by the authors, and talked to colleagues. We have come to recognize, for example, a greater need on the part of many students for more practice to reinforce conceptual understanding than we originally believed necessary. However, in fundamental ways, our vision from six years ago has provided a continuing philosophical framework such as the belief that concept development can occur through problem solving and that practice of skills properly follows that development. The butterfly on the cover of this textbook suggests the metamorphosis we have undergone as we developed these materials. We hope that students also experience a metamorphosis in their thinking and attitudes towards mathematics as they use this text. We believe that the following descriptions reflect our philosophy and the curriculum materials we have developed.

The Course and its Audience

This book is the second in a series that provides curriculum materials for teachers who want to make changes in the way they teach algebra. We want to teach an algebra course that focuses on function and concepts rather than on rote skills, but we found the materials available did not meet our needs for several reasons. Sometimes they were written for more mathematically mature students than were enrolled in college developmental algebra courses. Other materials did not fully integrate the tools we believe should become as automatic as pencil and paper to enhance the student's ability to explore mathematics. Most important, many of the materials seemed in conflict with mathematics education research on the teaching and learning of algebra. This book is not for the faint hearted. Teachers who have piloted these materials report that the use of these materials has changed the way they think about mathematics and the way they teach. Knowing something about the background and philosophy of these materials may help those contemplating the use of non-traditional materials in their classroom.

The National Council of Teachers of Mathematics (NCTM) *Curriculum and Evaluation Standards for School Mathematics* provided a vision and gave direction to change the way mathematics is taught. With the inclusion of the use of technology as an essential feature, the emphasis on problem solving, communication, reasoning, and connections empowered students in the field of mathematics and gave hope to frustrated teachers. Perhaps there was a way to make mathematics accessible to all students who were disenfranchised because of their lack of previous long-term success. Documents from the National Research Council, the Mathematical Sciences Education Board (MSEB), and the Mathematical Association of America (MAA) reinforced the need for change. We began to develop a set of materials that reflect our interpretation and agreement with ideas coming from research and working conferences. The American Mathematical Association of Two-Year Colleges (AMATYC) *Crossroads in Mathematics: Standards for Introductory College Mathematics Before Calculus* serves as a benchmark for reform materials in developmental mathematics. We have been gratified by the number of people familiar with our materials who feel that we have met that benchmark.

The changes called for in these documents focus attention on the need for restructured curriculum materials coupled with changes in instructional philosophy and strategy. Change occurs when teachers have a plan to achieve that change and transition materials to help all students make the change confidently. The materials must address appropriate technology use, learning styles, classroom management, assessment, and core curriculum.

An important aspect is the need for relevant content. Developmental algebra courses that meet the needs of students and industry can no longer be based solely on algorithmic techniques. Material must be presented in context. The sequence of ideas must be carefully considered. What is valued should be assessed but not necessarily graded. More emphasis should be placed on algebraic reasoning. A long-term immersion in language and structure, acquired and used over time, becomes an essential element of studying algebra. This kind of change does not occur magically. Teachers must have support and sufficient background to make these changes. Carefully planned in-service and on-going training to refocus and support teachers in acquiring a broader view of algebra is essential.

Approach

This text promotes a pedagogical approach based on a constructivist perspective of how mathematics is learned. A student learns mathematics by working in a social context to construct mathematical ideas, and to reflect on these constructions, all as a means of making sense out of problem situations. Each section in a chapter of this book begins with an Investigation of a problem situation. After gathering data, students work collaboratively on small tasks based on the Investigation activities. The tasks help students reflect on mental constructions they have made as a result of working with the calculator and/or computer.

Having students talk with the members of their group as they work on the task is another way to encourage them to reflect on the problems and the solutions, whether discovered by themselves or presented by someone else in class (such as a member of another group or even the instructor).

Students are expected to answer each Investigation in the text—*they write* portions of the book. A discussion in the text summarizes essential mathematical ideas. The instructor orchestrates intergroup and class discussions of the Investigation and the tasks. Explorations are included to reinforce the knowledge students are expected to have constructed during the first two steps of the cycle.

There is a significant difference between our approach and that of traditional materials: we avoid assigning exercises until *after* there is a reason to believe that students have constructed the relevant knowledge. When one is learning something new, there is a tendency for early interpretations to be inappropriate as students overgeneralize. Working with too many similar examples could cast these misinterpretations in stone. With our approach, we hope to reinforce understandings and not *mis*understandings.

Distinguishing Features of the Text

Making Connections

This intermediate algebra book is based on multiple representations of the function concept. Approaching functions from a data analysis perspective leads to a richer content than usually found in intermediate algebra. This edition begins each section with a Connections paragraph to emphasize how the material is tied to what came before or what follows.

Discovery-Based Learning

There are no illustrated examples followed by exercises that repeat the examples. The investigations tend to ask more questions than the subsequent Discussion answers. There are interactive investigations on calculator and computer designed to help students make mathematical discoveries, to support small group work both in and out of class, and to reduce the teacher's role as lecturer.

Perhaps the most innovative feature of the text is that, very often, students are asked to explore mathematical ideas on the calculator and/or computer before these ideas are discussed in class. Although this is a departure from standard practice, both faculty and students who have used the materials feel that this is an important component of our approach to helping students learn mathematical concepts and develop confidence in their ability to direct their own learning. Not only does such a style give students an opportunity to discover mathematical ideas on their own, but even if they don't succeed in making the discovery, the activities form an experiential base that helps them understand an explanation presented in class by another student or by the instructor.

Student Discourse A conversation between two students, Pete and Sandy, occurs regularly throughout the book. Pete is a student currently enrolled in the course and Sandy is a student who completed the course last term. Their conversations focus on key points of confusion and on making connections among ideas as students proceed through the course.

Glossary Students are expected to create their own glossary of terms and definitions. A blank Glossary appears at the end of the book. Each set of Explorations begins by asking students to write definitions for the new words and phrases encountered in the section. Each new word or phrase is printed in boldface italic type to set it apart.

Reflection A concluding feature of each section is a Reflection that asks students to write a short narrative discussing important concepts within a section and between sections.

Concept Map A concluding feature of each chapter is a Concept Map. The Concept Maps are centered on key ideas in the chapter and provide students with a chance to brainstorm about the ideas and draw connections between various aspects of the concepts. Detailed instructions and an example appear at the end of Chapter 1.

Student Contribution Another distinctive feature is that the materials include space to write in answers and observations to the Investigations in the text. Students develop organizational skills and an ownership in their text as they actually contribute to the writing of the Investigation sections during the term.

Unique Chapter Review Each chapter concludes with a review section. In addition to a wide variety of review problems to investigate, students are asked to reflect on the important ideas of the chapter. Suggestions for possible Concept Maps are listed.

Technology Focus

Students use graphing calculators regularly. The graphing calculator may be viewed as a function machine that has the capability of displaying both input and output on the screen simultaneously so that students can obtain immediate feedback, discover patterns, and identify their previously learned misconceptions. It is recommended that students have access to a computer algebra system during the course. Investigations done on a computer algebra system allow students to discover symbolic patterns and formulate rules. The use of technology provides opportunities for students to understand and interpret data in terms of the use of tools and techniques that are appropriate to the job. The emphasis is on the development of decision making and problem solving. Students are encouraged to make good choices about the techniques that make the most sense in a particular situation and to check answers for reasonableness and accuracy. While TI-83 and TI-92 views appear in the text, the materials are not dependent on a specific calculator or on specific computer software.

Skill Development

A key issue in algebra is the development of appropriate skills for later courses. However, we believe that problem-solving skills that encourage independent learning are more important than rote manipulation skills taught out of context. We cover most skills traditionally included in an algebra course plus advanced skills because students confront interesting problems that require these skills. The focus is on interesting problems that require certain skills in the process of investigating the problem. We prefer not to cover skills out of context. We want to emphasize connections among mathematical ideas and skills. This requires a major change in sequencing. If student learning is connected, the chance for retrieval of the knowledge in other problem situations is much greater. Based on constructivist principles, students will develop reasoning from a numerical standpoint by looking for patterns. By generalizing these patterns, they will create mathematics that is their own—this holds true for both concepts and skills.

Instructor Support

A support system must be in place for those who choose to implement materials that are radically different. We were awarded an NSF grant to aid us in developing workshops and an on-going network for faculty who use this text. Publisher-supported workshops may be available upon request.

Acknowledgments

There are many people to whom we need to say "Thank you." We would like to acknowledge specifically Frank Demana and Bert Waits for their pioneering efforts in promoting the widespread acceptance and use of technology to help students explore and visualize mathematics and who served as models for us.

Since the basis for much of the development of this text is grounded in the mathematics education research literature, we would also like to acknowledge the contributions of Jim Kaput and the members of the Algebra Working Group of the National Center for Research in Mathematical Sciences Education. Other researchers who have had a major impact on our work include Jere Confrey, Robert B. Davis, Ed Dubinsky, Jim Fey, Eddie Gray, Carolyn Kieran, Keith Schwingendorf, Richard Skemp, David Tall, Pat Thompson, Zalman Usiskin, and Sigrid Wagner. To this dedicated, articulate group of individuals, and to all those whose research guided our efforts and shaped our vision about what algebra should be, we say "Thank you."

We also need to acknowledge the major contribution of all those who have field-tested draft versions of this text, along with their students. The insightful recommendations for improvement, editorial comments, and energetic efforts of these pioneer field testers and students have been invaluable in making the text more useful and student friendly.

Preliminary Edition Class Testers

Calvin College

Central Washington University

Cerritos College

City University of New York La Guardia Community College

Clarke College

College of America

College of Lake County

Edison Community College

Fresno City College

Indiana University East

J. Sargeant Reynolds Community College

Lakeland Community College

Massachusetts Bay Community College

Minneapolis Community and Technical College

National-Louis University

Naugatuck Valley Community–Technical College

Northeast State Technical Community College

Northwestern Michigan College

Phoenix College

Point Loma Nazarene College

Richland Community College

San Diego Community College-Mesa College

San Jose Christian College

Siena Heights College

Tomball College

Unity College

University of Houston-Downtown

Walters State Community College

Western Connecticut State University

William Rainey Harper College

A remarkable group of colleagues, Lana Taylor, Gail Johnson, Nancy Rice, and Colette Currie, deserve a special mention. They were willing to test a very early draft of these materials in their classrooms without first seeing a completed project, receiving a chapter at a time during that first semester back in spring, 1992. There are no words that adequately acknowledge the gift of trust and support they gave to us in the initial stages of this writing project.

We greatly appreciate the detailed feedback and commentaries provided us by all the reviewers of the various drafts of this text. The reviews addressed specific points of concern, validated our vision of the direction we chose, and helped us clarify content issues.

Reviewers

Kathleen J. Bavelas, *Manchester Community Technical College*

Mary E. Clarke, *Cerritos College*

Elaine J. Dinto, *Naugatuck Valley Community–Technical College*

Tracey Hoy, *College of Lake County*

Marveen J. McCready, *Chemeketa Community College*

Joy McMullen, *Lakeland Community College*

Paula J. Mikowicz, *Howard Community College*

Mika Moteki, *Cardinal Stritch College*

Susan Parsons, *Cerritos College*

Debra A. Pharo, *Northwestern Michigan College*

John R. St. Clair, *Motlow State Community College*

Lana Taylor, *Siena Heights College*

Tom R. Williams, *Rowan-Cabarrus Community College*

Finally, we also wish to acknowledge our present editorial staff Jason Jordan, Kari Heen, Audra Connolly, Michelle Fowler, and Sara Peterson, along with our former editors, Anne Kelly, Greg McRill, and Karin Wagner, for their support during the development of this text. Their courage to allow this text to be developed with minimal editorial influence has resulted in what we believe to be truly a reform curriculum. Our production staff, Susan Carsten, Kathy Manley, and Joe Vetere should be recognized for their design and technical expertise in preparing this text. We also wish to thank our marketing staff, Andy Fisher, Liz O'Neil, and Mark Harrington for their support in the promotion of this text.

We invite you to join the on-going discussion among colleagues who are using these materials. We encourage you to share your experiences, both successful and not so successful, with others who are attempting to change not only the content but their instructional practices as well. We welcome your comments and suggestions for improving this text. Let us know what works and what doesn't so that we can continue to improve these materials. Our e-mail addresses are listed below. Contact us if you would like to become part of the network.

Sincerely,

Phil DeMarois
pdemaroi@harper.cc.il.us

Mercedes McGowen
mmcgowen@harper.cc.il.us

Darlene Whitkanack
darlene155@aol.com

1 Working with Data

SECTION 1.1

Data: Collecting, Organizing, and "Averaging"

Purpose

- Collect data.
- Introduce functions and relations to define relationships.
- Calculate measures of central tendency: mean, median, and mode.

A change in one thing regularly results in changes in other things. Think about everyday examples: If you work more hours, how does your paycheck change? When you hit the brakes in a car, how does your speed affect how far you travel before you stop? If you consume more calories, how does your weight change? We focus on describing the relationships between changing quantities throughout this book. Such descriptions involve the notions of relation and function, two of the key ideas in all of mathematics. We also focus on analyzing data. We begin the text by creating some data and discussing ways to describe the data. In the process, you will learn a little about statistics and you will see how functions can be used to illustrate the relationships between data and the description of data.

As you work through the text, write answers in the book to the questions titled Investigations. Make your best attempt at an answer for each question and discuss your thoughts with others in your class. It is okay if you can't answer a question immediately. Talk with others, look ahead to the section titled Discussion if you are stuck, and try to describe your struggles. You will make use of tables to organize information, explore the quantities in the relationships, and designate input and output. Enjoy the spirit of investigation. Do not worry if you make a mistake. We learn much from our mistakes when we reflect on the thought process that led to the error and figure out why it didn't work. After completing an investigation, study the discussion that follows. The discussion will help clarify the ideas that you explored in the investigation. The discussion and explorations will help you connect these ideas to mathematics you've already learned. As you work through the investigations and discussions, you will find many new mathematical terms—sometimes defined, sometimes not. To be successful in this class you must learn to speak the language of mathematics, so fill in the Glossary at the end of the book and learn to use the terms accurately.

Enough introduction. Let's take a test to generate a little data and see what we can discover.

1. **Test with No Questions:** You are taking a 10-question multiple-choice test with possible answers **a, b, c,** or **d** to each question. There are no questions, just answers. Although there are no questions, choose one letter for each answer and record your choice in Table 1.

TABLE 1 Your First Test

Question	Answer
1	
2	
3	
4	
5	
6	
7	
8	
9	
10	

2. Your instructor has received the correct answers from an ***oracle.*** Exchange papers. Check your answers with those supplied by your instructor. (No cheating!)

   Record the total number of correct answers on the board or the overhead as directed by your instructor.

3. With your classmates, decide as a group how to organize the class data.

   a. Organize and record the class data.

   b. Compare your organization to that of another group.

4. These data will be used to compare your class with other classes taking this test. If there are 50 such sets of data, what criteria would you use to determine which class has the best guessing ability?

5. How many ways could you describe the class performance using a single number? What are they? Do you know the name given to each number?

6. What number would you use? Why? What does this single number represent? Is your number actually in the data? How close is your number to the middle of the set of data?

Investigation 6 asked you to use a single number to describe the data. Three possible numbers are the *mean*, *median*, and *mode*.

The number usually referred to as the "average" is called the ***mean.*** (This term is an example of how mathematical terms are often common words that change their meaning when used in mathematics. The "mean" doesn't have anything to do with not being nice. It comes from a French word for "middle." You may not have used the term before, but the process of adding your test scores and then dividing by the number of tests to find the average is probably familiar. If you think of taking points from the high scores and adding them to the low scores so they are all about the same, you can see why the mean is a middle score.)

The number that is in the "middle" of the list after the numbers are arranged in numerical order is called the ***median***. Half the numbers are the same or larger than this score and the other half are the same or smaller.

The number that occurs most frequently in a list of data is called the ***mode***.

The three values—mean, median, and mode—are called the ***measures of central tendency***. This term indicates that we are trying to create one number that "measures" where the "center" of the data "tends" to be.

Let's investigate these three numbers in more detail.

7. Look at the class data from the **Test with No Questions**.

a. What is the mean of this data? Write a set of instructions that a friend could follow for finding the mean.

b. What is the median of this data? Write a set of instructions that a friend could follow for finding the median.

c. What is the mode of this data? Write a set of instructions that a friend could follow for finding the mode.

8. Which measure of central tendency best describes the class performance? Why?

The things we collect, such as the test scores, are called ***data***. If the data are in the form of numbers, they are called ***numerical data***. We frequently use a single number, such as the mean, median, or mode, as the best representation to describe a list of numerical data. This number is an example of a ***statistic***.

For our purposes, a ***statistic*** is a relation that receives as input a list of data and returns as output a number that describes the list of data. If there is exactly one output for each input, the relation is called a ***function***.

But what is a relation? What is a function?

A ***relation*** is a relationship between two quantities that change. For example, when you work more hours, your paycheck is larger. If the relation produces exactly one output for each input, the relation is called a function. One quantity is referred to as the ***input*** of the function and the other is called the ***output*** of the function. In this example, the number of hours worked is the input and the amount of the paycheck is the output.

There are many different relationships, some functions and some not, that use lists of data as input and whose output is a number representative of the entire list. Two of the three statistics referred to as ***measures of central tendency*** are functions—the mean and the median. For each set of data, there is only one mean and only one median. The mode is a relation that is not a function. Some sets of data have more than one mode. The mode may output more than one number for one input. For example, the modes of the list $\{1, 5, 5, 6, 6, 11\}$ are 5 and 6. One list as input produces two modes as output.

Given a value for the input, a function performs a process on the input resulting in *exactly one value* for the output. We visualize functions by using ***function machines*** (Figure 1).

FIGURE 1

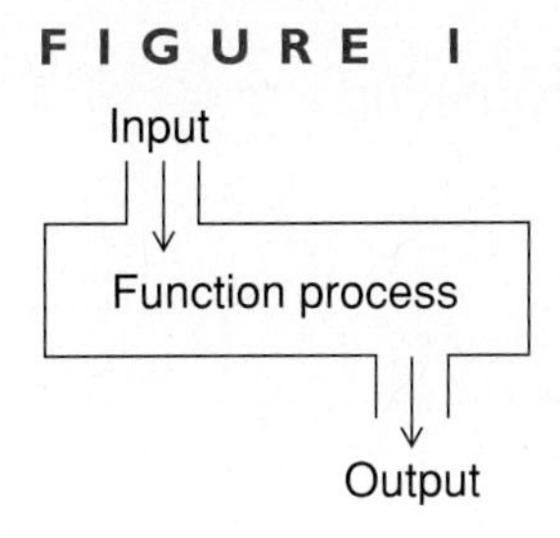

9. Consider the process of applying your brakes to stop your car. How does the speed you are traveling affect the distance it takes for the car to stop? What is the input? What is the output? Does this describe a relation? Does this describe a function? Explain your reasoning.

10. Refer to Investigation 7. Let's create function machines or relation machines for the statistics discussed inside each machine. Write a brief description of the following:

a. The process for computing the mean of a list of data.

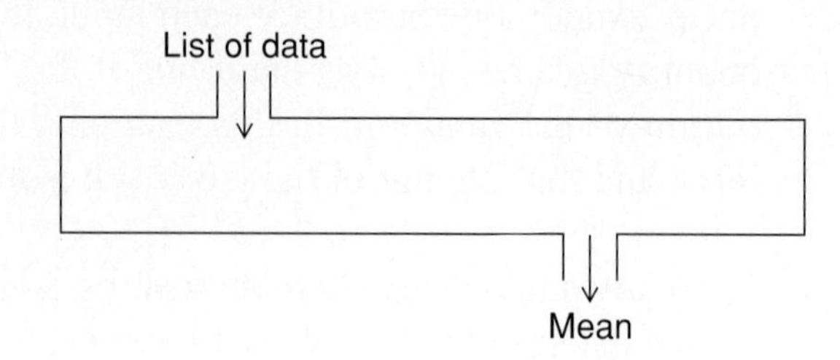

b. The process for computing the median of a list of data.

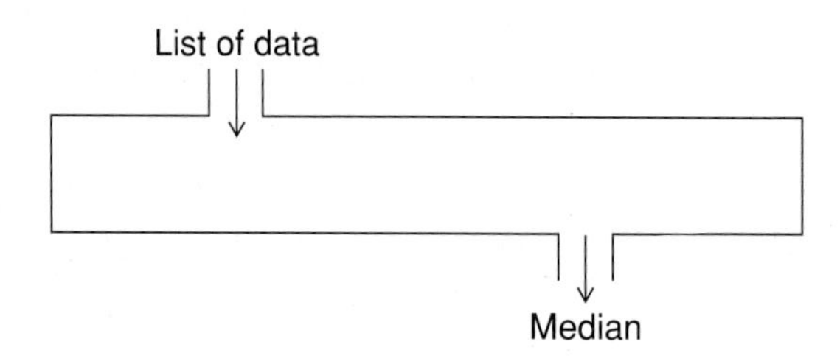

c. The process for computing a mode of a list of data.

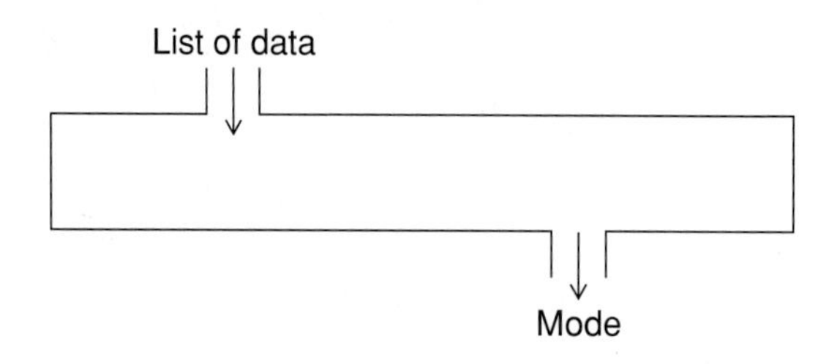

d. Will any one of the measures of central tendency ever have more than one output for a given list of data used as input? Justify your answer.

11. Identify the mode for each list. Write a sentence stating your reasons for your answer.

a. $\{0, 1, 1, 1, 2, 2, 4, 4, 4, 5, 5, 5, 5, 6, 7, 7, 9, 9, 9, 10\}$

b. $\{2, 3, 3, 4, 4, 4, 5, 11, 23, 3, 7, 3\}$

c. $\{0, 1, 2, 3, 3, 3, 3, 4, 4, 5, 5, 5, 5\}$

d. $\{3, 3, 4, 4, 5, 5, 8, 8, 1, 1\}$

12. Create an input data set representing temperatures at noon in February in which the mode is 17° Fahrenheit. Does knowing the mode tell you much about high or low temperatures for the month?

13. Create a set of data representing temperatures at noon in February in which the median is 17° Fahrenheit. Let the high be 40° and the low be −6°.

14. Create a set of data representing the temperatures at noon in February in which the mean is 17° Fahrenheit. Let the high be 40° and the low be 6°.

Let's discuss the data collected from the **Test with No Questions**. Since the data vary from class to class, we'll use a set collected from a very smart class. This class was told about the test beforehand and spent all night studying for it. The scores are listed as follows in increasing order.

$$\{0, 1, 1, 1, 2, 2, 4, 4, 4, 5, 5, 5, 5, 6, 7, 7, 9, 9, 9, 10\}$$

Find the *mean* by summing the data and dividing by the number of numbers in the list. The sum is 96 and there are 20 data values.

The mean of $\{0, 1, 1, 1, 2, 2, 4, 4, 4, 5, 5, 5, 5, 6, 7, 7, 9, 9, 9, 10\}$ is

$$\frac{96}{20} = 4.8.$$

Notice that the mean, 4.8, is not one of the numbers in the original list.

The *mean* is a function whose input is a list of data and whose output is a single number. The function machine for mean is shown in Figure 2.

The input is the list of numbers and the output is the mean.

FIGURE 2

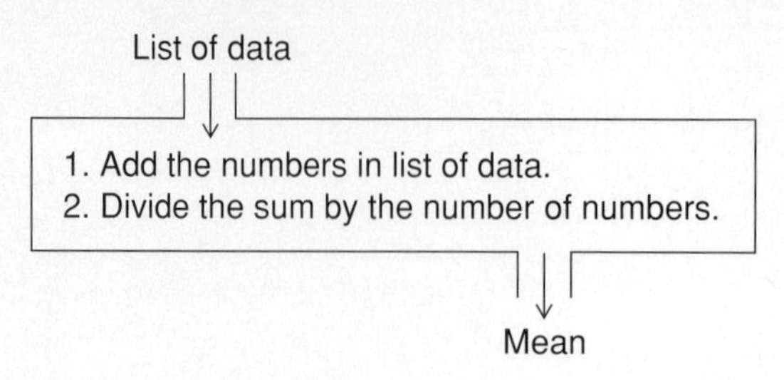

The next measure of central tendency is the median. To find the median, ***sort*** the scores in either increasing order or decreasing order. The way we sort is often determined by the data itself. In the following discussion, we will use an increasing sort. This is an arbitrary choice.

The list {0, 1, 1, 1, 2, 2, 4, 4, 4, 5, 5, 5, 5, 6, 7, 7, 9, 9, 9, 10} is already sorted.

After you sort the list, you can find the median by marking off the top and bottom (or first and last) scores in pairs until only one or two scores remain. If one score remains, that score is the median. If two scores remain, the median is the mean of the two scores.

Cross out top and bottom pairs.

$$\{\not 0, \not 1, \not 1, \not 1, \not 2, \not 2, \not 4, \not 4, \not 4, 5, 5, \not 5, \not 5, \not 6, \not 7, \not 7, \not 9, \not 9, \not 9, \not 1 0\}$$

Since two 5's remain, the median is probably 5. If you follow the procedure given, take the numbers that remain, add them, and divide by 2 to get the *median,* which is 5.

$$\frac{5+5}{2} = 5$$

Given input {0, 1, 1, 1, 2, 2, 4, 4, 4, 5, 5, 5, 5, 6, 7, 7, 9, 9, 9, 10}, the median of the list is 5.

The median is a function with input a list of data and output a number. The function machine for median is shown in Figure 3.

The input is the list of numbers and the output is the median.

Let's look at other examples.

FIGURE 3

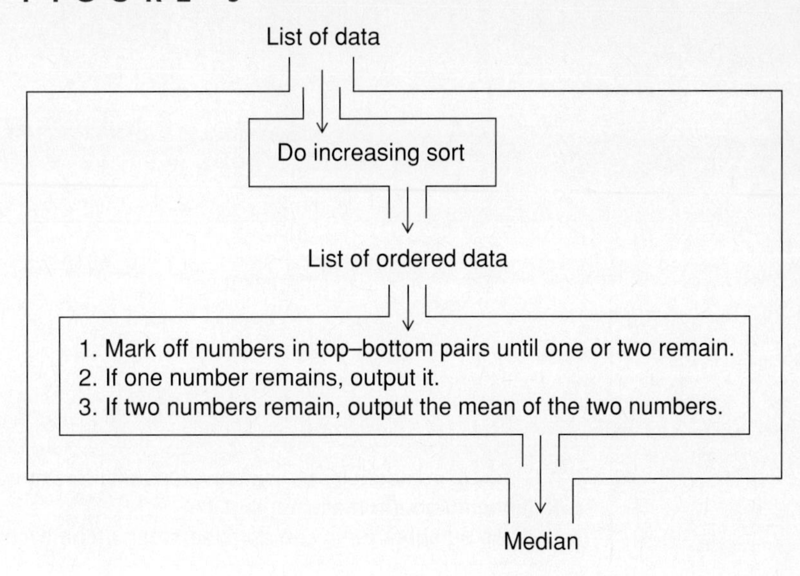

Consider a list of data that contains an odd number of elements, such as the list {19, 4, 40, 7, 3} (Figure 4).

FIGURE 4

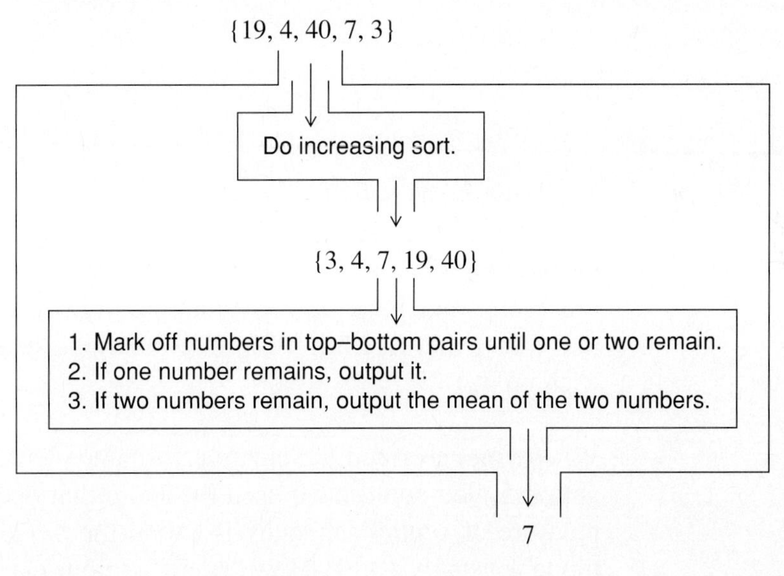

If the input is {19, 4, 40, 7, 3}, then the output of the median is 7.

Notice that the median is an element of the input list.

Consider a list of data that contains an even number of elements, such as the list {8, 4, 7, 40, 3, 19} (Figure 5).

FIGURE 5

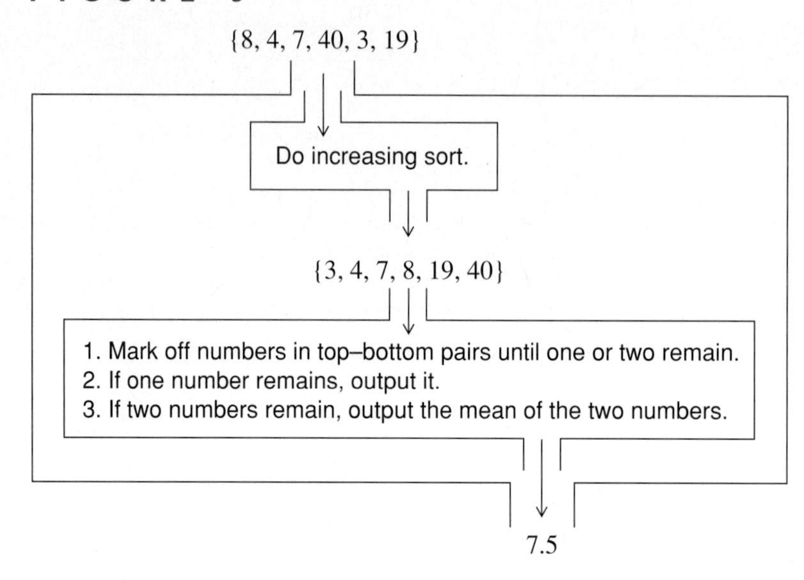

The median is the mean of the two middle numbers, 7 and 8, in the sorted list.

$$\frac{7+8}{2} = 7.5$$

The input is the list {8, 4, 7, 40, 3, 19} and the output is 7.5. In this case the median is not an element of the input list. (Notice that mathematicians could have defined the median in this set as the two middle numbers, 7 and 8, but this would not have been a function. Whenever it makes sense, we try to make sure that a relationship is a function—that is, there is one output for each input.)

When you described the class performance with a single number in Investigation 5, you could have used the score that occurred most frequently. This measure of central tendency is called the *mode*. For the test you took, the mode is usually 1 or 2. How "smart" is your class by comparison?

For the list {0, 1, 1, 1, 2, 2, 4, 4, 4, 5, 5, 5, 5, 6, 7, 7, 9, 9, 9, 10}, the data value 5 occurs four times, more than any other data value. So the mode of this class data is 5.

For a given list of data, the mode may not be unique.

For example, the list {0, 1, 2, 3, 3, 3, 3, 4, 4, 5, 5, 5, 5} has two modes, 3 and 5. Therefore, this list of data is said to be ***bimodal***. In general the modes of a list of data are the numbers in the data that occur with the greatest frequency. If all numbers in the list are distinct or if all numbers in a list occur with the same frequency, *there is no mode*. Such a list is not a legal input to the mode.

Mode is an example of a ***relation*** that is not a function. A function must have exactly one output for any given input. A relation may have several outputs for a given input.

For example, the mode of the list {2, 3, 3, 4, 4, 4, 5, 11, 23, 3, 7, 3} is 3. If the list {2, 3, 3, 4, 4, 4, 5, 11, 23, 3, 7, 3} is the input to *mode*, there is a single output, which is 3.

But the modes of the list {0, 1, 2, 3, 3, 3, 3, 4, 4, 5, 5, 5, 5} are both 3 and 5. If the list {0, 1, 2, 3, 3, 3, 3, 4, 4, 5, 5, 5, 5} is the input to *mode*, there are two outputs, 3 and 5. This is not acceptable for a function, because a function can have only one output for a given input. Since there are multiple outputs, we have a relation machine (Figure 6).

FIGURE 6

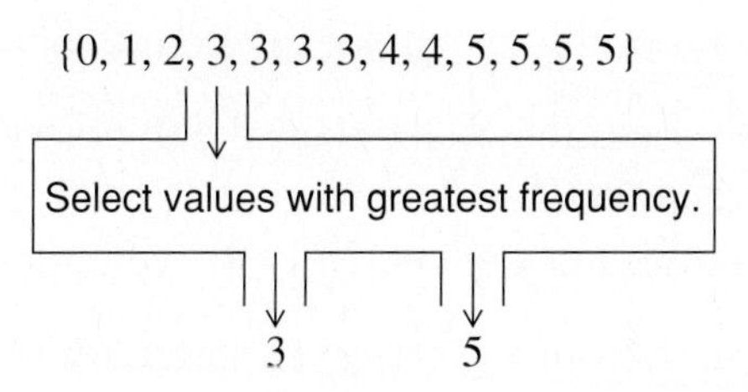

The mode of the list {3, 3, 4, 4, 5, 5, 8, 8, 1, 1} does not exist. No number occurs more frequently than any other number. There is no output for the input {3, 3, 4, 4, 5, 5, 8, 8, 1, 1} using the *mode* relation. The list {3, 3, 4, 4, 5, 5, 8, 8, 1, 1} is not a legal input to the *mode* relation.

The mode, as a measure of central tendency, is useful when a certain number or quality occurs often. Notice that the mode, if it exists, is always a number in the data list. Usually the mean is not a number in the data list and sometimes it is not even the same type of number. For example, all the test scores were ***whole numbers***, but the mean was probably a ***fraction or decimal*** number (***rational***). The median may or may not be a number in the data list. We'll leave it to you to decide when it is.

When you collect data, it is important to organize it and do calculations that help you interpret the data, just as was done with the class test data. If one number is a good representative of all the data collected, use this number to influence decisions. Often this number is one of the measures of central tendency.

STUDENT DISCOURSE

Meet Pete and Sandy. Pete is a student who is currently taking a course using this book. Sandy is a friend of Pete's who used this text in her course last term. Sandy enjoyed the course. She worked hard and did very well. You will find conversations between Pete and Sandy throughout the book that reflect a student's view of the material that includes discussions about what is ahead, that help to organize what was just covered, and that review some topics that often cause students problems. Pete and Sandy welcome you as you work through the course. We hope you find their discussions helpful throughout the course.

Pete: So, let's see if I have this straight. I know how to find the mean of some numbers. I just add them up and divide by how many numbers there are. I always called that the average.

Sandy: It seems like you have it, but let's check your understanding. What's the mean of $\{2, 0, 4\}$?

Pete: That's easy. Two plus 4 is 6, divided by 2 gives a mean of 3.

Sandy: Wait a minute! What happened to 0?

Pete: Zero is nothing so it doesn't count.

Sandy: Let's see if that makes sense. Suppose the list indicated the hours worked on three consecutive days. Would the average hours worked each day be 3? Try to even out the hours each day.

Pete: Well, if I took two hours from the four hours on the third day, I could say I worked two hours each day. I guess they will only pay me for 3×2 or 6 hours, not $3 \times 3 = 9$ hours. I guess I'd have to include the 0 some way, but zero doesn't change the sum.

Sandy: That's right. Zero is a legal data value. There are three data values, so you need to divide by 3, not 2.

Pete: Okay, I'll be careful with that. Using the same list, $\{2, 0, 4\}$, the median would be the middle number. Let's see—that's 0.

Sandy: I don't think so. You forgot to do something to the data first.

Pete: Oh, that's right. The data has to be sorted. If I write the list as $\{0, 2, 4\}$, then the median is 2.

Sandy: Very good.

Pete: Finally, I don't have to do any computations for mode—just find the value that occurs most often.

Sandy: What's the mode of $\{0, 2, 4\}$?

Pete: Hmmmm. They all occur most often. Is it all the numbers?

Sandy: The answer depends on how mode is defined. In this section, the mode is the value(s) that occurred more frequently than any others in the list. Is there such a number in $\{0, 2, 4\}$?

Pete: No. So there is no mode for $\{0, 2, 4\}$?

Sandy: That's right. According to the definition of mode in this section, the list $\{0, 2, 4\}$ does not have a mode.

Pete: One other question—how does all this function stuff tie in with this?

Sandy: That you'll learn about in the next two sections. Be patient. You'd better work on some Explorations first. You'll soon find out what I did—function is a really important idea. I use it all the time in the course I'm in now.

Pete: Okay. Thanks for your help. Talk with you later.

1. Create your own personal dictionary of vocabulary words. List and define the words in this section that appear in ***italic boldface*** type using your own words. Record definitions in the Glossary at the end of the book. An example may be given to clarify your definition, but it is not sufficient to define a mathematical term.

2. Indicate which definitions you knew and which definitions you found in an outside source. List your source (i.e., dictionary, friend, old math book, etc.).

3. For each given data list, find the mean, median, and mode.

a. The in-state tuition per semester hour at selected Chicago area colleges: \$30, 32, 125, 150, 76, 81, 30, 109, 125, 30.

b. My checkbook balances at the end of six consecutive days: \$2.53, 0, –5.62, –10.47, –38.16, 52.03.

c. The portion of a 12-slice pizza eaten by four children: $\frac{1}{3}, \frac{1}{6}, \frac{1}{4}, \frac{1}{6}$.

4. Refer to the data given in Exploration 3.

a. For the checkbook balances, what number did you divide by to get the mean? Why?

b. For the pizza data, how many slices of the pizza remain? Justify your answer.

5. When data are collected, the list is often rather messy. What does it mean to organize the data?

6. In each situation, which measure of central tendency would be most helpful in making a decision? You might want to create sample data.

a. The manager of a shoe store is looking at the sizes of shoes sold last month to decide what stock to order.

b. Sue is trying to decide which of two job offers to accept. Each company has provided salary data for the employees in the department in which Sue would be working.

7. Under what conditions will the median be a number in the original list?

8. Why do you think the term *measures of central tendency* is used to describe the mean, median, and mode?

9. Collect the following data from each member of the class and decide which measure of central tendency would be appropriate to describe the class.

a. The distance driven to school.

b. The hours you studied math this week.

c. The hours you worked this week.

Why are different relations or functions (mean, median, and mode) used to represent data? Are these three numbers ever the same for a given set of data? What are the advantages and disadvantages of each?

SECTION 1.2

Functions and Order of Operations

Purpose

- Explore the concept of function by making and analyzing tables, function machines, and equations.
- Investigate basic operations as functions.
- Analyze order of operations and decide whether it is a mandate or common sense.

In the previous section we investigated measures of central tendency. The mean and the median are examples of functions that use lists of data as inputs. The outputs are real numbers. The mode was a relation but not a function.

In this section we investigate mathematical functions with inputs consisting of a single real number rather than lists of real numbers. In the process, we visit the basic arithmetic operations and discuss order of operations.

Functions are important because they are mathematical structures for representing relationships between changing quantities. Such relationships occur often in our daily lives. Let's look at one such problem.

Selling Costume Jewelry: Meet Steady Sue and Erratic Ernie. They have formed a partnership. Ernie, being the creative one, makes costume jewelry. Sue, being business minded, markets the jewelry Ernie makes. In a recent month, they spent \$100 on raw materials to make 50 pieces of jewelry. Sue was able to sell several pieces of the costume jewelry for \$5.00 each. The state they work in has no sales tax. Their ***net profit*** (income minus expenses) depends on the number of pieces of jewelry they sell.

1. There are two ***constants*** (numbers that do not change) given in the problem situation. What are they and what do they represent?

2. What role does the number 50 play in the problem? Is it used in calculating the profit?

3. There are two ***variables*** (quantities that change) given in the problem situation. What does each represent? Choose a letter to represent each.

4. One of the variables is dependent on the other variable.

a. Which quantity represents the independent variable (input)? This is the variable whose value is independently determined. Write a reason for your answer.

b. What quantity represents the dependent variable (output)? Write a reason for your answer.

The **Selling Costume Jewelry** problem situation includes two constants: the fixed cost of materials ($100) and the price of each piece ($5.00). The relationship between the number of pieces of costume jewelry sold and the net profit involves two changing quantities. Mathematically, each changing quantity is identified by a variable name. If the value of the variable is chosen arbitrarily, that variable is called an ***independent variable***. The value of the other variable, the ***dependent variable***, is determined by the value of the independent variable.

In this problem situation, the net profit is determined by the number of pieces sold. This means that the variable for the number of pieces of costume jewelry sold is the independent variable. This variable is the input to the relationship. Let's use the letter n to represent the number of pieces of costume jewelry sold. We should note that 50 is the largest value of n because Ernie has only enough material for 50 pieces of jewelry. The net profit is the output of the function. The variable for net profit is the dependent variable. Let's use the letter P to represent the net profit measured in dollars.

5. Label the columns of Table 1 so that the left column represents values of the input and the right column represents values of the output. Choose five numbers as input (values of the independent variable). Calculate the output (values of the dependent variable).

TABLE 1 Selling Costume Jewelry

6. What kinds of numbers make sense for the input? For example, can you use positive numbers, negative numbers, fractions, decimals, etc. as inputs? State the restrictions, if any, on input and give a reason for your answer.

7. What numbers make sense for the output? For example, can you get positive numbers, negative numbers, fractions, decimals, etc. as output? State the restrictions, if any, on output and give a reason for your answer.

8. State in words the calculation that is done each time a new input is chosen.

We often use tables to display input and output. The left column usually represents input and the right column represents output. The values in the input column are freely chosen. The values in the output column are determined by the input.

If the relationship is such that a value for the independent variable (an input) determines *exactly one value* of the dependent variable (an output), the relationship between the two variables is called a ***function***.

A function represents a process that receives ***input*** (a value of the independent variable) and returns exactly one ***output*** (a value of the dependent variable). There must be exactly one output for each input.

The **Selling Costume Jewelry** problem situation can be visualized using a function machine (Figure 1). We must determine the process that goes in the function machine. We'll do that later in the section.

FIGURE 1

Number of pieces of costume jewelry sold n

Net profit P

Table 2 is a sample table for the costume jewelry problem.

TABLE 2 Selling Costume Jewelry

Number of Pieces of Costume Jewelry Sold (n)	Net profit ($) ($P$)
0	−100
5	−75
20	0
30	50
35	75

A table is often referred to as a ***numerical representation of the function***.

The set of "legal" inputs to a function is called the ***domain*** of the function. The word "legal" can have several meanings. If a process cannot be performed on a given number, that number is not a legal input to the process. For example, if the process is to divide 1 by the input, the input cannot be 0. Division by 0 is undefined. So 0 cannot be used as input for the process "divide 1 by the input." A legal input is one that makes sense for the situation and can be used in the calculation.

In the **Selling Costume Jewelry** problem situation, the inputs representing the number of pieces of jewelry sold must be whole numbers less than or equal to 50. Recall, the whole numbers consist of 0, 1, 2, 3, 4,... Negative numbers, fractions whose denominator is not 1, and decimals are not whole numbers.

The domain of this problem situation is the set of whole numbers less than or equal to 50 or $\{0, 1, 2, 3, 4, \ldots, 50\}$.

The set of outputs to a function is called the ***range*** of the function. In the **Selling Costume Jewelry** problem situation, the outputs represent the net profit. The smallest net profit is –100. The largest is 150, the profit if all 50 pieces of jewelry are sold. With each piece of jewelry sold, the profit increases by 5. The range is $\{-100, -95, -90, \ldots, 150\}$.

This set can be described as the multiples of 5 between –100 and 150 inclusive. The word "inclusive" means that both –100 and 150 are members of the set.

So we say the range of this problem situation is the set of multiples of 5 between –100 and 150 inclusive.

One reason we study mathematics is to learn ways to use its symbolism to describe the world around us. While the table view of the function is valuable, the table provides only a partial description of the function—we see only a finite number of input–output pairs in the relationship. The function can be described more generally by stating the process that is performed on an input to get an output. We saw this in Section 1.1 with the function machines for the mean and for the median.

In the **Selling Costume Jewelry** problem situation, we can compute the net profit from the number of pieces sold using the following process. The number of pieces sold is multiplied by 5. This gives us the income. To obtain net profit, we subtract the fixed cost of $100 from the income.

This process is composed of two basic operations: multiplication and subtraction. Before going further, let's take some time to think about the basic operations and the order of operations on them.

9. An operation that uses two numbers to produce an answer is called ***binary***. List the binary operations you know.

10. An operation that uses one number to produce an answer is called ***unary***. List the unary operations you know.

11. Irma decides to buy some packages of pencils and pens.

a. She buys four packages of pencils and three packages of pens. Each package costs $5. Write a mathematical expression that shows how to compute her total cost.

b. She buys one large package of pencils for $4 and three packages of pens for $5 each. Write a mathematical expression that shows how to compute her total cost.

12. How would you calculate $4 + 3(5)$? Why? Which situation from Investigation 11 does this expression represent? Does your answer agree with the answer on your calculator?

13. What do you think when you see -5^2? Why? Does your answer agree with the answer on your calculator? Compare it to the value of $(-5)^2$.

14. What do you think when you see $-a$?

15. List all the ways the symbol "–" is used in mathematics.

16. There is agreement in the mathematical and scientific community about the order in which operations are performed when there are several operations in a problem. List the order of operations as you know them. Compare it with the lists of others in your group.

The four basic binary operations (addition, subtraction, multiplication, and division), each of which receives two inputs and has one output, can be thought of as functions and displayed using a function machine (Figure 2).

FIGURE 2

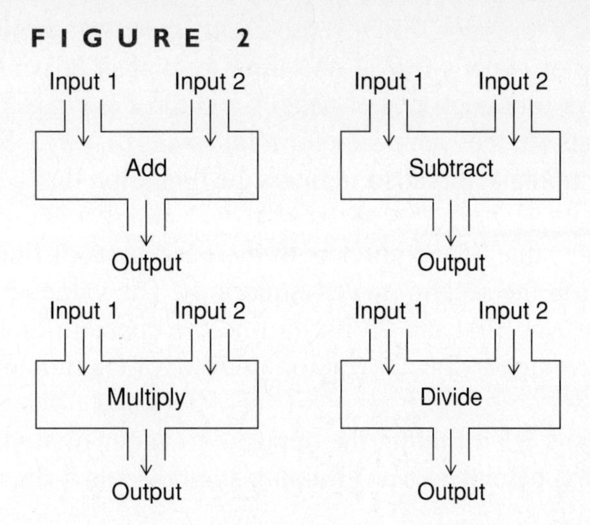

There are several unary operations that we will also use. They include finding the opposite of a number, finding the reciprocal of a number, ***exponentiation*** (raising an input to a power) and ***root extraction*** (finding a root of a given number). These functions each have one input and return one output (Figure 3).

FIGURE 3

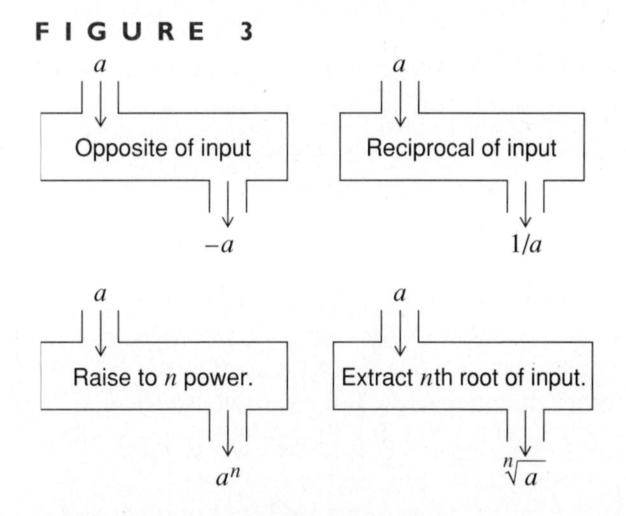

There are many different machines for exponentiation and root extraction depending on the choice of n.

With all these operations, it is not uncommon for several to occur in the same problem. For example, $4 + 3(5)$ includes both addition and multiplication and -5^2 includes both finding the opposite and exponentiation. The numerical result depends on which operation occurs first.

The expression $4 + 3(5)$ equals 19 if you do the multiplication first and 35 if you do the addition first. The correct result is 19 according to the convention of order of operations.

The expression $4 + 3(5)$ represents the total cost if Irma buys one large package of pencils for $4 and three packages of pens for $5 each. But if Irma buys four packages of pencils and three packages of pens and each package costs $5, the expression for total cost is $(4 + 3)5$. We must use parentheses as grouping symbols to indicate the operation that is performed first.

The value -5^2 requires both the operations of finding the opposite and raising to the second power (squaring). The value -5^2 equals –25 if you do the squaring first and 25 if you find the opposite first. The correct result is –25 according to the convention of order of operations.

If you want finding the opposite to occur first, you need to indicate this by using parentheses as grouping symbols and write $(-5)^2$.

The operation of finding an opposite is symbolized by "–." Be careful! This symbol (–) is used in three different ways:

- To indicate the ***binary operation*** of subtraction (it is binary because it operates on two numbers at a time) such as 5 – 7, read "5 minus 7." In this case, the "–" is a process performed on two numbers in a specific order.

- To indicate the unary operation of oppositing such as –5, read "the opposite of 5." In this case, the "–" is a process performed on one number.

- To indicate a negative number such as –5, read "negative 5." In this case, the "–" is part of the notation for a number. This number –5 is the ***additive inverse*** of the positive number 5.

These last two are difficult to distinguish. When in doubt, think opposite rather than negative. For example, $-a$ is read "opposite of a." Until we know the value of a, we don't know if $-a$ is positive or negative.

When doing a computation involving several operations, you need to recall order of operations. The order of operations is a convention agreed upon by scientists and mathematicians so there would be agreement about what mathematical expressions mean. Keep in mind that you can control the order in which the operations are done by using grouping symbols.

In the absence of grouping symbols, first remember the following rule:

Unary operations take precedence over binary operations.

The conventions for order of operations are listed in the following table.

Order of Operations

1. Operations inside grouping symbols. The ***grouping symbols*** include parentheses, fraction or division bars, and radicals. If there are several operations within a grouping symbol, the operations are done in the order specified in Steps 2–4.
2. Exponentiation from left to right.
3. Finding the reciprocal, then multiplication and division. Operations at the same level are done left to right.
4. Finding the opposite, then addition and subtraction. Operations at the same level are done left to right.

So -5^2 means that you must first square 5 and then take the opposite of the answer (Figure 4a).

However, $(-5)^2$ means that you take the opposite of 5 first and then square the result (Figure 4b).

FIGURE 4

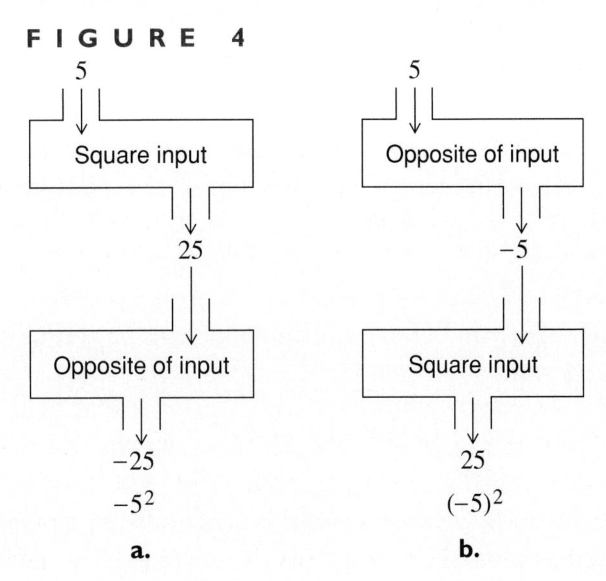

To conclude this section, let's look at function machines.

17. Algebraic expressions can be thought of in terms of function machines. The number of variables in the expression represents the number of different inputs. The process is determined by the operations on the variables. The output is the value of the algebraic expression. Draw a function machine for the expression $2x + 3$. Show what happens when the input is 2, 0, –3, and b.

18. Draw a function machine that clearly shows the input, the process, and the output for the **Selling Costume Jewelry** problem situation.

19. Refer to the **Selling Costume Jewelry** problem situation. Use your function machine to write an equation that shows algebraically how to compute the net profit given the number of pieces of jewelry sold.

20. Suppose that the profit in the **Selling Costume Jewelry** problem situation is $130.

a. Write an equation in one variable using the function written in Investigation 19 that expresses this fact.

b. Use your answer to part **a** to determine how many pieces of jewelry were sold if the profit is $130.

c. Explain how you could use the function machine to find the number of pieces of jewelry that were sold if the profit is $130.

21. Do you believe all functions can be written as equations? Why or why not?

In algebra, we use variable expressions to represent generalized relationships between numbers. Students in a math class were asked to come up with a rule for numerical operations on a variable x. Tom was the first to volunteer. Tom's rule was "Add 3 to the product of 2 and x" where x is the input and the output is determined by the evaluation of the ***expression*** $2x + 3$. We visualize Tom's function in Figure 5.

The function machine in Figure 5 is composed of two basic machines: a multiplication machine followed by an addition machine. The purpose of using two "machines" for this function is to emphasize how multistep function

FIGURE 5

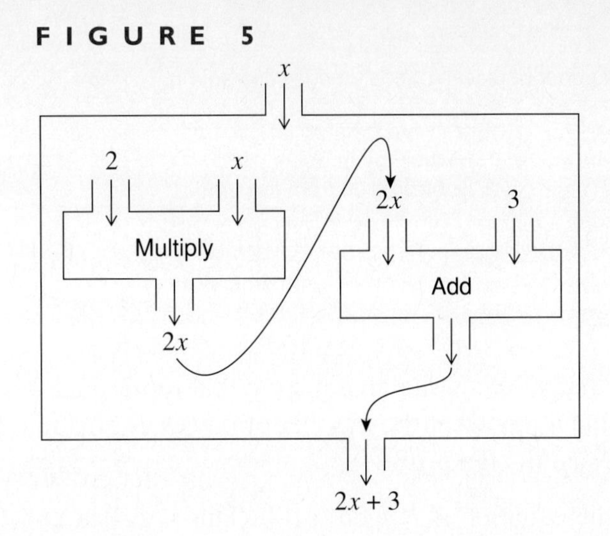

machines are constructed from basic function machines. A more compact way to represent this function is shown in Figure 6.

FIGURE 6

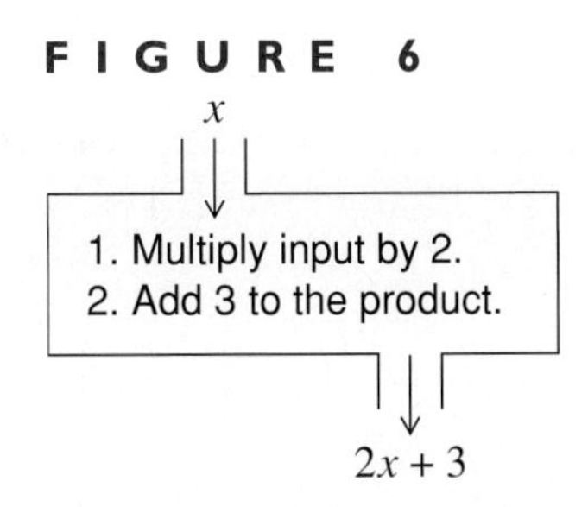

Notice that the defined process of each function machine establishes a relationship between input and output. For the machine in Figure 6, if x is 5, we input 5 to the machine. The machine returns 13 as output (Figure 7).

FIGURE 7

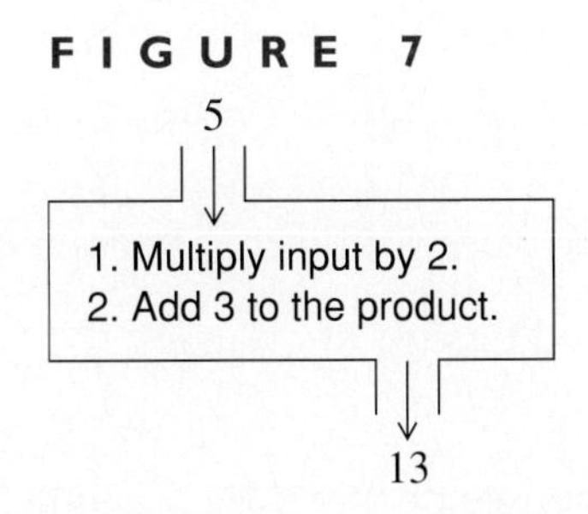

Returning to the **Selling Costume Jewelry** problem situation, we can compute the net profit from the number of pieces sold using a function machine. The number of pieces sold is multiplied by 5. This gives us the income. To obtain net profit, we subtract the fixed cost of \$100 from the income (see Figure 8 on next page).

FIGURE 8

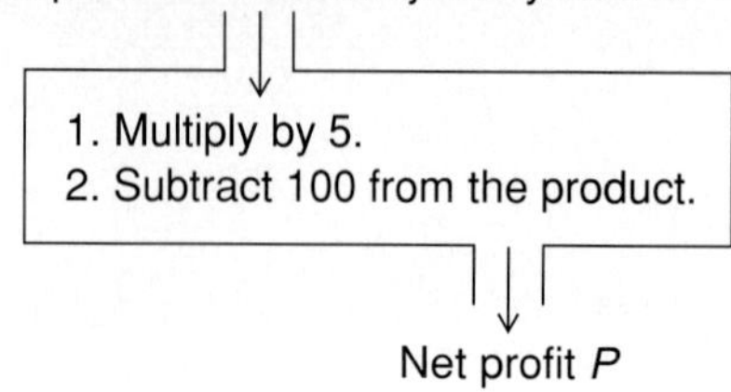

Finally, let's write an equation that represents this relationship. Let the variable n represent the number of pieces of costume jewelry sold and let P represent the net profit.

The equation is $P = 5n - 100$. This is called an ***algebraic representation of the function***.

Recall that the domain of this function is the set of all whole numbers less than or equal to 50 (these are the "legal" choices for n). The range is the set of integer multiples of 5 that are between –100 and 150 inclusive. These numbers are the only possible outputs of the function.

If the profit is $130, we are given the output of the function $P = 5n - 100$ and asked to find the input. Replacing P with 130, we get

$$130 = 5n - 100.$$

Let's solve this equation for n.

Add 100 to both sides:	$130 + 100 = 5n - 100 + 100$
Simplify:	$230 = 5n$
Divide both sides by 5:	$\frac{230}{5} = \frac{5n}{5}$
Simplify:	$46 = n$

So 46 pieces of jewelry were sold if the profit is $130.

We could have answered this question using the function machine. Given the output, we find the input by reversing the operations in the machine.

In the function machine, we multiply by 5 and subtract 100.

Reversing the machine, we add 100 and divide by 5. Adding 100 to the output, 130, we get 230. Dividing by 5, we get 46, the same answer as before.

While the function that mathematically describes the **Selling Costume Jewelry** problem situation was not difficult to write as an equation, there are many functional relationships that cannot be expressed in an equation form.

For example, refer to the median function in Section 1.1. While we specified a procedure for finding the median of a list, we did not and will not try to write this procedure as an equation.

STUDENT DISCOURSE

Pete: It seems to me that there are lots of ways to look at a function.

Sandy: What do you mean?

Pete: Well, the Selling Costume Jewelry problem described a relationship between sales and profit. Isn't that a function?

Sandy: Sure, the relationship is expressed in words. It's like a verbal description.

Pete: Okay, then we created a table.

Sandy: Right. That gives you a list of input–output pairs.

Pete: By the way, I noticed that the input appeared in the left column of the table and the output in the right column. Is that always the case?

Sandy: Not always, but it's a convention I noticed is used unless a problem states otherwise.

Pete: Okay. So I have a verbal description and a table of pairs of numbers. But then I created a function machine.

Sandy: That's right. While the verbal description and the table focus on input and output, there is another focus in the function machine. What is it?

Pete: It seems like the process. You have to describe the steps needed to move from input to output.

Sandy: Exactly. Are there any other ways to represent a function?

Pete: Sure. From the function machine we wrote an equation. The equation gives us the way to calculate the output from any input.

Sandy: Hey, that's pretty good. It took me a long time to figure all that out.

Pete: Are there any other ways to describe a function?

Sandy: Sure. One major way is geometrically, using a graph, but you will look at that later.

Pete: Yeah, I'm not ready to deal with graphs yet. I have enough to study already.

Sandy: I'll let you get to work. If you need any help, give me a call. I have all my notes from last term.

Pete: Thanks a lot. You'll probably be hearing from me.

1. List and define the words in this section that appear in ***italic boldface*** type. Write your definitions in the Glossary in the back of the book.

2. **a.** Create a function machine for $y = x^2 - 3x + 5$.

b. Find the outputs for the function in part a if the inputs are 4, –5, and $\frac{2}{3}$.

c. When you use –5 as input, do you write it as $y = -5^2 - 3(-5) + 5$ or as $y = (-5)^2 - 3(-5) + 5$? Does it make a difference? Why or why not?

3. Since we are going to discuss several functions, we will give each a name. Consider the function machine for the function *Tom* in Figure 9.

FIGURE 9

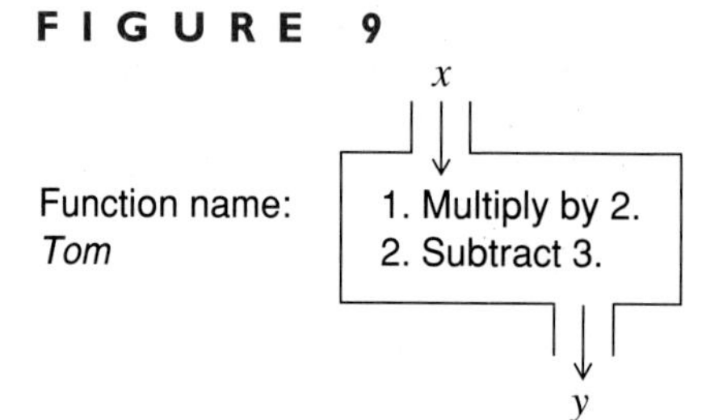

a. Create a table of five input–output pairs.

b. Write the algebraic representation (equation).

c. Find the output if the input is 7.

d. Write an equation that displays the case when the output is –13.

e. Find the input if the output is –13. Justify your work.

f. Write an inequality that represents the case when the outputs are greater than 5. Solve the inequality.

4. Consider the function machine for the function *Ann* in Figure 10.

FIGURE 10

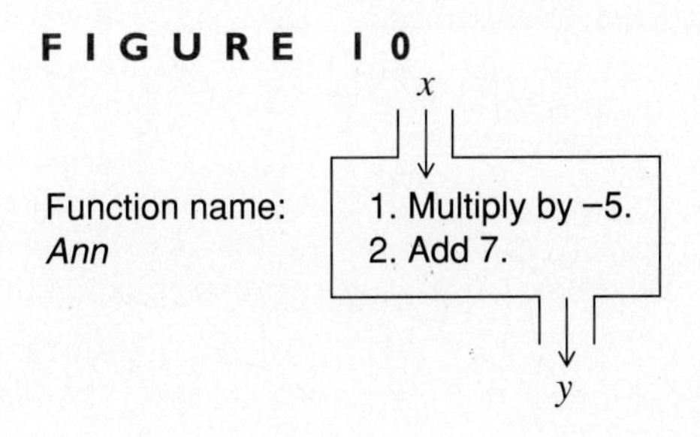

a. Create a table of five input–output pairs.

b. Write the algebraic representation (equation).

c. Find the output if the input is 2.6.

d. Write an equation that displays the case when the output is –13.

e. Find the input if the output is –13. Justify your work.

f. Write an inequality that represents the case when the outputs are less than or equal to 2. Solve the inequality.

5. Consider the expression $\dfrac{5+2\,(4^3-6)}{\sqrt{8+1}+7} - 2$.

a. What are the grouping symbols in the expression? Identify the expressions that each groups.

b. List the operations in the order they would be performed.

c. What is the final answer?

6. Consider the function machine in Figure 11.

FIGURE 11

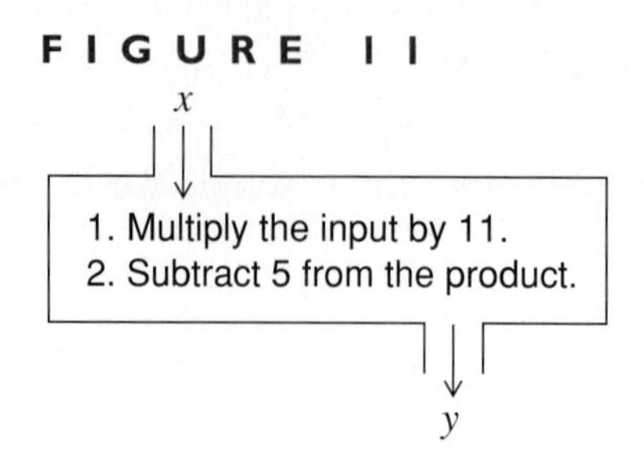

a. Write an algebraic representation of the function.

b. Find the output if $2t$ is input.

c. Find the input if the output is $c + 4$.

7. Create an example of:

a. a function. Justify your answer in words.

b. a relation that is not a function. Justify your answer in words.

8. How do you decide whether:

a. given information constitutes a relation?

b. a given relation is a function?

9. Is it true that all:

a. functions are relations? Justify your answer.

b. relations are functions? Justify your answer.

10. Carrie's Job Offer: Carrie has been offered a job with a starting salary of $30,000. She estimates that she will receive, on the average, a 5% increase each year. The data in Table 3 displays her annual salary for the next 10 years

a. Find the mean, median, and mode for Carrie's annual salary.

b. Which measure of central tendency best describes Carrie's annual salary? Why?

c. In this problem what are the independent and dependent variables?.

TABLE 3 **Carrie's Annual Salary by Year of Employment**

Year	Salary ($)
0	30,000.00
1	31,500.00
2	33,075.00
3	34,728.75
4	36,465.19
5	38,288.45
6	40,202.87
7	42,213.01
8	44,323.66
9	46,539.85
10	48,866.84

11. Each of the following equations is an example of a function set equal to a numerical output. Identify the function and the output.

   a. $4x + 9 = -23$ b. $5 = 2 - 9x$

12. Solve each of the following equations or inequalities.

   a. $5x - 2 = 17$ b. $3t - 7 < -8$

   c. $4p - 9 = 3 - 7p$ d. $1 - x > 4x + 7$

REFLECTION

Explain the idea of a mathematical function in your own words. Give an example of a function in everyday life. Is a variable necessary to describe a function? Clearly identify your input and output.

SECTION 1.3

Function Notation

Purpose

- Examine the role of notation in mathematics.
- Gain experience using and interpreting function notation.
- Explore composition of functions.

MAKING CONNECTIONS

We studied functions in some detail in the last section. Commonly accepted notation that is used regularly in the context of function plays a role in understanding functions. In this section function notation is introduced and investigated. This notation is used throughout the rest of the text.

Let's begin by exploring your understanding of some mathematical phrases.

1. What do you think of when you see each of the following?

a. $5x$

b. $x5$

c. 45

d. $5(x)$

e. $x(5)$

f. xy

g. yx

h. $y(x)$

It is likely that you identified most of the expressions (all except the number 45) in the previous investigation with the operation of multiplication. When we see two mathematical symbols juxtaposed (next to each other), we think, "Multiply." Often, we can only interpret notation if we understand the context in which it is used.

Could the other expressions mean anything but multiplication? For most of them, no. But for $x(5)$ and $y(x)$, there is another interpretation. Each of these expressions could be ***function notation***.

You probably are wondering: "What is function notation?"

That's what we look at now.

We think of a function as a mathematical process on an input that produces a single output. Recall Tom's rule from Section 1.2: Multiply by 2 and add 3. We could give the process a name, such as *Tom*. This is the name of the function (Figure 1).

FIGURE 1

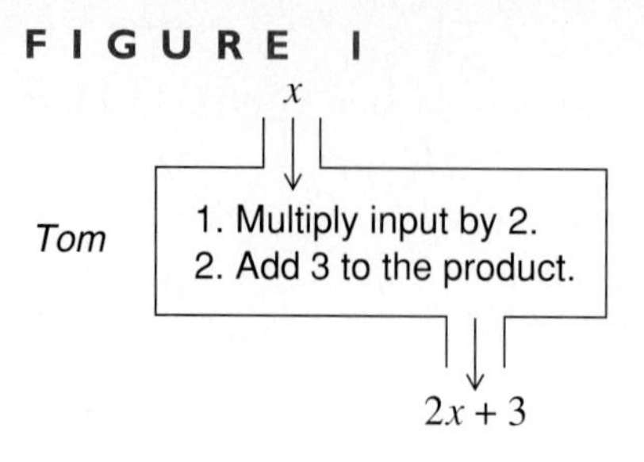

The variable x is the input. We say *Tom* is a function of the variable x.

This is written in function notation as $Tom(x)$, and read "*Tom* is a function of x."

Note that this notation does not mean multiplication here. The quantity inside the parentheses represents the input. The ambiguity of mathematical notation is illustrated when we write $Tom(x)$ and mean function, not multiplication. Furthermore, *Tom* may represent both the name of the process and the resulting output. Symbolically,

$$Tom(\text{input}) = \text{output}.$$

If we know Tom's rule for the function process, we can define the function algebraically using the equation

$$Tom(x) = 2x + 3.$$

This is function notation for writing the function symbolically using an equation. Notice that $Tom(x)$ has two meanings when used in the context of functions.

$Tom(x)$ means the output of process *Tom* when x is the input.

$Tom(x)$ means the process that is performed when x is the input.

What happens if the input is 5? This is represented using function notation as $Tom(5)$. To calculate the function value, we substitute 5 for x in the rule.

$$Tom(5) = 2(5) + 3 = 13 \qquad \text{or we write } Tom(5) = 13,$$

which means that the output *Tom* is 13 when the input x is 5.

Figure 2 is a function machine that provides a more visual interpretation of function notation.

FIGURE 2

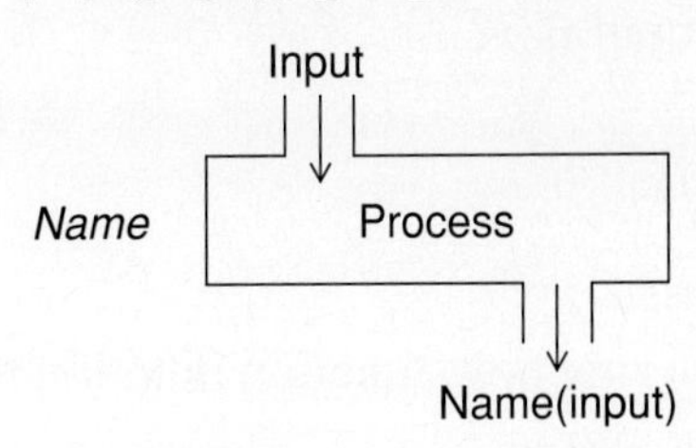

Notation, such as function notation, is an important part of studying mathematics. Analyze new notation by addressing the following standards.

Standards for Mathematical Notation

- The notation makes the process or concept easier to understand.
- The notation makes it easier to discuss the process and/or results with others.
- The notation is used consistently by everyone.
- The notation is efficient.

Let's think about function notation in light of these standards.

2. Analyze function notation by addressing how well the notation measures up to each standard of mathematical notation. In other words, does this notation make sense to you? How do the standards help explain why you have that feeling? How could mathematicians have done a better job inventing notation to use for functions?

3. Write at least three different interpretations of:

a. $x(5)$

b. $y(x)$

4. Consider the equation $f(t) = 4t - 9$.

a. What is the name of the function?

b. What is the variable that represents input?

c. Without computing the output, how would you represent the output when the input is 7?

We begin by considering how function notation measures up to the standards of mathematical notation.

- Some who are just learning function notation will argue that the notation makes the function harder, rather than easier, to understand. As you learn more about functions, your appreciation for the power of the notation will no doubt increase.

- Function notation can represent both the function's process and the function's output. It is common in mathematics for a single set of symbols to mean two different things, depending on the context. One of the keys to success in learning mathematics lies in developing flexibility in interpreting notation.

- Function notation is used consistently in the mathematical community. (It may take a little thought to determine when we have a function and when we have multiplication, however.)

- Function notation is efficient, but, in its efficiency lies some confusion. A very brief set of symbols carries a great amount of information that must be interpreted in the context of a given problem.

Let's look at the other problems in the investigation. There are several ways to interpret $x(5)$.

It might represent the *product* of variable x and constant 5. It might represent the *output* of the process name x when the input is 5. It might represent the *process* that function x performs on an input of 5.

Similarly, from the previous discussion, there are several ways to interpret $y(x)$. It might represent the product of variables x and y. It might represent the output of the process name y when the input is x. It might represent the process that function y performs on an input of x. While the output and the process may seem like the same thing, recognizing the difference between what you do to the input and the result of that action, the output, is important.

In $f(t) = 4t - 9$, f is the name of the process (function) and t is the input. If the input is 7, the output can be expressed as $f(7)$.

We now return to our statistical examples to see how we might use function notation with mean, median, and mode.

Tire Tread Wear: John wants to know when he needs to replace his tires. DMW Reports compiled the following information on tread wear as measured by the depth of the tread. Most new tires have a tread depth of $\frac{11}{32}$ inch. Tires need to be replaced when the tread depth is $\frac{2}{32}$ inch or less. Fifteen DMW Reports staff members recorded the number of miles they drove before their tires needed replacement. The distances given are 49,522, 37,805, 55,711, 74,907, 51,280, 62,053, 61,928, 65,416, 58,253, 31,605, 69,258, 62,415, 51,826, 72,398, 58,832.

5. Use a variable to represent the given list of numbers. Calculate the mean and median for the tire data. Write your answers using function notation. Use the function names *mean* and *median* and use the variable you selected for input.

6. When should John replace his tires? Which measure of central tendency, if any, is most helpful in making a decision? Why?

Let's name the list {49,522, 37,805, 55,711, 74,907, 51,280, 62,053, 61,928, 65,416, 58,253, 31,605, 69,258, 62,415, 51,826, 72,398, 58,832} using the letter *L*.

So *L* = {49,522, 37,805, 55,711, 74,907, 51,280, 62,053, 61,928, 65,416, 58,253, 31,605, 69,258, 62,415, 51,826, 72,398, 58,832}.

To find the *mean*, we sum the numbers in list *L*. The sum is 863,209. There are 15 numbers in the list.

$$mean(L) = \frac{863{,}209}{15} \approx 57{,}547$$

Note that *mean*(*L*) is function notation representing the output of the *mean* function when list *L* is the input.

The first step in finding the median is to sort the tire data. Doing an increasing sort, we get the sorted list

{31,605, 37,805, 49,522, 51,280, 51,826, 55,711, 58,253, **58,832**, 61,928, 62,053, 62,415, 65,416, 69,258, 72,398, 74,907}.

By crossing off bottom–top pairs, we end up with the eighth number in the list, 58,832.

$$median(L) = 58{,}832$$

That was rather time-consuming, but we can duplicate the work we did in this discussion on some calculators. The next investigation gives you a chance to do so. If you don't know how to do the problems in the next investigation, ask a classmate, your instructor, or consult the calculator manual. Since you will use this procedure frequently in this text, you may want to make notes that will reflect how you used the calculator, not just answer the questions.

7. a. Enter the tire data as a list in your calculator. What is the name of the list you used?

b. Calculate the mean by applying the *mean* function to the list. How does the answer compare to the previous discussion?

c. Calculate the median by applying the *median* function to the list. How does the answer compare to the previous discussion?

Figure 3 displays the data (partial list) entered as L1, the mean, and the median. Notice the use of function notation.

FIGURE 3

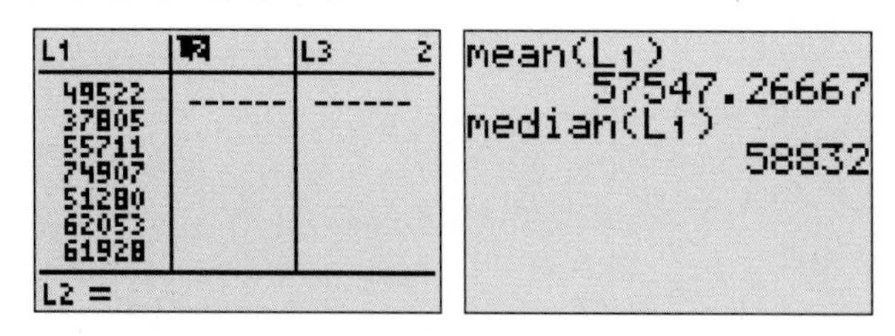

Let's explore the use of function notation in more detail.

You are given the function $Tony(x) = 7x^2 - 5x + 2$.

8. What is the name of this function?

9. What is the independent variable (input) for this function?

10. Find the output of *Tony* if –4 is input.

11. Find the output of function *Tony* for five additional input values. Record your answers in Table 1.

TABLE 1 Function *Tony*

Input	Output
–4	

12. Consider the function $Carol(x) = 3x + 5$. Interpret the meaning of the following by identifying the input and the output. Identify any operation that is performed on the output. Find the answer to each expression.

a. $Carol(t)$

b. $Carol(t + 1)$

c. $Carol(t) + 1$

d. $Carol(t + 1) - Carol(t)$

e. $Carol(-x)$

f. $-Carol(x)$

In the function statement $Tony(x) = 7x^2 - 5x + 2$, *Tony* is the name of the function, x represents the input, and the expression $7x^2 - 5x + 2$ represents the process to follow to obtain the output.

If the input is –4, then

$$\begin{aligned} Tony(-4) &= 7(-4)^2 - 5(-4) + 2 \\ &= 7(16) + 20 + 2 \\ &= 112 + 20 + 2 \\ &= 134. \end{aligned}$$

Thus $Tony(-4) = 134$. This says that the function's name is *Tony*, the input is –4, and the output is 134. It is important to note that the expression $Tony(-4)$ represents the output of function *Tony* when –4 is the input.

Now let's analyze some of the work you did with function *Carol*.

$Carol(t)$ represents the output of function *Carol* when t is input. To input t, substitute t for the input variable x.

Since $Carol(x) = 3x + 5$, we know that $Carol(t) = 3t + 5$.

$Carol(t + 1)$ represents the output of function *Carol* when $t + 1$ is input. Thus the input variable x must be replaced with the expression $t + 1$.

$$\begin{aligned} Carol(t + 1) &= 3(t + 1) + 5 \\ &= 3t + 3 + 5 \\ &= 3t + 8. \end{aligned}$$

$Carol(t) + 1$ represents two steps. First find the output of function *Carol* when t is input. Then add 1 to the output.

We know $Carol(t) = 3t + 5$.

So $Carol(t) + 1 = (3t + 5) + 1 = 3t + 6$.

$Carol(t + 1) - Carol(t)$ represents the difference of two outputs. Find the output of function *Carol* when $t + 1$ is input. Find the output of function *Carol* when t is input. Subtract to find the difference of the two outputs.

Since $Carol(t) = 3t + 5$ and $Carol(t + 1) = 3t + 8$,

$$\begin{aligned} Carol(t+1) - Carol(t) &= (3t+8) - (3t+5) \\ &= 3t + 8 - 3t - 5 \\ &= 3. \end{aligned}$$

$Carol(-x)$ represents the output when the input is opposite of x.

$Carol(-x) = 3(-x) + 5 = -3x + 5.$

$-Carol(x)$ represents the opposite of the output of *Carol*.

$-Carol(x) = -(3x + 5) = -3x - 5.$

Let's complete our initial study of function notation by returning to our intrepid costume jewelry salespersons, Sue and Ernie. Recall that their profit function was

$$P = 5n - 100,$$

where n is the number of pieces of jewelry sold and P is the number of dollars of profit.

We rewrite this equation in function notation as $P(n) = 5n - 100$. The letter P serves a dual role: It is the name of the function and the variable used for output. We conclude the section by building another function from the profit function.

Tax on Profits: Though Steady Sue and Erratic Ernie do not have to charge sales tax on their merchandise, they must pay federal income tax on their net profit. They decide to estimate what they owe by computing 28% of their net profits.

13. Complete Table 2 as follows. Select five different values for input n. Compute the net profits. Compute 28% of the net profits to obtain the outputs that represent the amounts of tax they would have to pay.

14. Draw a function machine for the tax function. Use net profits as input and tax as output.

TABLE 2 Computing Tax on Net Profits

Number of Pieces Sold, n	Net Profits ($)	Tax ($)

15. Let P represent the net profits and let T represent the tax. Both are measured in dollars. Write an equation that represents the relationship between net profit and tax. Use function notation.

16. Net profits were originally computed from the number of pieces of costume jewelry sold. How can you compute the tax if you know the number of pieces of costume jewelry sold?

17. Draw a function machine for the tax function. The input should be the number of pieces of costume jewelry sold and the output should be the tax.

18. Write an equation for a function whose input is the number of pieces of costume jewelry sold and whose output is the tax. Use the variables n for the input and T for the output. Write in function notation.

If we know the net profits, we can compute the tax by multiplying the net profits by 0.28.

So, if P is the input and T is the output, we get the algebraic representation $T(P) = 0.28P$.

The function machine appears in Figure 4.

FIGURE 4

Net profit P
Multiply by 0.28.
Tax $T(P)$

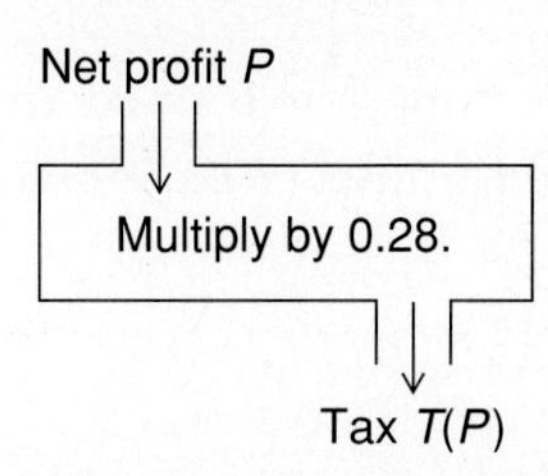

To create a function that computes tax from the number of pieces of jewelry sold, we must use the output of one function as the input to another function. This is called ***composition of functions***.

Let's look at the composite function machine first (Figure 5).

FIGURE 5

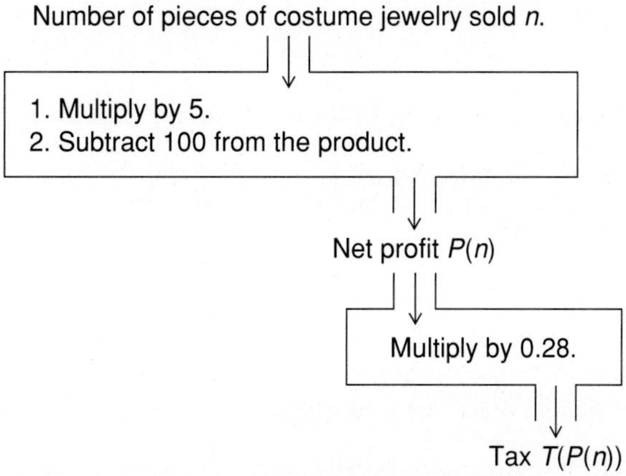

Number of pieces of costume jewelry sold n.
1. Multiply by 5.
2. Subtract 100 from the product.
Net profit $P(n)$
Multiply by 0.28.
Tax $T(P(n))$

Notice that the output of the first function is used as input to the second function.

Let's try and create one function rule that accomplishes this.

We know $P(n) = 5n - 100$ computes the profit from the number of pieces of jewelry sold.

We know $T = 0.28P$ computes the tax from the profits.

If we substitute the rule for P in the second function, we will have tax computed from the number of pieces of jewelry sold.

$$\begin{aligned} T &= 0.28P \\ &= 0.28(5n - 100) \\ &= 1.4n - 28. \end{aligned}$$

So $T(n) = 1.4n - 28$ is one function formed from the composition of two functions. In function language, this is T of P of n. Using function notation,

$$T(P(n)) = 1.4n - 28.$$

Composing functions in this way allows us to create more complex functions from basic functions. This is an important tool in the mathematician's arsenal.

One last comment on this problem. We realize that the tax computation will output a negative number if Sue and Ernie don't make a profit. That's okay since we'll consider it a reduction in their taxes if they lose money.

We'll continue to study statistical functions in the next section.

STUDENT DISCOURSE

Pete: One thing bothers me about function notation.

Sandy: What's that?

Pete: How do you distinguish it from multiplication?

Sandy: That's a good question. I had trouble with that too. It seems to have to do with the context the notation is used in.

Pete: What do you mean?

Sandy: Suppose I write $a(b)$. How would you interpret that?

Pete: I'd think that it meant a times b.

Sandy: That's probably the most common interpretation. However, suppose I told you a was a function of b. In other words, the value of a depends on the value of b. You might recall the area of a triangle does depends on its base. Then what would you think?

Pete: It seems like $a(b)$ would be function notation, expressing the dependency you just described.

Sandy: Okay. Why did you change your interpretation?

Pete: Because you gave me information that indicated I was to think in terms of functions.

Sandy: That's what I mean by context. If a notation is given within a certain context, its meaning is determined by the given setting. In the case of the triangle, knowing that area is not multiplied by the base but determined by the base might be the only clue we were discussing a function rather than multiplication.

Pete: Hmmm. I think I get it, but I need to think about this a bit more. I have to run. See you later.

Sandy: So long.

EXPLORATIONS

1. Define the words in this section that appear in ***italic boldface*** type. Write your definitions in the Glossary in the back of the book.

2. Suppose that y is a function of x.

a. What does $y(3)$ mean?

b. What does $y(x) = 5$ mean?

c. What does $y(x) = 3x - 5$ mean?

3. Given any function $Sue(x)$, describe the meaning in words of:

a. $Sue(t)$

b. $Sue(t + 1)$

c. $Sue(t) + 1$

d. $Sue(t + 1) - Sue(t)$

e. $Sue(-x)$

f. $-Sue(x)$

4. Given the function $Sue(x) = 10 - x$, evaluate the following if $t = 3$.

a. $Sue(t)$

b. $Sue(t + 1)$

c. $Sue(t) + 1$

d. $Sue(t + 1) - Sue(t)$

e. $Sue(-x)$ **f.** $-Sue(x)$

5. Given the notation $Joe(m) = p$, identify the input, the name of the function, and the output.

6. A herd of wild wallabies is introduced onto an island. The function $N(t)$ represents the number of wallabies still alive after t years, where

$$N(t) = -t^4 + 37t^2 + 400.$$

a. How many wallabies will be alive after two years? Write a justification for your answer.

b. Create a table showing how many wallabies will be alive at the end of each of the first five years. Describe your results.

c. How many years will it be until no wallabies remain on the island? Justify your answer.

7. Consider the function definition $Fun(x, y, z) = 3x - 5y + 2z$.

a. What is $Fun(-5, 2, 4)$?

b. What variables are used for input?

c. What is the name of the function?

d. Create a function machine.

8. Refer to the function machines in Figures 6 and 7.

FIGURE 6

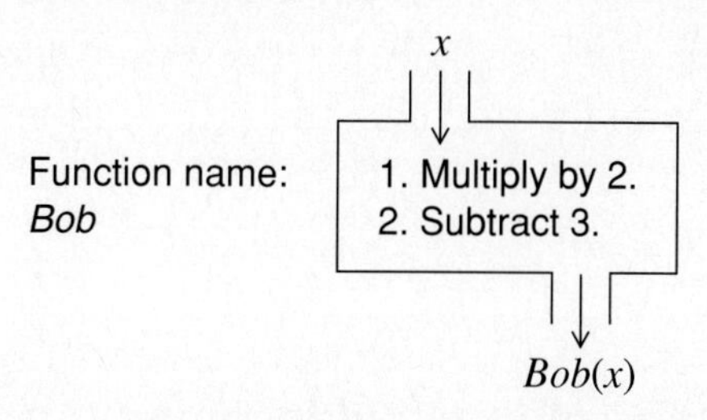

FIGURE 7

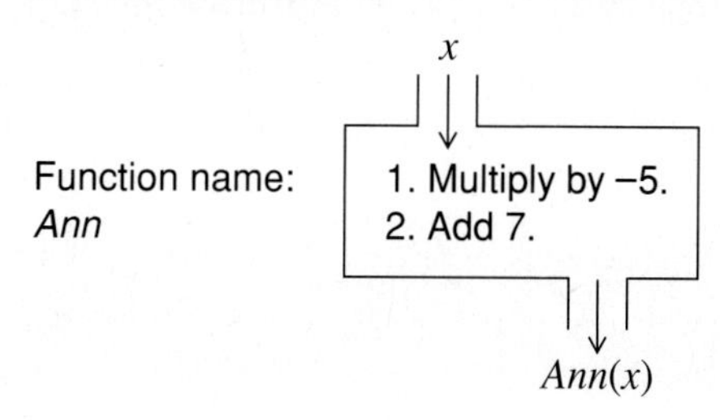

a. Create a composite function machine by using the output of function *Bob* as the input to function *Ann*.

b. Create a table of five input–output pairs for the function $Ann(Bob(x))$.

c. Write the algebraic representation (equation) for the function $Ann(Bob(x))$.

9. Refer to the function machines in Figures 6 and 7.

a. Create a composite function machine by using the output of function *Ann* as the input to function *Bob*.

b. Create a table of five input–output pairs for the function $Bob(Ann(x))$.

c. Write the algebraic representation (equation) for the function $Bob(Ann(x))$.

10. Refer to the previous two explorations. Are the functions $Ann(Bob(x))$ and $Bob(Ann(x))$ the same function? Why?

11. Median Household Income: You are working for a statistical firm that has the task of analyzing the median household income in the United States. A survey of 30 randomly chosen counties has been done. The following data are the median household incomes ($) of the populations of the counties sampled.

33,825	36,241	29,792	36,952	32,181	41,721
29,334	30,319	28,706	28,797	32,252	27,128
26,832	29,281	34,875	39,386	20,795	26,647
22,534	24,807	23,597	20,136	25,267	21,147
32,270	27,742	22,988	25,257	27,096	30,140

a. Calculate the mean, median, and mode of the data. Write answers using function notation.

b. Write a short report to your supervisor that includes your analysis of the data. Be sure to back up your statements with hard facts.

c. Suppose a co-worker informs you that the second data value, 36,241, was a mistake and that the real data value is 86,241. How would you modify your report to reflect the corrected data?

12. Consider the functions $f(x) = 3x - 7$ and $g(x) = 6x + 2$.

a. Calculate $f(-2.3)$

b. Calculate $g\left(\frac{7}{5}\right)$

c. Find x if $f(x) = 17$

d. Find x if $g(x) = -93$

e. Find $f(g(3))$

f. Write an equation that represents the case in which the outputs of $f(x)$ and $g(x)$ are the same.

g. Solve the equation you wrote in part **f**. Describe the meaning of the answer.

REFLECTION

A friend of yours is confused about function notation. Write a paragraph explaining function notation to your friend.

SECTION 1.4

More Analysis of Measures of Central Tendency

Purpose

- Analyze measures of central tendency in the context of functions and relations.
- Investigate the similarities and differences between functions and relations.
- Investigate the domain and range of both functions and relations.

The chapter began by introducing relations and functions using measures of central tendency. We have spent the last two sections developing a deeper understanding of relations and functions. We now return to our study of mean, median, and mode. We begin by looking at domain and range in a particular problem situation. Later in the section, you will be introduced to two new functions as a way of transforming data. Keep in mind that the focus is deepening the understanding of relations and functions. The focus of the next investigation is on domain and range.

Grading with Six Exams: Consider a class in which your course grade is determined by six exams, each with a possible grade consisting of *whole*

numbers up to 100 inclusive. Use your calculator, where appropriate, in the following investigations.

1. Create five different lists, each containing six exam scores. Record them in the left column of Table 1. Record the mean, the median, and the mode of each list in the appropriate column.

TABLE 1 Mean, Median, and Mode of Six Exam Scores

List of Exam Scores	Mean	Median	Mode

2. Were there any inputs for which there was no output for either the mean, the median, or the mode? If so, record the inputs and the function or relation involved.

3. Were there any inputs that resulted in more than one output for either the mean, the median, or the mode? If so, record the inputs, the outputs, and the function or relation involved.

You created five different lists in Table 1 that may be used as input to the mean function, the median function, and the mode relation. Each list is in the *domain* of the relation or function if the list produces an output. Each number in the mean, median, and mode columns is in the *range* of the corresponding relation or function. Use these examples to help you write a general answer to the next investigation.

4. Within the context of the **Grading with Six Exams** problem, write a sentence that describes the set of all possible inputs (domain) to the:

a. *mean* function.

b. *median* function.

c. *mode* relation.

5. Compare and contrast (find the things that are the same and the things that are different about) the domains of the *mean* function, *median* function, and *mode* relation.

6. Within the context of the **Grading with Six Exams** problem, write a sentence that describes the set of all possible outputs (range) of the

   a. *mean* function.

   b. *median* function.

   c. *mode* relation

7. Compare and contrast (find the things that are the same and the things that are different about) the ranges of the *mean* function, *median* function, and *mode* relation.

Remember that the set of all legal inputs to a relation or function is called the ***domain of the relation or function***. An input is legal if it produces at least one output in the case of a relation and exactly one output in the case of a function

The *mean*, *median*, and *mode* each receive input in the form of finite lists of numbers.

For the **Grading with Six Exams** investigation:

- The domain for both the mean and median functions is the set of all lists containing six whole numbers between 0 and 100 inclusive.

Some lists in the domain are {0,0,0,0,0,0}, {94, 83, 72, 98, 56, 67}, {100, 100, 100, 100, 100, 100}, and {87, 76, 87, 76, 87, 98}. The five lists you wrote in Table 1 are elements of the domain too.

- The domain of the mode relation is the set of all lists containing six whole numbers between 0 and 100 inclusive in which all numbers in the list do not occur with the same frequency.

  For example, the lists {54, 68, 79, 79, 84, 69} and {98, 89, 89, 98, 98, 98, 100} are in the domain of the *mode* relation, but the list {88, 76, 88, 76, 76, 88} is *not* in the domain of the *mode* relation. This list is not in the domain of the mode relation since no output is produced if the list is used as input. You may have had a list in Table 1 that did not have a mode. Such a list is not in the domain of the mode relation.

The set of all outputs of a relation or a function is called the ***range of the relation or function***. The *mean* and *median* produce a single number as output. However, *mode* is a relation since, for a given list of numbers as input, the mode may have more than one number as output. There is *exactly one* output for a given input in the case of a function. There may be more than one output for a given input in the case of a relation.

For the **Grading with Six Exams** investigation:

- The range of the *mean* function is a subset of the rational numbers between 0 and 100 inclusive. The elements of the range are the ratios of the sum of six whole numbers between 0 and 100 inclusive and the number 6.

- The range of the *median* function is the set {0, 0.5, 1, 1.5, 2, 2.5, 3, . . . , 99, 99.5, 100}.

- The range of the *mode* relation is the set of all whole numbers between 0 and 100 inclusive.

To investigate further the idea of domain and range, we return to the **Test with No Questions** from Section 1.1. We look at two functions that transform the original scores (data). Transformation of data is one of the many uses of functions. In this case, we will transform the original scores by adding a number to the original scores to get a transformed score.

8. Scaling Scores I: Recall the **Test with No Questions** from Section 1.1. The test scores were so low that we decided to scale the scores by adding 3 points to each possible score (0–10). Let's create a function called *add3* to perform this transformation on each possible test score.

a. Record in the left column all possible inputs (0–10) to this function. In the right column, record the output of the function *add3*.

TABLE 2 The *add3* Function

Input: Original Scores	Output: Transformed Scores

b. What is the domain of the *add3* function?

c. What is the range of the *add3* function?

d. Draw a function machine for the *add3* function.

e. If the variable s represents any score, use function notation to write an algebraic equation for the *add3* function.

In the next investigation, we use the random number function *Rand*. This function outputs a random number between 0 and 1. The input is a seed value internal to the calculator. See the calculator guru (the manual) if you want more information.

9. Scaling Scores RANDomly: Instead of adding 3 points to each student's score, we decided to use a random number generator to determine how many points to add to each score. We will use the units digit of the result of $10 \times Rand$ to identify the number of points added to a given score. (For example, if *Rand* generates 0.5147019505, then $10 \times Rand$ will give the number 5.147019505, and we will add 5.)

a. Record the possible original scores (0–10) in the Input 1 column. For each score, generate a random number and record the number of extra points in the Input 2 column. Record the transformed score (sum of Input 1 and Input 2) in the Output column.

TABLE 3 The *AddRand* Function

Input1: Original Score	Input2: Random Digit	Output: Transformed Score

b. The function defined in this problem has two inputs. How would you define the domain of the input Original score? How would you define the domain of the input Random digit?

DISCUSSION

For the *add3* function, the domain is the set of all possible test scores, or {0, 1, 2, 3, 4, 5, 6, 7, 8, 9, 10}.

The range is the set obtained by adding 3 to each number in the domain. The range is {3, 4, 5, 6, 7, 8, 9, 10, 11, 12, 13}.

For *add3*, the input is any whole number between 0 and 10 inclusive. The process is "add 3," and the output is a whole number between 3 and 13 inclusive.

One way to visualize *add3* using the input, 5, as an example is shown in Figure 1.

FIGURE 1

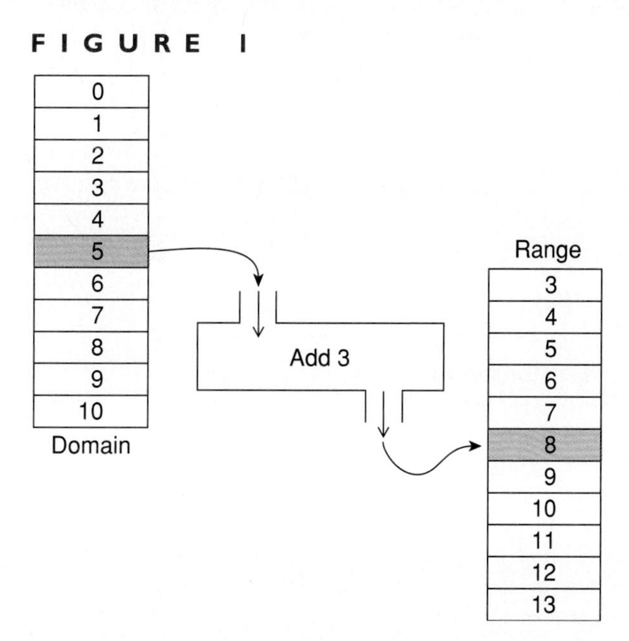

Symbolically, using function notation, the action depicted in Figure 1 is written as *add3* (5) = 8.

Using the form *function name* (input) = output, the *add3* function can be written as

$$add3(s) = s + 3,$$

where s represents the input (a test score) to the function and *add3* is the name of the function.

Another visualization is shown in Figure 2.

FIGURE 2

Domain	Add 3	Range
0	→	3
1	→	4
2	→	5
3	→	6
4	→	7
5	→	8
6	→	9
7	→	10
8	→	11
9	→	12
10	→	13

The *add3* function had one input and one output. Functions may have two or more inputs. The operation machines for addition, subtraction, multiplication, and division are examples of functions with two inputs. A general function machine that has two inputs is shown in Figure 3.

FIGURE 3

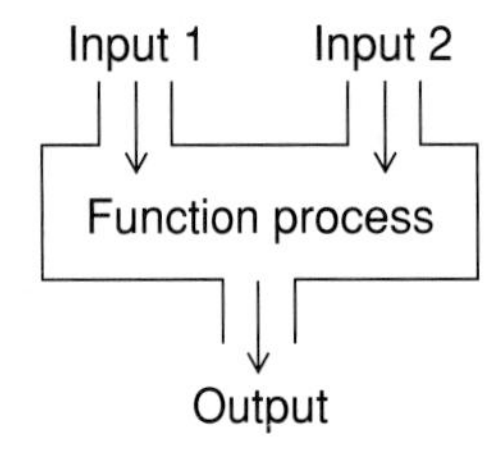

Symbolically, we could write

function name (Input1, Input2) = Output.

The *AddRand* function is a function that requires two inputs to return one output. Symbolically, we could write

AddRand (Input1, Input2) = Output.

For the *AddRand* function,

- the set of legal values for Input1 is the set of possible original scores {0, 1, 2, 3, 4, 5, 6, 7, 8, 9, 10}.
- the set of legal values for Input2 is the set of random digits {0, 1, 2, 3, 4, 5, 6, 7, 8, 9}.
- the set of outputs or range is the set of digits {0, 1, 2, 3, 4, 5, 6, 7, 8, 9, 10, 11, 12, 13, 14, 15, 16, 17, 18, 19}.

The process is visualized for the inputs (5, 7) in Figure 4.

FIGURE 4

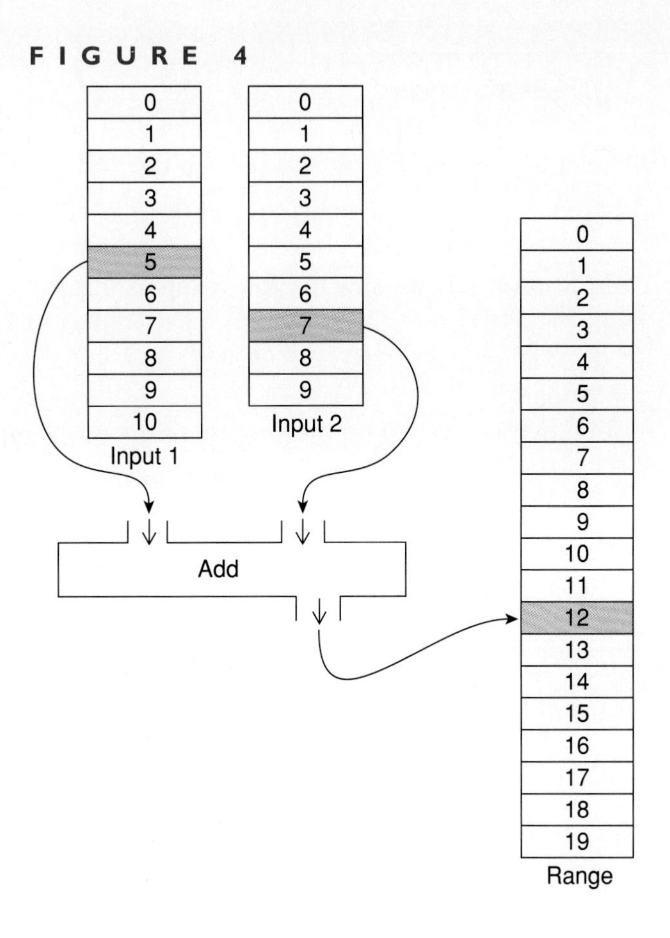

Symbolically, the action in Figure 4 is written as

$$AddRand(5, 7) = 12.$$

AddRand has two inputs, but, for each pair of inputs, there is exactly one output. Since the input always produces exactly one output, *AddRand* is a function.

The *mode* is an example of a relation that is not a function. While a function must have exactly one output for a given input, a relation may have several outputs for a given input.

A relation that has two outputs for a given input is shown in Figure 5.

FIGURE 5

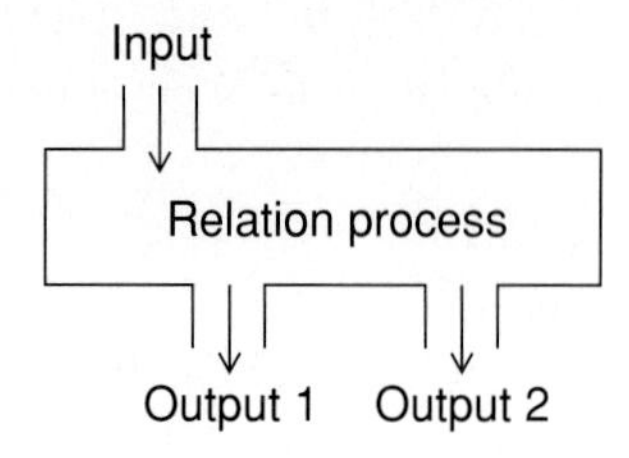

In Section 1.1, we saw the following example.

$$mode(\{0, 1, 2, 3, 3, 3, 3, 4, 4, 5, 5, 5, 5\}) = 3 \text{ or } 5.$$

Figure 6 is a visual representation of the mode relation for this example.

FIGURE 6

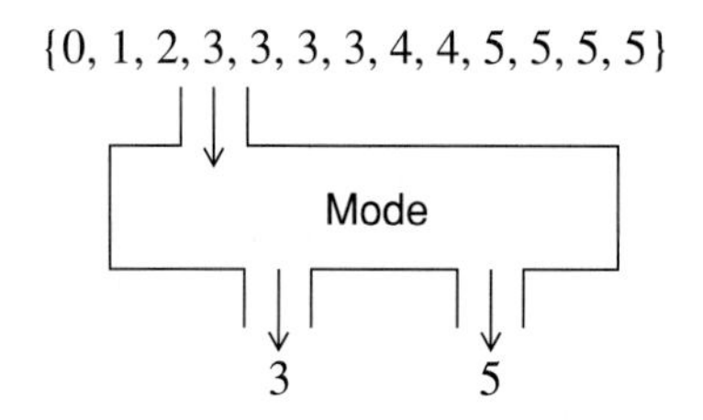

STUDENT DISCOURSE

Pete: I have a question about domain.

Sandy: I'm listening.

Pete: Suppose I use a number as an input to a function and there is no output. Is that number in the domain of the function?

Sandy: Give me an example.

Pete: Okay. . . . let's see. I have one. Suppose my process is $1/x$ and I use 0 as input. There is no output since I can't divide 1 by 0. Is 0 in the domain of this function?

Sandy: Let me think. A value is in the domain of a function if it is a "legal" input. "Legal" means that the process can be performed on the input to produce an output. So. . . .

Pete: Zero is not in the domain of the process $1/x$. I think that answers my question. If an input does not result in an output, the input must not be in the domain of the function.

Sandy: That seems right to me.

Pete: Okay, I think I have that straight. Thanks for the help. Just talking about this with someone else really clarifies things.

1. List and define the words in this section that appear in ***italic boldface*** type. Write your definitions in the Glossary in the back of the book.

2. Answer the given statement: always, sometimes, or never. Write a sentence or two justifying your answer. Include examples.

   a. The outputs of the *mean* function are elements of the input list.

   b. The outputs of the *median* function are elements of the input list.

   c. The outputs of the *mode* relation are elements of the input list.

3. Create a list of data in which the mean, median, and mode are all different. Justify your answer by calculating the mean, median, and mode for your list.

4. Create a list of data in which the mean, median, and mode are all the same. The list you create should have at least three different numbers. Justify your answer by calculating the mean, median, and mode for your list.

5. One finite list is not in the domain of the *mean* function and the *median* function. What is it? (*Hint*: Think about the computation for the mean. When will this computation not be possible?)

6. If a list of data contains only integers, what types of numbers are output by:

   a. the *mean* function?

   b. the *median* function?

   c. the *mode* relation?

7. Describe a problem situation in which the list of data consists of:

a. whole numbers.

b. integers.

c. rational numbers.

8. Assume you are in a class where your grade is determined by six exams, each worth 100 points. Create a list of five test scores, each above 60. Suppose you miss the sixth exam, earning a zero.

a. How does the missed test affect your mean score for the class?

b. How does the missed test affect your median score for the class?

c. Suppose you contact your instructor and make up the test. If your score for that test is 50, how is your mean score for the class affected? How is your median score for the class affected?

9. Consider the function $Joe(t) = 3t - 5$.

a. Create an input–output table where the inputs go from 1 to 6 in increments of 1.

b. What is the value of $Joe(t + 1) - Joe(t)$? Find the relationship between this value and the numbers in your table.

c. Describe, in words, the domain of function *Joe*.

d. Describe, in words, the range of function *Joe*.

10. Create another function, similar to *add3* or *AddRand*, that you could apply to the data from the **Test with No Questions**.

a. What is the domain of your function?

b. What is the range of your function?

c. Draw a function machine for your function.

d. Write an equation, if possible, in function notation for your function.

11. Consider the function $f(x) = 3x^2 - 5x + 2$. Find:

a. $f(4)$

b. $f(-7)$

c. $f(0)$

d. $f(4p)$

e. $f(k-2)$

f. $f(2t+5)$.

12. Each of the following equations is an example of a function set equal to a numerical output. Identify the function and the output.

a. $7x - 5 = 12$

b. $-8 = 3 + x$

13. Solve each of the following equations or inequalities.

a. $2 - 3x < 11$

b. $5 + 2x = -15$

c. $4 + 7a = 2a - 1$

d. $5t - 7 < t + 2$

14. Consider the function $Joe(x) = 7x - 9$. Find x if:

a. $Joe(x) = -2$

b. $Joe(x) > 4$

c. $Joe(x) < -3$

d. $Joe(x) = 0$.

REFLECTION

Write a paragraph that explains to a friend what is meant by both the domain of a relation or function and the range of a relation or function. Be sure to include examples to help your friend understand.

SECTION 1.5

Data: Graphical Analysis

Purpose

- Visualize data.
- Construct histograms and scatter plots with a calculator.

We have investigated relations and functions numerically using tables, with function machines, and symbolically using equations where possible. The most common visual method for investigating relations and functions involves graphing. Graphs provide a method to display data visually. Statistics uses some special graphs that may be unfamiliar to you. The first type of graph involves a single list of data. Later in the section, we look at graphs of paired lists of data. This second type of graph will be used extensively throughout the rest of the text.

We begin by constructing a ***frequency plot*** for the data from the **Test with No Questions**.

1. Look at the results of the **Test with No Questions** in Section 1.1.

a. Label the tick marks on the number line in Figure 1 with units corresponding to possible scores on the test).

b. Place a dot or an X for each test score above the number line in Figure 1 at the appropriate unit.

FIGURE 1

2. Is it easier to summarize the data from the **Test with No Questions** by looking at the actual data or by looking at the frequency plot? Defend your answer.

We use the list of sample test data to create an example of the frequency plot. The class data is

$$\{0, 1, 1, 1, 2, 2, 4, 4, 4, 5, 5, 5, 5, 6, 7, 7, 9, 9, 9, 10\}.$$

In Figure 2, the number of X's above each tick mark correspond to the number of times the given score occurred.

FIGURE 2

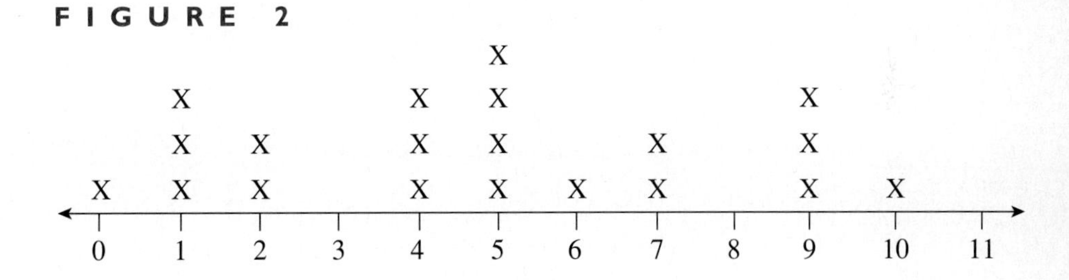

The frequency plot displays the data visually. We can see various trends in the test scores simply by glancing at the frequency plot. It is easy to tell what score occurred most often (the mode) and what scores occurred least.

A frequency plot is a method of representing one list of data visually. In algebra, it is more common to graph two lists—the list of inputs and the list of corresponding outputs. The result is a two-dimensional graph with an *input (horizontal) axis* and an *output (vertical) axis*. Let's investigate three such graphs. We investigate each of these using the same data.

A ***histogram*** is a bar graph consisting of connected rectangles. The set of inputs consists of values, such as the hours of a day. The outputs are frequency of occurrence of some phenomenon—for example, the number of cars leaving campus at a given hour.

Campus Vehicle Congestion: College officials are concerned about traffic patterns on their two campuses. To analyze the situation, they record the number of cars that leave each campus within fifteen minutes of a given hour. The data appear in Table 1.

Officials want to compare the distribution of cars leaving each campus.

TABLE 1 Cars Leaving Campus by Time of Day

Time of Day	Number of Cars Leaving Green Campus	Number of Cars Leaving Concrete Campus
8 A.M.	159	207
9 A.M.	341	223
10 A.M.	467	301
11 A.M.	537	310
12 noon	551	298
1 P.M.	509	254
2 P.M.	411	238
3 P.M.	257	219

3. Construct a histogram for the number of cars leaving the green campus by hour. Let the horizontal axis represent the hour and let the vertical axis represent the number of cars leaving the green campus.

4. Construct a histogram for the number of cars leaving the concrete campus by hour. Let the horizontal axis represent the hour and let the vertical axis represent the number of cars leaving the concrete campus.

5. Use your calculator to construct two histograms on the same grid. The two histograms should represent the number of cars leaving the green campus and the number of cars leaving the concrete campus. (Ask a friend, your instructor, or consult the manual if you don't know how to do this.)

a. What is an advantage of displaying both histograms on the same grid?

b. What is a disadvantage of displaying both histograms on the same grid?

Figure 3 shows the two separate histograms along with the overlapping histograms.

FIGURE 3

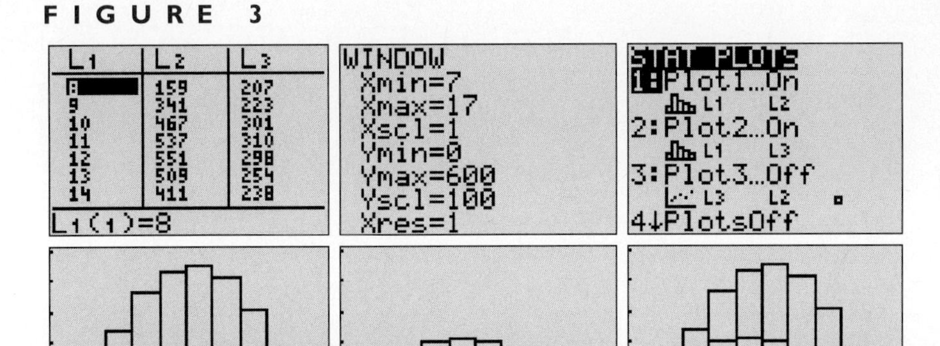

Green campus Concrete campus Combined

It is easy to tell from the histogram

- when most cars left campus,
- what times were busier than 2 P.M.,

and many other pieces of information. The combined graph is useful for making comparisons between the two sets of data, but it can also be confusing because of the overlap. Different colors are helpful in such a case.

The final two graphs we discuss are the ***scatter plot*** and the ***line plot***. We think of the data as consisting of ordered pairs for both of these graphs.

We mark the points corresponding to each ordered pair to create the scatter plot. This is a plot of individual, or ***discrete,*** points. Notice how the points are not connected by lines in a scatter plot.

We use the scatter plot to create the line plot. We create the line plot by connecting the successive points of the scatter plot with straight-line segments.

Use the data from the **Campus Vehicle Congestion** problem (Table 1) for this Investigation.

6. Let time of day (use a 24-hour clock) be the input and let number of vehicles leaving the green campus be the output.

a. Write the ordered pairs implied by the table of data.

b. Construct a scatter plot where the number of cars leaving the green campus is the output.

c. Construct a line plot where the number of cars leaving the green campus is the output.

7. Let time of day (use a 24-hour clock) be the input and let number of vehicles leaving the concrete campus be the output.

a. Write the ordered pairs implied by the table of data.

b. Construct a scatter plot where the number of cars leaving the concrete campus is the output.

c. Construct a line plot where the number of cars leaving the concrete campus is the output.

8. Write three observations about the data using either the scatter plot or line plot.

Scatter plots are used whenever we wish to graph paired data. We use scatter plots to identify trends and to model data with algebraic functions. Line plots can be used similarly. The line segments can be used to analyze the rate of change of the output with respect to the input for two consecutive points. We'll look at this idea in detail in Section 2.1. Now look at the scatter plots and line plots for the **Campus Vehicle Congestion** problem.

The scatter plot and line plot for the number of cars leaving the green campus appear in Figure 4.

FIGURE 4

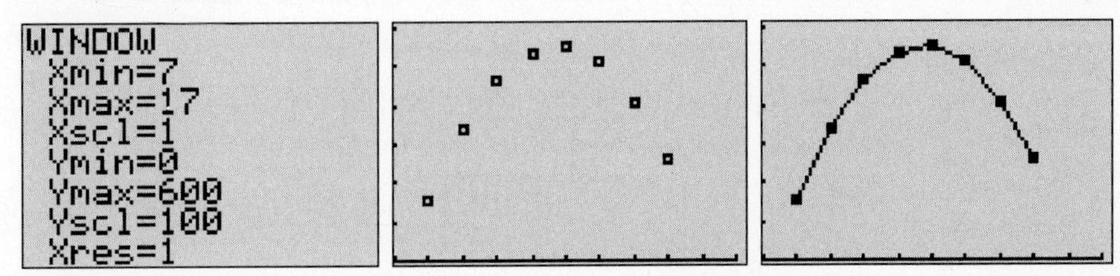

The scatter plot and line plot for the number of cars leaving the concrete campus appear in Figure 5.

FIGURE 5

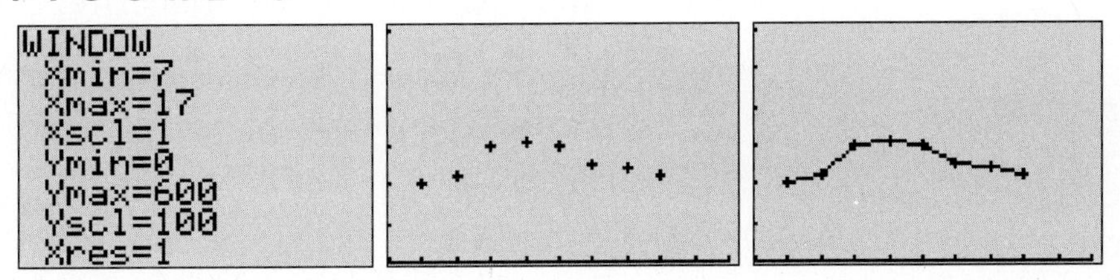

Comparing the two plots reveals much greater vertical change (increase, then decrease) in the cars leaving the green campus as compared to the cars leaving the concrete campus.

Graphs are a visual way of studying functions. As we proceed through the book, we will often use graphs such as those shown in Figures 4 and 5 to analyze problem situations. Get to feel comfortable using and interpreting graphs. They give you a tremendous amount of information in a compact format.

Pete: So a scatter plot just consists of the points that are given by the ordered pairs in the table, is that correct?

Sandy: Yes. It's a way to visualize the individual data points.

Pete: But a line plot isn't a graph of a line, is it? I can see that from the line plots in this section.

Sandy: That's right. A line plot consists of several connected line segments.

Pete: And I get those by connecting the points from the scatter plot?

Sandy: By connecting successive points.

Pete: Okay. Does the line plot give more information than the scatter plot?

Sandy: Not necessarily. It may be easier to spot trends with the line plot, but that depends on how you see graphs.

Pete: I can see that, but I do have one last question. How would we know that we should connect successive points with a line segment? Couldn't the actual data be more erratic?

Sandy: Certainly, you really make an assumption about the trends when you draw the line segments.

Pete: Sure, one assumption is that there are data points between the ones given.

Sandy: That's right and in many cases there are not. What else?

Pete: Well, the data could act erratically between the two points.

Sandy: Exactly. These are just some things to be aware of when you interpret line plots.

Pete: Thanks for helping me clarify my thinking. I needed to talk about it to make sure I understood how to look at the graphs.

1. In the Glossary, define the words in this section that appear in ***italic boldface*** type.

2. Choose one of the graphs from the investigations in this section and list four important features of the graphs that are necessary to understand the message in the graph.

3. For each of the following questions, describe how you would collect the data and what kind of graph would be appropriate to portray the data.

a. I am having trouble getting all my work done. Perhaps if I look at how I spend my time this Wednesday, I can see how to manage my time more efficiently.

b. My dad is really concerned about the credit card bill. I pay it off every month, but he thinks too much of my income has been spent on impulsive purchases with the credit card and he thinks it is getting worse. (He may be right since I find each month it is a little more difficult to pay my bills.) I wonder how my monthly charges have changed since I got the card 18 months ago.

c. I think it is important to read from a wide variety of genres. I subscribe to a number of book clubs and also buy books in various bookstores. How can I check to see if my reading is balanced?

4. Carrie's Job Offer: Carrie has been offered a job with a starting salary of $30,000. She estimates that she will receive, on the average, a 5% increase each year. The data in Table 2 display her annual salary for the next 10 years.

TABLE 2 **Carrie's Annual Salary by Year of Employment**

Year	Salary ($)
0	30,000.00
1	31,500.00
2	33,075.00
3	34,728.75
4	36,465.19
5	38,288.45
6	40,202.87
7	42,213.01
8	44,323.66
9	46,539.85
10	48,866.84

a. Construct a scatter plot of the paired data.

b. What do you observe about the shape of the graph? Relate your answer to Carrie's annual salary.

5. Compare the use of a table with the use of a histogram to present the data. Which is easier to read? From which do you get the most information? Why?

6. Jerome's Quiz Scores: The graph in Figure 6 displays Jerome's weekly quiz scores (out of 20) in his Chaos class. Write three observations you can make about Jerome's progress in the class.

FIGURE 6

20
15
10
5
Number correct
1
2
3
4
5
6
7
8
9
Week

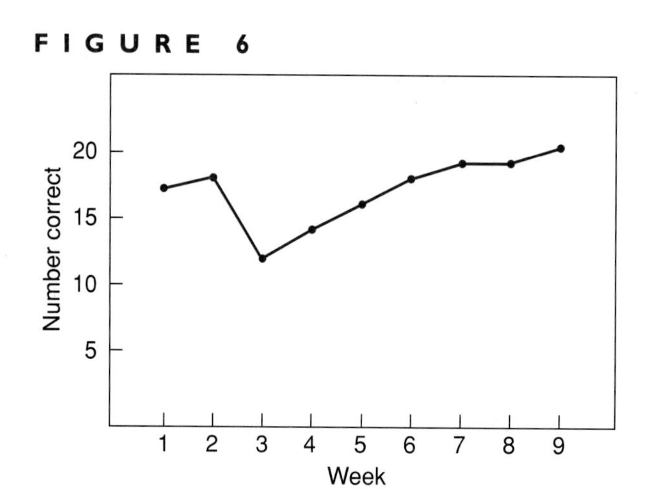

7. Ginny's Budget: The ***circle graph*** (***pie chart***) in Figure 7 displays Ginny's spending habits over a two-week period. Write three observations about Ginny's financial management based on the graph.

FIGURE 7

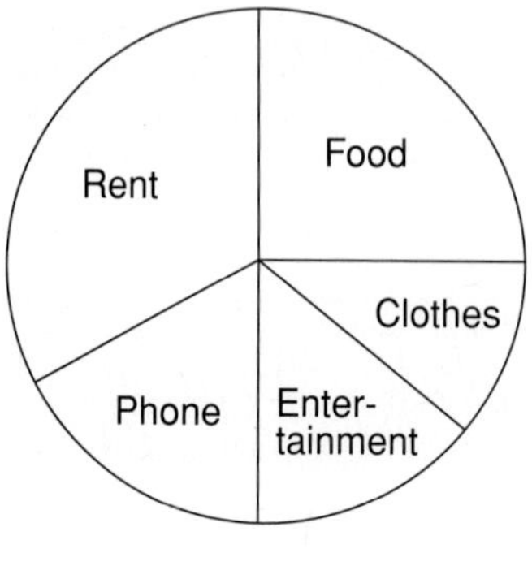

Rent
Food
Clothes
Phone
Enter-
tainment

8. Barb's Chaos Project: The graph in Figure 8 displays Barb's scores on three of the four projects required in her Chaos class. The average grade on the projects is displayed also.

FIGURE 8

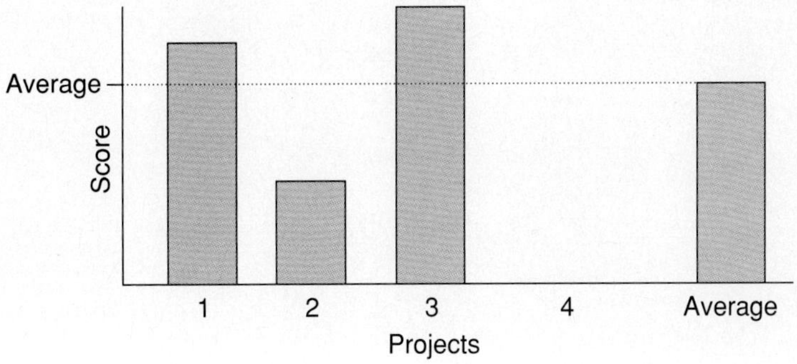

Average
Score
1 2 3 4 Average
Projects

a. Draw a rectangle indicating the score Barb needs on the fourth project so that the mean score on her four projects equals the class average.

b. Write three questions that can be answered by the graph in Figure 8.

9. Given the function machine in Figure 9,

FIGURE 9

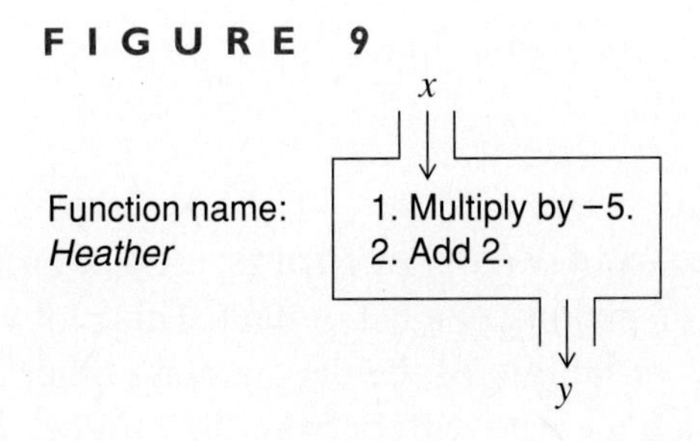

a. Create a table of five input–output pairs.

b. Write the algebraic representation (equation).

c. Find the output if the input is –4.

d. Write an equation that displays the case when the output is 27.

e. Find the input if the output is 27. Justify your work.

10. Consider the functions $f(x) = 5x + 11$ and $g(x) = -3x + 1$.

a. Calculate $g(7.9)$

b. Calculate $f\left(-\frac{9}{7}\right)$

c. Find x if $f(x) = -4.1$

d. Find x if $g(x) = -\frac{2}{3}$

e. Write an equation that represents the case in which the outputs of $f(x)$ and $g(x)$ are the same.

f. Solve the equation you wrote in part **e.** Describe the meaning of the answer.

11. Solve each of the following equations or inequalities.

a. $-5 = 3 + 7y$

b. $3p - 5 = 11p$

c. $-19 < 12 - 17x$

d. $4 - 2x > 11x + 7$

12. Box-and-Whisker Plots: A box-and-whisker plot is a visual method of displaying *one* list of data. This plot visually displays the median and the variability of the data, among other things. Variability refers to how much the data differs from the middle. The ***statistical range***, the difference between the largest and smallest score is a simple measure of variability. Figure 10 displays a sample box-and-whisker plot.

The difference between the first quartile and the third quartile is called the ***interquartile range***. This difference is another measure of the variability of the data.

FIGURE 10

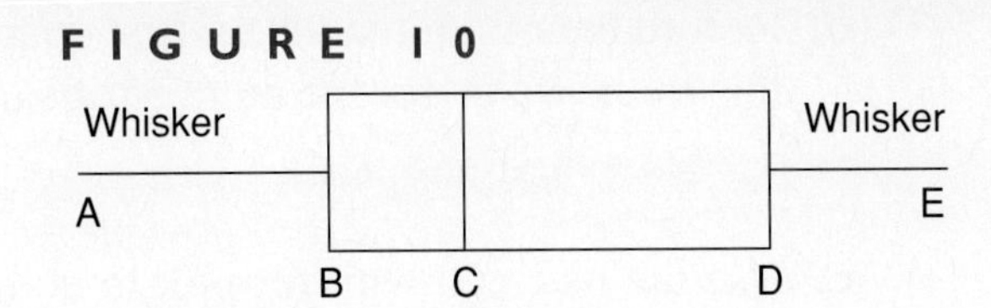

A: the minimum data value.

B: the ***first quartile***—25% of the data values are less than this value.

C: the median—50% of the data values are less than this value.

D: the ***third quartile***—75% of the data values are less than this value.

E: the maximum data value.

For the Campus Vehicle Congestion data listed in Table 3, use your calculator to construct two box-and-whisker plots: one for the number of cars leaving the green campus; the other for the number of cars leaving the concrete campus. (Ask a friend, your instructor, or consult the manual if you do not know how to do this.)

TABLE 3 Cars Leaving Campus by Time of Day

Time of Day	Number of Cars Leaving Green Campus	Number of Cars Leaving Concrete Campus
8 A.M.	159	207
9 A.M.	341	223
10 A.M.	467	301
11 A.M.	537	310
12 noon	551	298
1 P.M.	509	254
2 P.M.	411	238
3 P.M.	257	219

a. Record the minimum value, first quartile, median, third quartile, and maximum value for the number of cars leaving the green campus.

b. Record the minimum value, first quartile, median, third quartile, and maximum value for the number of cars leaving the concrete campus.

c. Use the box-and-whisker plot to compare and contrast the two sets of data. Write at least three different observations about the data.

REFLECTION

Write a paragraph or two comparing and contrasting various methods of representing data graphically. How are the graphs in this section similar to the graphs of functions you have done in a previous algebra course? How are they different?

SECTION 1.6

Working with Data: Review

Purpose

- Reflect on the key concepts of Chapter 1.

In this section you will reflect upon the mathematics in Chapter 1—what you've done and how you've done it.

1. State the most important ideas in this chapter. Why did you select each?

2. Identify all the mathematical concepts, processes, and skills you used to investigate the problems in Chapter 1.

There are a number of really important ideas that you might have listed including mean, median, mode, measures of central tendency, function, input, output, domain, range, function notation, function machines, graphs, frequency plots, histograms, scatterplots, and line plots.

1. Mathematicians use rules to make a math process more efficient. Write a rule to find the median of 100 pieces of data quickly.

2. A company offers a mean annual salary of \$45,000, a median annual salary of \$20,000, and a modal annual salary of \$10,000. Create a list of data that represents this situation.

3. a. Create a function machine for $y = x^2 - 5x + 4$.

b. Find the outputs for the function in part a if the inputs are 4, –5, and $\frac{2}{3}$.

c. When you use –3 as input, do you write it as $y = -3^2 - 5(-3) + 4$ or as $y = (-3)^2 - 5(-3) + 4$? Does it make a difference? Why or why not?

4. Given the function machine in Figure 1,

FIGURE 1

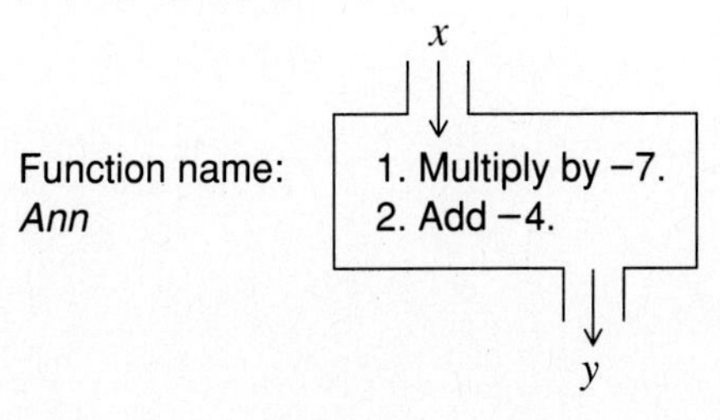

a. Create a table of five input–output pairs.

b. Write the algebraic representation (equation).

c. Find the output if the input is 3.9.

d. Write an equation that displays the case when the output is –11.

e. Find the input if the output is –17. Justify your work.

f. What are the inputs if the output is less than 9?

5. How do you decide if:

a. given information constitutes a relation?

b. a given relation is a function?

6. Consider the function definition $Fun(x, y, z) = 5x - 2y + 3z$.

a. What is $Fun(-4, 3, 5)$?

b. What variables are used for input?

c. What is the name of the function?

d. Create a function machine.

7. Consider the function $f(x) = 2x^2 - x + 6$. Find:

a. $f(3)$

b. $f(-5)$

c. $f(0)$

d. $f(2p)$

e. $f(k-1)$

f. $f(3t-5)$

8. For each of the following questions, describe how you would collect the data and what kind of graph would be appropriate to portray the data.

a. I have noticed that the cost of buying a new computer fluctuates both in terms of where you purchase it and when you purchase it. I would like to give a friend information that includes the range as well as the "average" cost of the computer.

b. I am becoming concerned about my weight. Although frequent weighing is not recommended, I get on the scales to check my weight every hour from 6 A.M. until 11 P.M. If I kept the data as ordered pairs (time, weight) for a full week, I wonder if I would discover a pattern of weight fluctuation versus time.

9. The following expression is equal to an important year, 1997. Change the order of operations in the following expression by adding parentheses to get that result. (Try this on your calculator and make a table of all the outputs you get when you change the location of the parentheses.)

$$-53^2 + 4/2 \times 26 - 72 \times 6$$

10. Tim has kept track of the first digit in the Pick Three lottery drawing for the last 4 weeks. Find the measures of central tendency. Show his data on a frequency plot, a histogram, and a box-and-whisker plot. If you were picking a number tomorrow, what number would you choose for your first digit? Explain your strategy. Which presentation of the data was most helpful?

5, 8, 7, 4, 6, 1, 0, 6, 2, 9, 2, 2, 1, 2, 0, 4, 2, 3, 1, 1, 0, 5, 2, 3, 5, 7, 5, 6

11. If $P(x) = 2x - 3$ and $M(x) = 5 - x^2$, draw a function machine to show $M(P(x))$. Evaluate $M(P(-2))$.

12. A cookout to celebrate the final problem in Chapter 1 was planned for the whole Algebra class. Joel bought the hamburger, of which he had to choose among several package sizes. He found that a 1 pound package cost $1.50, a 3 pound package cost $4.50, a 5 pound package cost $7.50, and a 2½ pound package cost $3.75. Use this information to make a table and write an algebraic expression that shows the relation between the weight and the cost of the hamburger. Identify the input and output. Is this a function? Make a scatter plot using the data.

13. Each of the following equations is an example of a function set equal to a numerical output. Identify the function and the output.

a. $2 - 11x = -23$ **b.** $15 = 3p + 2$

14. Solve each of the following equations or inequalities.

a. $5 - 3t = -24$ **b.** $5p + 7 = -8p - 9$

c. $-12 > 7k - 9$ **d.** $5 - x < 4 + 15x$

15. Consider the function $g(t) = 4t + 11$. Find t if:

a. $g(t) = 0$ **b.** $g(t) > 4$

c. $g(t) < -7$ **d.** $g(t) = -9$.

Pete: Reviewing what we did in Chapter 1 wasn't bad, but the next thing in the book is called a Concept Map. Sounds scary. How can a Concept Map—whatever that is, help me learn math?

Sandy: Actually, Concept Maps helped me a lot last term. Think of a concept map as a visual map of how you are thinking about what you've learned. They helped me to see connections between the concepts and procedures that I was learning. I really improved my understanding of terms and definitions as I had to review the terms and definitions that I was unclear about.

Pete: But how do I get started?

Sandy: One way to begin a Concept Map is to write down the main topic at the top of a piece of paper—sort of like a title. Then list all of the words that come to mind when you think about that topic. Each time you write down a word, try to think of something that you associate with that word or with the overall topic—it can be another idea, a procedure, or your feelings about the topic.

Pete: Ok, but I'm not certain how to get started on the Map part.

Sandy: A Concept Map shows how all the words you've written down are related. It really helps to have some self-stick notes. Here's a check list to help you get started.

- On a piece of paper, write down the main topic of the concept map, then list all of the words that you associate with that topic. Concepts, procedures, your feelings about the topic, other concepts and visual representations like graphs or tables can also be added to your list.

- Think in terms of making an outline, with a main idea, supporting ideas, and details. Making your list of concepts and procedures is kind of like making an outline of a book you've been assigned in English class.

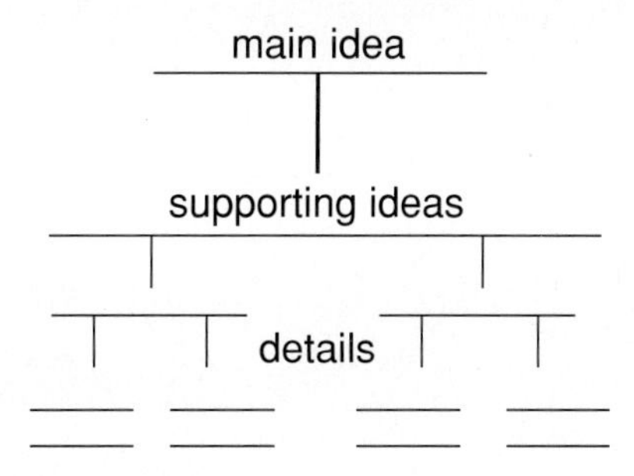

- Working from your list, use a highlighter and identify what you think are the important supporting ideas for your topic. Identify what words, representations, and procedures you associate with each of your main supporting categories.

- After you have finished analyzing your list, write each word on a separate self-stick note

- Post the main topic in the middle of a piece of paper. Place a self-stick note for each key word or idea you've listed around the main topic.

- Near each key word, place self-stick notes with words, representations, and procedures you associate with that key word. Writing the words on stick-on notes allows you to move the words around until you have placed them near other words that you think relate to each other.

- Build out from the main topic in a way that makes sense to you.

- When you've posted all of your words, draw in the linkages and connections between words and between groups of words. Use arrows to indicate the direction of each link. Wherever possible, write in the relationship along the connecting link.

- Sometimes, you have a word that you connect with more than one key word. Arrange your key words so that the shared connecting word is located between them so you can draw connections to both key words.

- Be creative and personalize your map.

Pete: I think that what I really need is an example.

Sandy: Well, you might get the idea from one of my early Concept Maps.

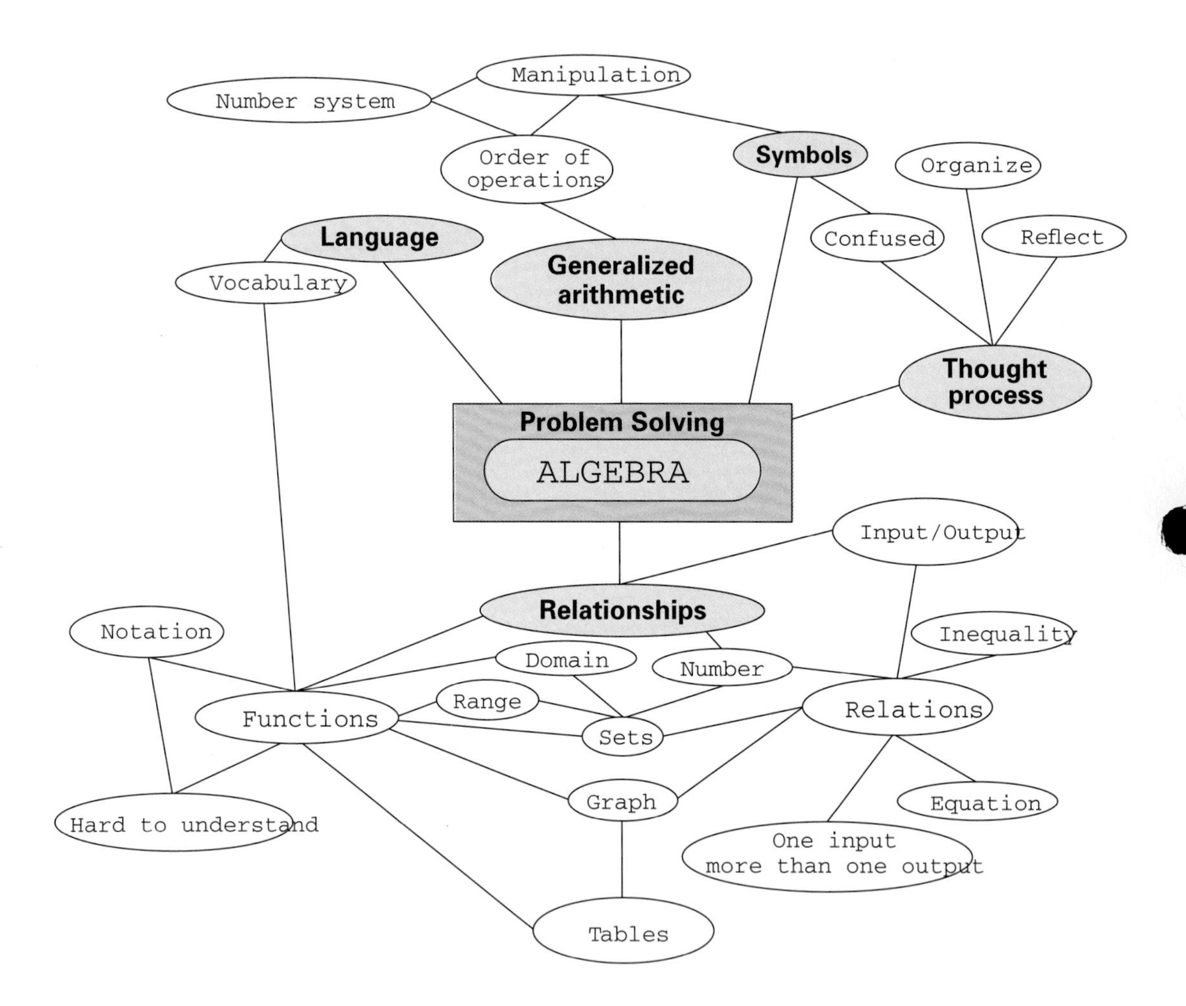

Sandy: Once you have listed your ideas, arranged the self-stick notes, and drawn your Concept Map, it will probably look very different from mine. Remember that no two maps will look exactly the same, even if the list of ideas is identical, because we all think about things a little differently and make different connections.

CONCEPT MAP

Construct a Concept Map centered on one of the following:

a. statistic

b. order of operations

c. function

d. measures of central tendency

e. mathematical notation

f. graph

REFLECTION

Write a paragraph describing the most interesting mathematical idea in this chapter. Include sentences that begin with the phrases:

Before studying this chapter, I used to think....

After this section, I now think....

because....

2

Constant and Accelerated Change

SECTION 2.1

Rates of Change

Purpose

- Explore rates of change for paired data presented graphically.
- Explore rates of change for paired data presented numerically.
- Explore rates of change for a function presented symbolically.

One focus in Chapter 1 that continues in this chapter is the mathematical description of data. Initially in Chapter 1, we discussed single numbers that could be used to describe a list of data. In Chapter 2, we look at paired lists of data and attempt to analyze the relationship between the pairs. One list of data provides inputs and the other list provides outputs.

Most of the relations and functions we studied in Chapter 1 used lists as inputs. The functions we study in Chapter 2 will have single numbers as inputs. Many real-life functions are functions of time. This means that a time value is the input to the function. Some quantity that depends on the time value is the output. The table of values for **Carrie's Job Offer** from Chapter 1 is an example of such a function. The input is the years that Carrie worked at the job. The output is Carrie's salary during that year. We will investigate many such functions in this chapter.

Remember that relations and functions provide a way to measure mathematically how change occurs. Several common types of functions occur naturally. This causes us to focus on developing an understanding of certain major classes of functions. We looked at one such class in Chapter 1—statistics is a class of relations and functions. We will begin to study two other major classes in this chapter.

One important idea that distinguishes one class of functions from another is ***rate of change***, the ratio of the change in one variable with respect to change in another variable. A common example of rate of change is speed, which is the ratio of change in distance to change in time. Before we look at some new classes of functions, we focus on the concept of rate of change.

Barb's Bicycle Ride: What a beautiful day! Barb left home on her bicycle at 8 A.M. intent on enjoying the day to its fullest. The relationship between the time of day (input) and Barb's distance from home (output) at that time are marked as points in Figure 1.

FIGURE 1

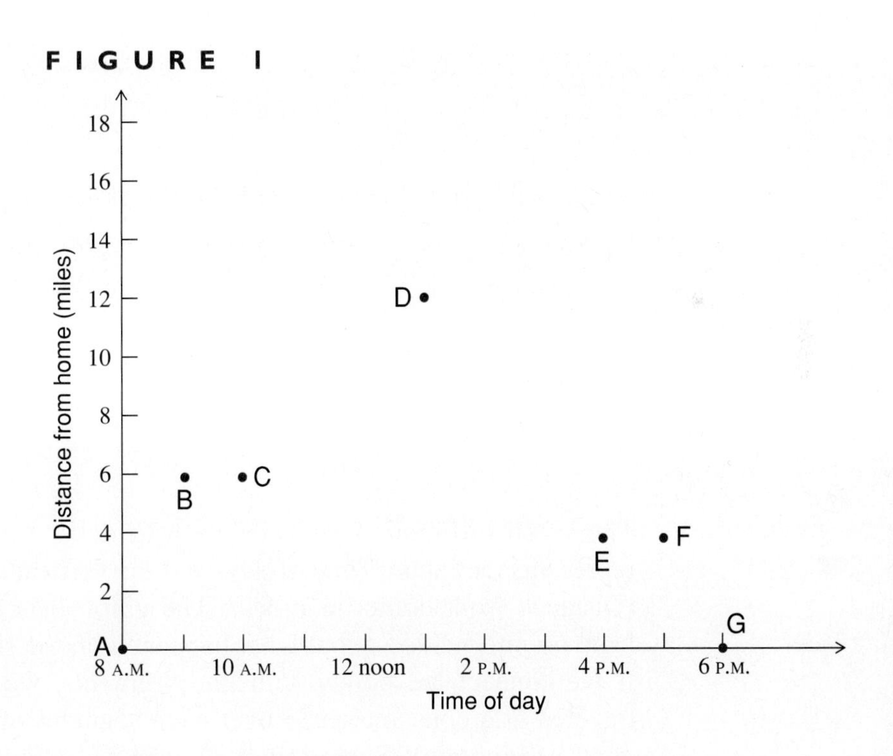

1. Write a brief story describing the graph in terms of Barb's travels.

2. Identify two consecutive points when Barb's ride was the

a. easiest. Why?

b. hardest. Why?

3. What is different about Barb's ride between 10 A.M. and 1 P.M. as compared to her ride between 1 P.M. and 4 P.M.?

4. How did the outputs change from

a. 8 A.M. to 9 A.M.?

b. 10 A.M. to 1 P.M.?

c. 1 P.M. to 4 P.M.?

5. How would you calculate Barb's average speed from

a. 8 to 9 A.M.?

b. 10 A.M. to 1 P.M.?

The graph in Figure 1 provides us with data visually. The horizontal axis represents the input (Time of day) and the vertical axis represents the output (Distance from home, in miles). The graph lists only discrete points. We have no information about what happened on the ride between these points. If we connect the points with line segments, we are assuming that Barb traveled at a constant speed over each segment of the trip (probably not a good assumption). Some things we can say about Barb's trip follow. She increased her distance from home between 8 and 9 A.M. She was the same distance from home at 9 and 10 A.M. She increased her distance from home between 10 A.M. and 1 P.M., then moved toward home until 4 P.M. Barb was the same distance from home at 4 and 5 P.M., and then returned home at 5 and 6 P.M.

We need to discuss the idea of finite differences to address Investigations 4 and 5. A ***finite difference*** is an amount and direction of change in two different values of a variable.

To compute the finite differences of time, subtract two values of time. To compute the finite differences of distance from home, subtract two values of distance. Be sure to *subtract the values in reverse order of occurrence*: second value minus first value.

Let's introduce two variables for Barb's ride. Let t represent the time of day. Thus, t is the variable for input. Let d represent the distance from home in miles. Thus, d is the variable for output.

If t represents time, a finite difference of time is represented by the symbol Δt (the symbol Δ (delta) is used to represent "change in"). Similarly, if d represents distance from home in miles, a finite difference of distance from home is represented by the symbol Δd.

For example, consider point A represented by the ordered pair (8, 0) and point B represented by (9, 6).

The finite difference in times is

$$\Delta t = 9 - 8 = 1$$

The finite difference in distances is

$$\Delta d = 6 - 0 = 6$$

Investigation 4 asks how the outputs change between two given times. We are asked for Δd in each case.

Between 10 A.M. and 1 P.M., $\Delta d = 12 - 6 = 6$.

Between 1 P.M. and 4 P.M., $\Delta d = 4 - 12 = -8$.

The ratio of change in distance to change in time provides information about Barb's average speed. Just knowing the change in time or the change in distance does not give a complete picture of how fast Barb was traveling. Notice that Δt represents a change in input and Δd represents a change in output. The ratio $\frac{\Delta d}{\Delta t}$ represents

$$\textbf{\textit{rate of change}} = \frac{\text{Change in output}}{\text{Change in input}}.$$

Points A through G provide us with data points. In this chapter, we look at paired data—that is, data in which there is a relationship between two quantities. An ultimate goal is to create a ***mathematical model***—a function that describes the data symbolically and that can be used to answer questions about the data. Before doing this, let's investigate the relationship between changes in the input quantity (time of day) and changes in the output quantity (distance from home). We will make comparisons using finite differences by subtracting values of distance from home and values of time. The ratio of the differences provides a measure of how Barb's distance from home is changing with respect to time.

6. Complete Table 1 by recording the values from Figure 1 for time (the first coordinate of the ordered pairs) and for distance from home (the second coordinate of the ordered pairs). Compute the finite differences for the column headed time and the column headed distance from home. The first two are already done as examples.

TABLE 1 Finite Differences in Time and Distance for Barb's Trip

Label	Finite Difference, Δt	Input Time of Day, t	Output Distance *d* from Home, (miles)	Finite Difference, Δd
A	9 – 8 = 1	8	0	6 – 0 = 6
B	10 – 9 = 1	9	6	6 – 6 = 0
C		10	6	
D				
E				
F				
G				

7. Complete Table 2 on page 87 by finding the rate of change $\frac{\Delta d}{\Delta t}$ for each finite difference in Table 1. The first one is an example.

8. Interpret in words the meaning of $\frac{\Delta d}{\Delta t}$ with respect to Barb's trip.

TABLE 2 Rates of Change in Distance to Change in Time

Endpoints	Δt	Δd	Rate of Change, $\frac{\Delta d}{\Delta t}$
A to B	1	6	$\frac{6}{1} = 6$
B to C			
C to D			
D to E			
E to F			
F to G			

9. Interpret in words what Barb is doing when the ratio $\frac{\Delta d}{\Delta t}$ is

a. positive.

b. negative.

c. zero.

10. Write a note to your friend Izzy describing the procedure for finding the rate of change between two data points.

11. Using the graph of Figure 1 on page 83, decide if the given assumption is valid. Write a statement to justify your answer.

a. The trip from A to B is uphill.

b. By traveling from D to E, Barb moves closer to home.

c. The trip from D to E is downhill.

d. Barb is traveling at a slower speed from F to G than from D to E.

12. What assumptions about speed are we making when we connect the labeled points in Figure 1 with line segments? Are these assumptions reasonable in context of this problem?

You investigated the change in each dimension as you moved from one point to another for Barb's trip. The rates of change from Table 2 are recorded on Figure 2.

FIGURE 2

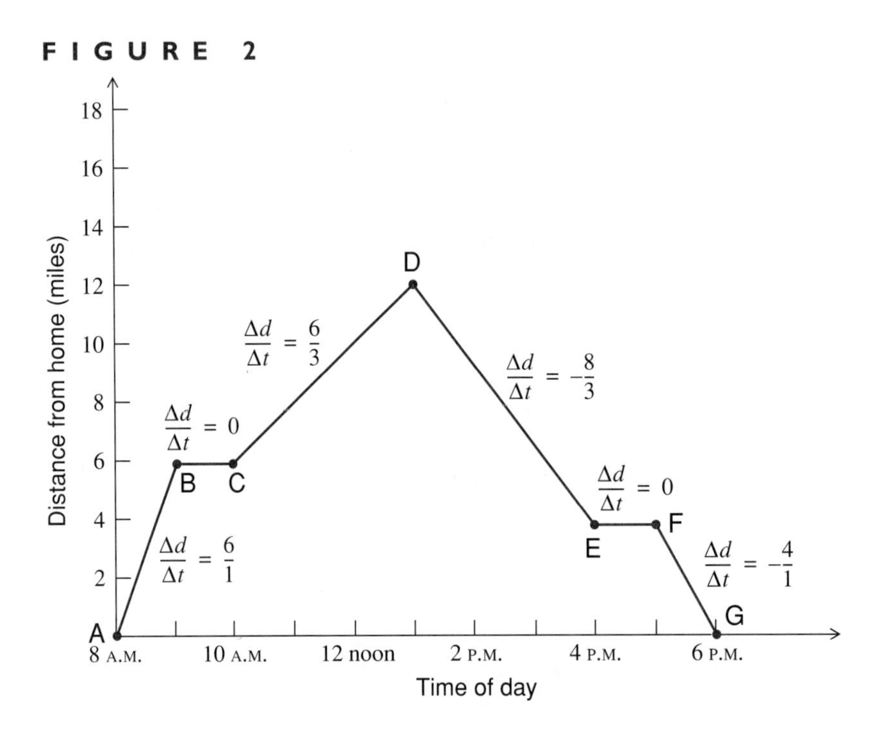

Line segments have been drawn between successive points. Drawing line segments implies two implicit assumptions.

First, we assume that the inputs (time of day) and the outputs (distance from home) are ***continuous*** quantities—that is, they may take on all values between the whole number quantities that correspond to the labeled points. This assumption seems reasonable in this problem.

A second assumption is that Barb rides at a steady speed between labeled points. This most likely is not true, so the use of line segments to connect points is a way of representing Barb's average speed between points, but not her actual speed at a specific location between the points. For example, between any two points Barb might stop at a stop sign and then pedal hard to get up to a comfortable speed.

The change in time, Δt, varies from 1 hour to 3 hours while Barb is moving between points.

The change in distance from home, Δd, varies from –8 to 6 miles while she is moving from point to point.

It is necessary to consider both changes to measure Barb's velocity (change in distance as compared to change in time).

Average velocity $\frac{\Delta d}{\Delta t}$ represents the rate of change of distance from home with respect to change in time. Ignoring the sign, the larger this value, the faster Barb's pace. The smaller the value, the slower her pace. If the ratio is positive, she is moving away from home. If the ratio is negative, she is moving toward home.

Slope is the ratio of change in output to the change in input along a straight line. Slope is a special case of rate of change. Each rate of change is the slope of the line segment connecting the two data points. For example, the slope of the line segment connecting points D and E is $-\frac{8}{3}$. The word "slope" is used when the rate of change is a constant.

Rate of change can be positive, zero, or negative. Let's interpret each in terms of Barb's ride.

A ***positive rate of change*** means that, as the input increases, the output increases. For Barb's ride, a positive rate of change indicates that, as the time increases, Barb's distance from home increases. This occurs, for example, from 8 to 9 A.M., when the rate of change is 6 miles per hour.

A ***negative rate of change*** means that, as the input increases, the output decreases. For Barb's ride, a negative rate of change indicates that, as the time increases, Barb's distance from home decreases. This occurs, for example, from 1 to 4 P.M., when the rate of change is $-\frac{8}{3}$ miles per hour.

A ***zero rate of change*** means that, as the input increases, the output remains the same. For Barb's ride, a zero rate of change indicates that, as the time increases, Barb's distance from home remains the same. For example, from 9 to 10 A.M., the rate of change is 0 miles per hour.

Investigation 11 asks about assumptions that can be drawn from the graph. It is impossible to make any assumptions about whether Barb is riding uphill or downhill during any time interval. The graph indicates Barb's distance from home at a given time, but gives no information about the terrain. However, the graph does indicate that Barb is moving closer to home from D to E. This is exactly the information that the vertical axis tells us. Barb's rate of change from F to G indicates a faster average speed than that from D to E.

Summarizing, given two input–output pairs, we find the average rate of change between these two points by dividing the difference of the outputs by the difference of the inputs. It is important that the order of subtraction of the outputs and the order of the subtraction of the inputs be the same. Otherwise, a sign error will result.

We use rate of change as a way to understand the behavior of the outputs of a function as the inputs increase. For Barb's Bicycle Ride, we began with a graph that we interpreted as a function. We recorded the information in a table and calculated rates of change from the table. We could also begin with a function defined by an equation. Let's investigate rate of change when a function is given symbolically.

The Cold Front: A cold front arrived at midnight. The temperature is T degrees Fahrenheit n hours after midnight. This means that n is 0 at midnight. It takes 12 hours for the cold front to pass. If n is the input and T is the output, the relationship is given by $T(n) = 0.1(400 - 40n + n^2)$.

13. What is the domain of the temperature function? That is, what are the legal values for n based on the problem restrictions?

14. Find Δn from

a. 1 A.M. to 4 A.M.

b. 3 A.M. to 8 A.M.

c. midnight to noon.

15. Find ΔT from

a. 1 A.M. to 4 A.M.

b. 3 A.M. to 8 A.M.

c. midnight to noon.

16. Find the average rate of change $\frac{\Delta T}{\Delta n}$ when n changes from

a. 1 A.M. to 4 A.M.

b. 3 A.M. to 8 A.M.

c. midnight to noon.

17. Interpret the meaning of $\frac{\Delta T}{\Delta n}$ in this problem situation.

Since it takes 12 hours for the cold front to pass, the function can be used as a model from midnight until noon. This means the input n is restricted to values between 0 and 12 inclusive. Since time is essentially a continuous quantity, n can take on any real number value between 0 and 12 inclusive.

To answer Investigation 14, we need to subtract values of n. To answer Investigation 15, we need to subtract values of T. These values must be computed using the equation and substitution (Figure 3).

FIGURE 3

X	Y1
0	40
1	36.1
3	28.9
4	25.6
8	14.4
12	6.4

Y1=.1(400−40X+X…

Table 3 displays the finite differences in inputs n and outputs T over the time intervals requested in the investigation. The rates of change are calculated in the rightmost column.

The rates of change in Table 3 measure the average rate at which the temperature is changing per hour. The negative in each rate indicates that the temperature is dropping as time passes. (I guess that's why it's a cold front.)

For example, from 1 to 4 A.M. the temperature fell at a rate of 3.5° Fahrenheit per hour. From 3 to 8 A.M., the temperature fell at a rate of 2.9° Fahrenheit per hour. Since these two rates are different, the temperature is not falling at a constant rate. In fact, the temperature fell faster from 1 to 4 A.M. than from 3 to 8 A.M.

TABLE 3 Selected Rates of Change for Cold Front Problem

Time Interval	Change in Time, Δn	Change in Temperature, ΔT	Rate of Change in Temperature with Respect to Time, $\frac{\Delta T}{\Delta n}$
1 to 4	3	$25.6 - 36.1 = -10.5$	$\frac{-10.5}{3} = -3.5$
3 to 8	5	$14.4 - 28.9 = -14.5$	$\frac{-14.5}{5} = -2.9$
midnight to noon	12	$6.4 - 40 = -33.6$	$\frac{-33.6}{12} = -2.8$

Before concluding the section, let's look at one more example of computing rates of change for functions.

18. Consider the function $f(x) = \frac{3}{2}x - 5$.

a. Complete Table 4.

TABLE 4 Numeric Values of $f(x)$

x	$f(x)$
−8	
6	
10	
−2	
4	

b. Compute the rate of change for each successive pair of points in Table 4. What do you notice?

c. Look at the equation that defines the function. Do you see the rate of change in the equation? Where?

Given $f(x) = \frac{3}{2}x - 5$, the rate of change between any two points is always $\frac{3}{2}$. This is an example of a function with a *constant rate of change*. The constant rate of change $\frac{3}{2}$ is called the *slope*. In this case, the slope is the coefficient of the x term. We will consider this important case in more detail later in the chapter.

The rate of change of output per unit change in input is one of the key concepts in mathematics. We will use change in inputs, change in outputs, and rates of change to create several important classes of functions from data. Before we end the section, consider the following idea. We measured average rates of change in temperature per hour in the **Cold Front** problem. What if we wanted to know how fast the temperature was changing at a given time, say 5 A.M.? Now we want to know how the temperature is changing at an instant rather than over a period of time. How would you compute such a value? The precise answer is found when you study calculus. However, there is a way to approximate the answer using the ideas of this section. One of the explorations gives you a chance to explore this idea. Watch for it.

STUDENT DISCOURSE

Pete: A real basic thing confuses me—what is meant be the term "rate"?

Sandy: Good question. I had some trouble with that myself. Basically, rates come from comparing two quantities using division. Often the quantities have different units.

Pete: What do you mean?

Sandy: Well, you probably deal with rate every day when you talk about the speed you drive.

Pete: You mean like 50 miles per hour?

Sandy: That's right. The 50 is a rate formed by dividing the number of miles by the number of hours. Notice that the units—miles and hours—are different in this case.

Pete: Does the word "per" indicate that a division was performed?

Sandy: In a way.

Pete: Okay, when I computed rates of change in this section, I did divisions. For the bicycle ride, the division was distance from home divided by time. In the **Cold Front** problem, the division involved temperature and time. Where does the word "change" fit into this?

Sandy: Well, you didn't divide distance from home by time or temperature by time, did you?

Pete: No, I really divided finite differences.

Sandy: And what do finite differences measure?

Pete: Hmmm. I have it—change. So, when computing a rate of change, I am dividing a change in one thing by a change in the other.

Sandy: In particular, it's change in output divided by change in input.

Pete: Thanks, that helps me understand the words better. Now that I know what we are talking about, it makes a lot more sense.

1. In the Glossary, define the words in this section that appear in ***italic boldface*** type.

2. Take one segment of the graph in Figure 1. Describe what Barb's trip might have been like along this segment. Draw a more realistic graph of this segment.

3. Selling Costume Jewelry: Remember Erratic Ernie and Steady Sue? The last we looked, they were selling costume jewelry in which their net profit was defined by the function $P(n) = 5n - 100$, where P dollars of profit was realized when n pieces of jewelry were sold.

a. Find the rate of change in net profit per unit change in the number of pieces of jewelry sold when n changes from 10 pieces to 15 pieces.

b. Find the rate of change in net profit per unit change in the number of pieces of jewelry sold when n changes from 15 pieces to 16 pieces.

c. Regardless of the change in n, what is the rate of change of P per unit change in n? Does this value for rate of change appear in the equation? Where?

4. When finding the finite differences in time, arithmetic on a 12-hour clock must be performed (this is called ***modular arithmetic***).

a. When you find Δt from 12 noon to 4 P.M., what adjustments must you make?

b. When you find Δt from 8 A.M. to 10 P.M., what adjustments must you make?

5. Campus Vehicle Congestion: Recall the survey of cars leaving campus during certain times of day. The data appear in Table 5.

TABLE 5 Cars Leaving Campus by Time of Day

Label	Time of Day	Number of Cars Leaving Campus
A	8 A.M.	159
B	9 A.M.	341
C	10 A.M.	467
D	11 A.M.	537
E	12 noon	551
F	1 P.M.	509
G	2 P.M.	411
H	3 P.M.	257

Find the rate of change in the number of cars leaving campus at a given hour when the time changes from

a. 8 to 9 A.M.

b. 9 A.M. to 12 noon

c. 1 to 3 P.M.

6. The Cold Front: Recall the function $T(t) = 0.1(400 - 40t + t^2)$ expressing temperature as a function of time. Let's approximate how fast the temperature is changing at 5 A.M. While we can't find this exactly, we can approximate it by looking at the rate of change as time changes from 5 A.M. to 5 + (some very small time value). Calculate $\frac{\Delta T}{\Delta t}$ where time changes from 5 to

a. 5.1

b. 5.01

c. 5.001

d. Based on your approximations, how fast is the temperature dropping at 5 A.M.?

7. Any two distinct points define a line segment. We can find the slope of that line segment by calculating the rate of change between the two points. Find the slope of the line segment with endpoints

a. (4, –5) and (11, 6)

b. (–3, 0) and (–5, –7)

c. (13, 11) and (–24, 9)

d. (–5, 8) and (4, 8)

e. (–5, 8) and (–5, 4)

8. Credit Hour Fees: The fee structure in Table 6 on page 97 represents the total fees for a college student as determined by the number of credit hours the student takes.

Find the ratio of the change in total fees to the change in credit hours when the credit hours change from

a. 1 to 4. Write a sentence interpreting your answer.

b. 7 to 12. Write a sentence interpreting your answer.

c. 5 to 6. Write a sentence interpreting your answer.

9. Determine the rate of change for the following functions.

a. $g(x) = 3x - 2$

b. $h(x) = \frac{7}{3}x + 5$

c. $p(t) = -2.8t - 1.5$

d. $k(q) = \frac{2}{3} - \frac{5}{9}q$

TABLE 6 Total Fees as a Function of Credit Hours

Credit Hours	Total Fees ($)
1	57
2	89
3	121
4	153
5	185
6	217
7	249
8	281
9	313
10	345
11	377
12	409

10. Consider the graph of three lines given in Figure 4. Two of the lines are parallel and the other is perpendicular to the two parallel lines.

FIGURE 4

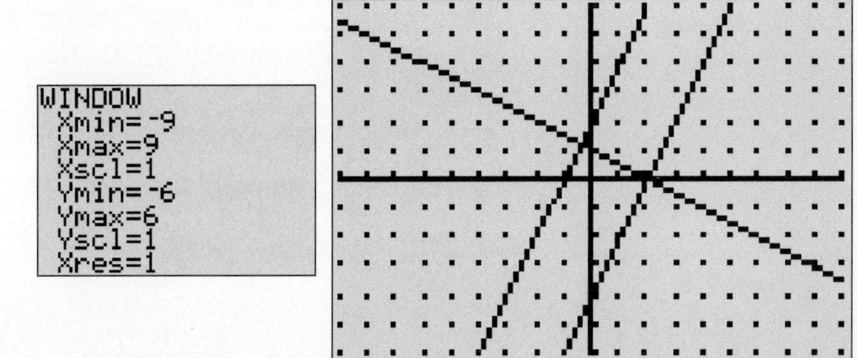

a. Find the slope of each line.

b. How do the slopes of the parallel lines compare?

c. How does the slope of the perpendicular line compare to the slope of the parallel lines?

11. Create a table for the function $f(x) = 6x^2 - x + 1$. Find the rates of change of f with respect to x as x increases from

a. 1 to 2

b. 2 to 3

c. 3 to 4

d. 4 to 5

e. 1 to 5

REFLECTION

Write a paragraph describing your understanding of *rate of change*. Specifically address what it means when a rate of change is positive, when a rate of change is negative, and when a rate of change is zero. Also discuss what it might mean when a rate of change is constant.

SECTION 2.2

Number Sequences

Purpose

- Explore number sequences as functions.
- Introduce arithmetic sequences.
- Introduce geometric sequences.

In Section 2.1, we investigated the general idea of rates of change. Finite differences are the key to finding rates of change. We continue studying the idea of change in this section by looking at number sequences and how they change from one number to the next number in the sequence. In addition to using finite differences to measure the change, we will use finite ratios between successive terms—finite ratios are found by dividing successive elements. Recall that we found finite differences by subtracting successive elements.

We have used the symbol Δ to mean "change in." The symbol Δ represents a finite difference and is computed using subtraction.

We will use the symbol ® to mean "ratio between." Thus the symbol ® represents a finite ratio and is computed using division.[1]

To investigate these ideas, let's explore the details of three different salary offers for a full-time job. We think of each as a function of time. The input is the number of years worked and the output is the salary for that year.

CDs-R-Us: The starting annual salary is $35,000 and you will receive a fixed raise of $2000 annually.

Super Dog Burger Bits: The starting annual salary is $25,000 and you will receive a 7% raise annually.

Virtual Reality Experiences: The starting annual salary is $30,000. You receive a fixed annual raise of $2000 on even years and a 7% annual raise on odd years.

1. At CDs-R-Us, the salary is $35,000 for year 1 and increases by $2000 annually.

a. Complete Table 1.

TABLE 1 CDs-R-Us Salaries

Year	1					
Salary ($)	35,000					

1. We use the symbol ® for finite ratio with permission from Jere Confrey, who first suggested the notation.

b. Explain to your friend, Izzy, how to compute the year's salary from the previous year's salary.

2. At Super Dog Burger Bits, the salary is \$25,000 for year 1 and increases by 7% annually.

a. Complete Table 2.

TABLE 2 **Super Dog Burger Bits Salaries**

Year	1					
Salary (\$)	25,000					

b. Explain to Izzy how to compute the year's salary from the previous year's salary.

3. At Virtual Reality Experiences, the salary is \$30,000 for year 1 and increases by \$2000 in even years and by 7% in odd years.

a. Complete Table 3.

TABLE 3 **Virtual Reality Experiences Salaries**

Year	1					
Salary (\$)	30,000					

b. Explain to Izzy how to compute the year's salary from the previous year's salary.

Each ordered list of salaries (the outputs in Tables 1–3) is an example of a ***sequence***. The ***elements*** of a sequence are the individual items in the list. The first six elements of these sequences are:

CDs-R-Us: 35,000, 37,000, 39,000, 41,000, 43,000, 45,000.

Super Dog Burger Bits: 25,000, 26,750, 28,622.50, 30,626.075, 32,769.90025, 35,063.79327.

Virtual Reality Experiences: 30,000, 32,000, 34,240, 36,240, 38,776.80, 40,776.80.

The processes used to generate each sequence might be as follows.

CDs-R-Us: Add 2000 to the salary from the previous year. For example, the salary in year 1 is 35,000. The salary in year 2 is found by computing 35,000 + 2,000.

Super Dog Burger Bits: Multiply the salary from the previous year by 0.07. Add the product to the salary from the previous year. For example, the salary in year 1 is 25,000. The salary in year 2 is found by first computing (25,000)(0.07) = 1750. Add this product to the salary from previous year to get 25,000 + 1750 = 26,750.

How could we make this a one-step process? Let's look at the processes again. The new salary is

$$\begin{aligned} 25{,}000 + 25{,}000(0.07) &= 25{,}000(1) + 25{,}000(0.07) \\ &= 25{,}000(1 + 0.07) \\ &= 25{,}000(1.07). \end{aligned}$$

We factored 25,000 from the two terms to move from the first step to the second step. It appears that a salary can be obtained by multiplying the previous year's salary by 1.07. Check it out. This is a shorter one-step process that is equivalent to the two-step process stated initially.

Virtual Reality Experiences: If the previous year number is odd, add 2000 to the salary from the previous year. If the previous year number is even, multiply the salary from the previous year by 1.07. (This is reworded so the salary used in the computation matches year.)

Since these lists of salaries are ordered, we can refer to a sequence element (a salary) by stating the element's position (the number of years). For example, the third element (year 3) of the sequence corresponding to CDs-R-Us salaries is $39,000.

Sequences can be thought of as functions where the input is the position of an element and the output is the sequence element in that position (Figure 1).

FIGURE 1

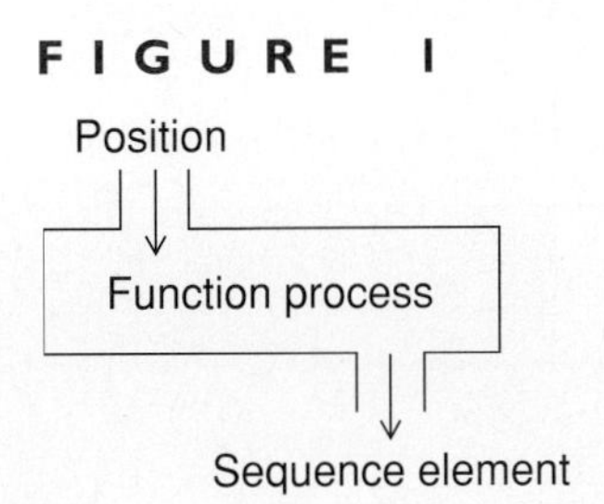

For the salary offers, the position corresponds to the year and the sequence element corresponds to the salary during that year. For example, at CDs-R-Us, for input (position or year) 3, the output is 39,000 (the sequence element or salary).

Since the position must be a natural number 1, 2, 3, 4, . . . , the ***domain of any sequence of numbers*** is the set of ***natural numbers***.

These ideas provide a good definition for a sequence. A ***sequence*** is a function whose domain is the natural numbers.

Let's investigate the functions that correspond to each salary offer. The input is the year and the output is the salary during that year. Understanding the behavior of these functions is closely tied to finding finite differences and finite ratios between successive outputs. Let's take a look. Use the variable n for year number (input) and S for salary (output) in dollars.

4. For each salary offer, complete the table. Next to the years (inputs), record the finite differences in successive inputs. Next to the salaries (outputs), record the finite differences in successive outputs and the finite ratios of the outputs. The first two are done for you.

a. CDs-R-Us: Complete Table 4.

TABLE 4 **CDs-R-Us Salaries**

Finite Differences of Years, Δn	Year, n	Salary, S	Finite Differences of Salaries, ΔS	Finite Ratios of Salaries, ®S
2 − 1 = 1	1	35,000	37,000 − 35,000 = 2000	$\frac{37,000}{35,000} \approx 1.057$
3 − 2 = 1	2	37,000	39,000 − 37,000 = 2000	$\frac{39,000}{37,000} \approx 1.054$
	3	39,000		
	4			
	5			
	6			

b. Super Dog Burger Bits: Complete Table 5.

TABLE 5 Super Dog Burger Bits Salaries

Finite Differences of Years, Δn	Year, n	Salary, S	Finite Differences of Salaries, ΔS	Finite Ratios of Salaries, ®S
2 − 1 = 1	1	25,000	26,750 − 25,000 = 1750	$\frac{26{,}750}{25{,}000} = 1.07$
	2	26,750		
	3			
	4			
	5			
	6			

c. Virtual Reality Experiences: Complete Table 6.

TABLE 6 Virtual Reality Experiences Salaries

Finite Differences of Years, Δn	Year, n	Salary, S	Finite Differences of Salaries, ΔS	Finite Ratios of Salaries, ®S
2 − 1 = 1	1	30,000	32,000 − 30,000 = 2000	$\frac{32{,}000}{30{,}000} \approx 1.067$
	2	32,000		
	3			
	4			
	5			
	6			

5. For each salary offer, complete the grid below, indicating whether the finite differences or ratios were or were not constant. One of them is done for you.

	Finite Differences in the Inputs	Finite Differences in the Outputs	Finite Ratios in the Outputs
CDs-R-Us	Constant		
Super Dog Burger Bits			
Virtual Reality Experiences			

Since the inputs to a sequence are natural numbers, the finite differences in successive inputs (successive years) are a constant equal to 1.

Thus, in the first column of Tables 4–6, $\Delta n = 1$, where n is the year number.

To analyze data in terms of finite differences in outputs or finite ratios in outputs, *it is important that the change in input be a constant.*

The patterns in the finite differences and the finite ratios for each salary offer are as follows.

CDs-R-Us: $\Delta S = 2000$ and ®S varies. The finite differences in outputs are constant and the finite ratios in outputs are not constant.

Super Dog Burger Bits: ΔS varies and ®$S = 1.07$. The finite differences in outputs are not constant and the finite ratios in outputs are constant.

Virtual Reality Experiences: Both ΔS and ®S vary. The finite differences in outputs are not constant and the finite ratios in outputs are not constant.

The cases in which either ΔS or ®S are constant will be our focus. These cases lead to two special types of sequences called ***arithmetic sequence*** and ***geometric sequence.***

If a sequence has finite **differences** in outputs which are constant, the sequence is called ***arithmetic***. This constant finite difference is called the ***common difference.***

If a sequence has finite ***ratios*** in outputs which are constant the sequence is called ***geometric***. This constant finite ratio is called the ***common ratio.***

Let's look at the three salary offers in light of these new ideas. In the process, we can develop some notation to represent the relationship between sequence elements.

6. Which of the salary offers is represented by

a. an arithmetic sequence? Why? What is the common difference?

b. a geometric sequence? Why? What is the common ratio?

7. Let a function named *CD* represent the sequence defined by the CDs-R-Us salary offer. So $CD(1)$ is the salary in year 1 or $CD(1) = 35{,}000$.

a. What is the value of $CD(4)$?

b. Write the function notation for the element preceding $CD(4)$.

c. What is $CD(4) - CD(3)$?

d. What is $CD(20) - CD(19)$?

e. What is $CD(n) - CD(n-1)$? What does this mathematical statement represent in terms of salaries?

f. Use your answer to **e** to complete the following mathematical statement: $CD(n) = CD(n-1) + \square$.

8. Let a function named *SD* represent the sequence defined by the Super Dog Burger Bits Salary offer. So $SD(1)$ is the salary in year 1 or $SD(1) = 25{,}000$

a. What is the value of $SD(3)$?

b. Write the function notation for the element preceding $SD(3)$.

c. What is $\frac{SD(3)}{SD(2)}$?

d. What is $\frac{SD(20)}{SD(19)}$?

e. What is $\frac{SD(n)}{SD(n-1)}$? What does this mathematical statement represent in terms of salaries?

f. Use your answer to **e** to complete the following mathematical statement: $SD(n) = SD(n-1) \times \boxed{}$.

The previous investigation was designed to help you state explicitly, using algebraic notation, the relationship between elements of both arithmetic and geometric sequences.

Since the finite differences in outputs for CDs-R-Us salaries are always 2000, the CDs-R-Us salaries form an arithmetic sequence with $CD(1) = 35{,}000$ and common difference $d = 2000$.

Since the finite ratios in outputs for Super Dog Burger Bits salaries are always 1.07, the Super Dog Burger Bits salaries form a geometric sequence with $SD(1) = 25{,}000$ and common ratio $r = 1.07$.

Neither the finite differences in output nor the finite ratios in output for Virtual Reality Experiences salaries are constant. The sequence for these salaries is neither arithmetic nor geometric.

Let's focus on the salaries at CDs-R-Us and Super Dog Burger Bits. Use the function *CD* to represent the sequence defined by CDs-R-Us salaries and the function *SD* to represent the sequence defined by Super Dog Burger Bits salaries.

For CDs-R-Us, a salary is computed from the previous year's salary by adding 2000 to the previous year's salary.

Symbolically, $CD(n) = CD(n-1) + 2000$. Note that if n is a given year, then $n-1$ is the year before.

The sequence given by CDs-R-Us salaries is generalized by

$$CD(1) = 35{,}000$$

and

$$CD(n) = CD(n-1) + 2000, \quad n = 2, 3, 4, \ldots.$$

This formula is an example of a ***recursive definition of the sequence***. Recursive means that an element (salary) is computed using a previous element (previous year's salary).

Figure 2 contains a function machine representation for CDs-R-Us salaries.

FIGURE 2

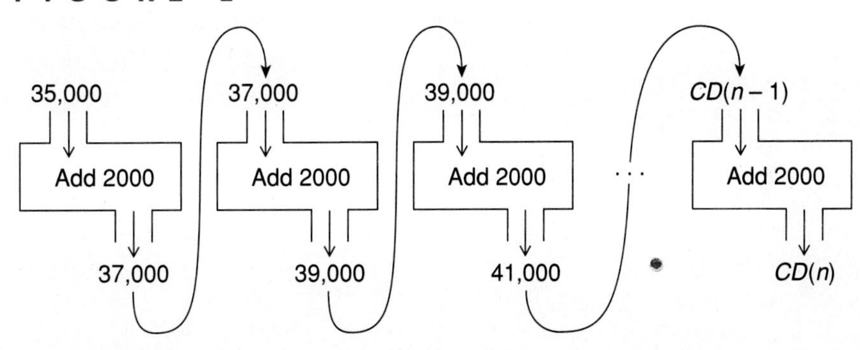

For Super Dog Burger Bits, a salary is computed from the previous year's salary by multiplying the previous year's salary by 1.07.

Symbolically, $SD(n) = SD(n-1) \times 1.07$. If n is a given year, then $n-1$ is the year before.

The sequence given by Super Dog Burger Bits salaries is generalized by

$$SD(1) = 25{,}000$$

and

$$SD(n) = SD(n-1) \times 1.07, \quad n = 2, 3, 4, \ldots$$

This formula is another example of a recursive definition of a sequence.

Figure 3 contains a function machine representation for Super Dog Burger Bits salaries.

FIGURE 3

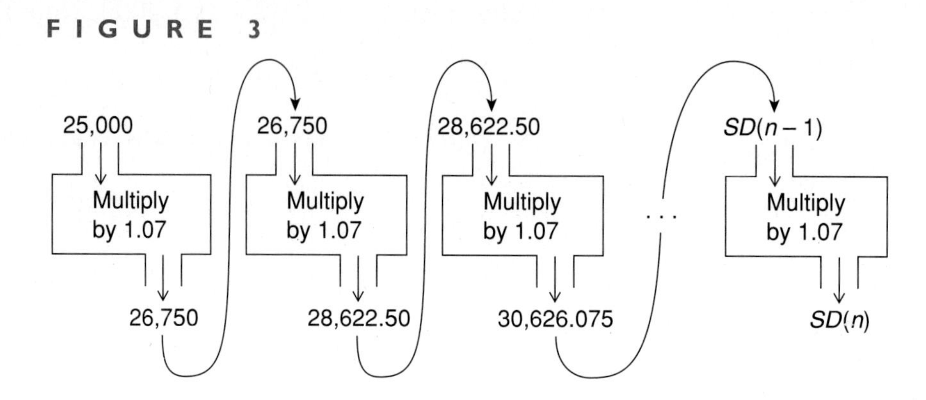

The recursive formulas tell us how to compute a salary from the previous year's salary. This is inconvenient since finding the 30th year salary requires that we compute the salaries for the previous 29 years. We'd like to find a method of computing the salary from the year rather than from the previous salary. We focus on this in the next section.

Before closing this section, we can use ***sequence mode on the calculator*** to both generate and graph sequences. This mode can be used to generate the elements of either an arithmetic or geometric sequence.

9. Use the sequence mode to generate the first 20 elements of the sequence represented by

a. CDs-R-Us salaries.

b. Super Dog Burger Bits salaries.

10. Use your calculator to graph the sequence generated in

a. Investigation 9a.

b. Investigation 9b.

11. Compare and contrast the two graphs obtained in Investigation 10.

In the *sequence* mode on your calculator, a letter, such as *u*, *v*, or *w*, is used in place of the function name. So $u(n)$ represents the sequence element in position n. Some calculators use subscript notation instead of function notation. In ***subscript notation***, a ***subscript*** is used for the position. Thus u_n and v_n represent elements of the sequence (in this situation, they represent salaries). In the following discussion, use letter u in place of CD and letter v in place of SD.

For example, $CD(4)$ may be expressed as either $u(4)$ or u_4.

The sequence defined by CDs-R-Us salaries is defined recursively as

$$u(1) = 35{,}000$$

and

$$u(n) = u(n-1) + 2000.$$

Remember this is function notation and does not mean multiplication of u and $(n-1)$, which may be why some calculators use subscripts.

The function definition, the graph of the first 20 elements, and the window setting appear in Figure 4.

FIGURE 4

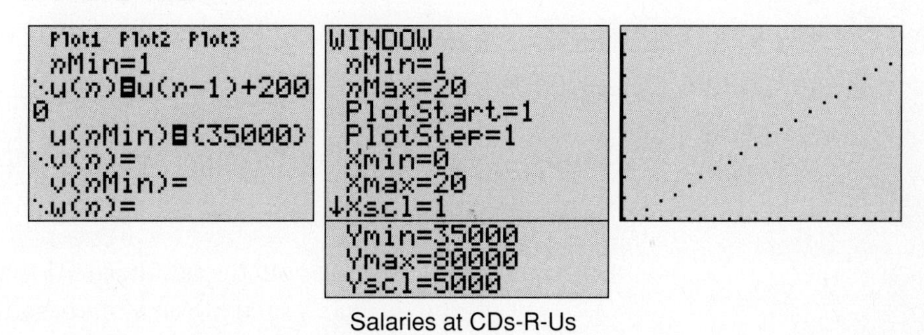

Salaries at CDs-R-Us

Notice that the sequence elements appear to lie in a straight line consistent with the idea of a constant rate of change.

The sequence defined by Super Dog Burger Bits salaries is defined recursively as

$$v(1) = 25{,}000$$

and

$$v(n) = v(n-1) \times 1.07 .$$

The function definition, the graph of the first 20 elements, and the window setting appear in Figure 5.

FIGURE 5

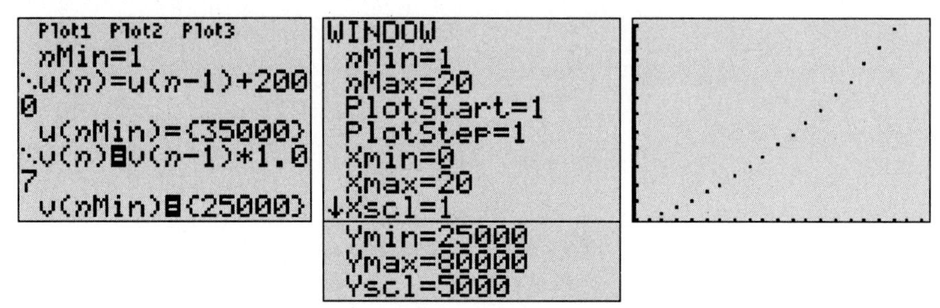

Salaries at Super Dog Burger Bits

Notice that the graph of the sequence elements appears to curve upward. This is consistent with the idea of an increasing rate of change.

Pete: I have to ask a question about Super Dog Burger Bits salaries.

Sandy: Fire away.

Pete: I found the salary for a given year using two steps: multiply by 0.07 and then add the previous year's salary to the product. Isn't that correct?

Sandy: Sure.

Pete: But then where did the 1.07 come from?

Sandy: Well, what does 0.07 times the previous year's salary represent?

Pete: The amount of the raise.

Sandy: Right. Think about the number 1.07 that we multiplied by the previous year's salary. You already said that the 0.07 represented the raise. What do you think the 1 represents?

Pete: Hmmm. Let me see. The previous year's salary?

Sandy: That's right.

Pete: So 1.07 can be thought of as 1 + 0.07 where the 1 is the previous amount and the 0.07 is the percent increase?

Sandy: Correct. What if we invested money in an account that was losing money at a rate of 4% annually. How would you figure out the money you had left in a given year using the preceding year's amount?

Pete: Let's see. I could use a two-step approach. Multiply the previous year's money by 0.04 and subtract the difference from the preceding year's money.

Sandy: Sure, but how could you do it in one step?

Pete: Well, I start with 1 and, hmmm, subtract 0.04 to get 0.96. Is that right?

Sandy: It sure is.

Pete: So I could multiply the year's amount of money by 0.96 to get the amount of money the following year.

Sandy: I think you have it. Good job!

1. In the Glossary, define the words in this section that appear in ***italic boldface*** type.

2. The Dying Town: Because of the scarcity of jobs, a town's population is decreasing at a rate of 3% every year. The population in 1996 was 6354.

a. Create a list of the town's annual population up to the year 2002.

b. Does the list in part a form an arithmetic sequence or a geometric sequence? Why?

c. Write a recursive definition for computing the town's population.

3. The Diet: Tyler has been told he must go on a diet or he will have a heart attack. Tyler goes on a diet in which he will lose 3 pounds per month. He weighs 254 pounds when he begins his diet.

a. Create a list of Tyler's weights for the first six months.

b. Does the list in part **a** form an arithmetic sequence or a geometric sequence? Why?

c. Write a recursive definition for computing Tyler's weight.

4. Credit Hour Fees: The fee structure in Table 7 represents the total fees for a college student as determined by the number of credit hours the student takes.

TABLE 7 **Total Fees as a Function of Credit Hours**

Credit Hours	Total Fees ($)
1	57
2	89
3	121
4	153
5	185
6	217
7	249
8	281
9	313
10	345
11	377
12	409

a. Is the sequence formed by total fees arithmetic, geometric, or neither? Why?

b. Use function notation, if possible, to write a recursive formula that describes the sequence.

5. The first three numbers in a sequence are 1, 3, 9. Write at least two different sets of numbers for the next three elements. Explain how you created each sequence.

6. Write recursive definitions for the following sequences.

a. The first element is 7 and each element is $\frac{4}{3}$ less than the preceding element.

b. The first element is $\frac{2}{3}$ and each element is $\frac{1}{2}$ of the preceding element.

7. Write the first five terms of each sequence. Identify if the sequence is arithmetic, geometric, or neither. Why?

a. $u(1) = 7$ and $u(n) = u(n-1) - 3$

b. $u(1) = 64$ and $u(n) = \dfrac{u(n-1)}{-2}$

8. Carrie's Job Offer: Carrie has been offered a job with a starting salary of \$31,500. She estimates that she will receive, on the average, a 5% increase each year. Table 8 displays her annual salary for the next 10 years.

a. Is the sequence formed by the annual salaries arithmetic, geometric, or neither? Why?

b. Use function notation, if possible, to write a recursive formula that describes the sequence.

TABLE 8 Carrie's Annual Salary by Year

Year	Salary ($)
1	31,500.00
2	33,075.00
3	34,728.75
4	36,465.19
5	38,288.45
6	40,202.87
7	42,213.01
8	44,323.66
9	46,539.85
10	48,866.84

9. Lottery Doubles: You have just won a lottery in which you will receive money on each of 30 consecutive days. The payoff structure is as follows: 1¢ the first day, 2¢ the second day, 4¢ the third day, 8¢ the fourth day, 16¢ the fifth day, and so on.

a. How much will you be paid on the tenth day? How did you find the answer?

b. Is the sequence defined by the daily payouts arithmetic, geometric, or neither? Why?

c. You paid $10 for the winning lottery ticket. Was it worth it? Why or why not?

10. Given $y = 7x - 2$, find the rate of change of y with respect to x as:

a. x increases from –3 to 4

b. x increases from 1 to 5

c. y increases from –1 to 3

d. y decreases from 7 to 0

11. Given $4x + 3y = 7$, assume that y depends on x. Find the rate of change of y with respect to x as x increases from 2 to 6.

12. Find the slope of the line between the points:

a. (2, –4) and (–3, –1)

b. (0, 5) and (–1, 7)

c.

x	y
4	7
–2	3

d.

x	–8	2
y	4	1.3

13. Consider the sequence 11, 21, 1112, 3112, 211,213, 312,213,... .

a. Determine the pattern and write the next three elements. (*Hint*: Read aloud digit by digit.)

b. Continue generating elements. Is there a position after which point on all the elements are the same? If so, what is that position and what is the element in that position?

14. The graphs of two sequences are displayed in Figure 6.

FIGURE 6

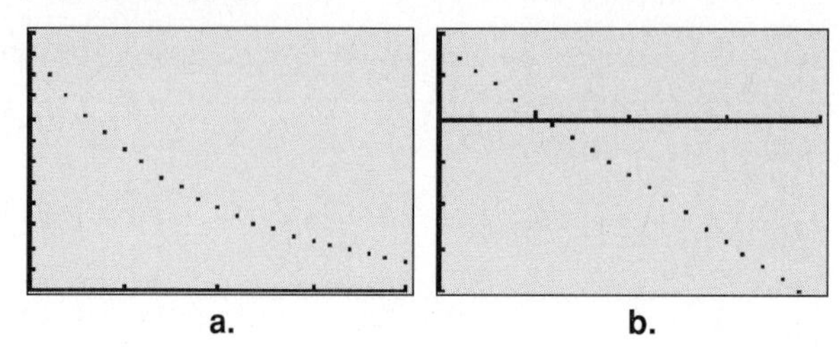

Which graph represents an arithmetic sequence and which represents a geometric sequence? Justify your answer.

15. Subtracting 1, 2, and 2 respectively from the first three elements of a particular arithmetic sequence produces the first three terms of a geometric sequence. The common difference of the arithmetic sequence is 4. Find the first three elements of the arithmetic sequence, the first three elements of the geometric sequence, and identify the common ratio of the geometric sequence.

16. Pascal's Triangle: The following array of numbers is called ***Pascal's Triangle***.

```
          1
        1   1
      1   2   1
    1   3   3   1
  1   4   6   4   1
1   5  10  10   5   1
```

a. Write the next three rows of the triangle. Describe how to create one row from the previous row.

b. Find at least five different sequences in the triangle. Identify them as arithmetic, geometric, or neither.

Where have you seen sequences in other contexts? Provide a real-life example of an arithmetic sequence. Provide a real-life example of a geometric sequence.

SECTION 2.3

Generalizing Sequences

Purpose

- Develop general formulas for arithmetic and geometric sequences.

In the previous section, we wrote recursive formulas that generate the annual salaries at CDs-R-Us and at Super Dog Burger Bits. The word *recursive* means that each salary was computed using the previous year's salary. This is an inefficient technique if we'd like to know the annual salary at year 30, for example.

In this section, we look at how to generate the salaries using the year as input. In other words, we'd like to identify the process for the function machine in Figure 1.

FIGURE 1

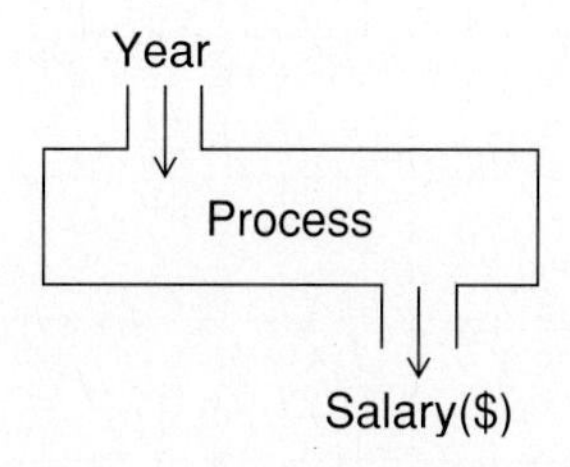

We create formulas (the processes) using the first year's salary and the finite difference or finite ratio in successive salaries as constants.

Before beginning, recall two of the three different scenarios involving annual salary and increases to salary.

CDs-R-Us: The starting annual salary is $35,000 and you will receive a fixed raise of $2000 annually.

Super Dog Burger Bits: The starting annual salary is \$25,000 and you will receive a 7% raise annually.

We determined that the salaries for CDs-R-Us formed an arithmetic sequence because the finite differences in successive outputs were constant. The salaries for Super Dog Burger Bits formed a geometric sequence because the finite ratios in successive outputs were constant.

When trying to generalize the output from the input, remember that the output (salary) depends on the input (year). This suggests that you must always keep one eye on the input as you generalize how to calculate output. Somehow the input must be used to find the output. Keep these ideas in mind as you work through the next investigation.

1. Complete Table 1 for the salaries at CDs-R-Us. Record the process (arithmetic computations), rather than the final answer, for the last two columns). The first two have been completed as examples. Use the last row to write a generalization.

TABLE 1 Generalizing CDs-R-Us Salaries

Year (Input)	Number of \$2000 Raises	Total Raise from Starting Salary Process	Salary (\$) (Output) Process
1	0	0(2000)	35,000 + 0(2000)
2	1	1(2000)	35,000 + 1(2000)
n			

2. If $CD(n)$ is the salary in year n, write a general function for $CD(n)$. Use the last line of Table 1 as a guide.

3. Write a sentence describing the formula you wrote in words.

4. Use the CDs-R-Us salary function to find the salary in year 35. State what you did.

5. Let $u(n)$ represent an arithmetic sequence. Let $u(1)$ be the first element and d be the constant finite difference in successive outputs. Using CDs-R-Us as an example, write a general function for $u(n)$.

When attempting to generalize, it is very important to concentrate on the *process* that you are following rather than on the answer. Table 2 displays the process required to develop a general rule for the sequence determined by CDs-R-Us salaries.

TABLE 2 Generalizing CDs-R-Us Salaries

Year (Input)	Number of \$2000 Raises	Total Raise from Starting Salary Process	Salary (\$) (Output) Process
1	0	0(2000)	35,000 + 0(2000)
2	1	1(2000)	35,000 + 1(2000)
3	2	2(2000)	35,000 + 2(2000)
4	3	3(2000)	35,000 + 3(2000)
5	4	4(2000)	35,000 + 4(2000)
6	5	5(2000)	35,000 + 5(2000)
7	6	6(2000)	35,000 + 6(2000)
n	$n-1$	$(n-1)2000$	$35{,}000 + (n-1)2000$

To use this approach, we need only know the first element and the common difference of the arithmetic sequence. The last row of Table 2 suggests the following algebraic statement for function CD.

$$CD(n) = 35{,}000 + (n-1)2000$$

The function machine appears in Figure 2.

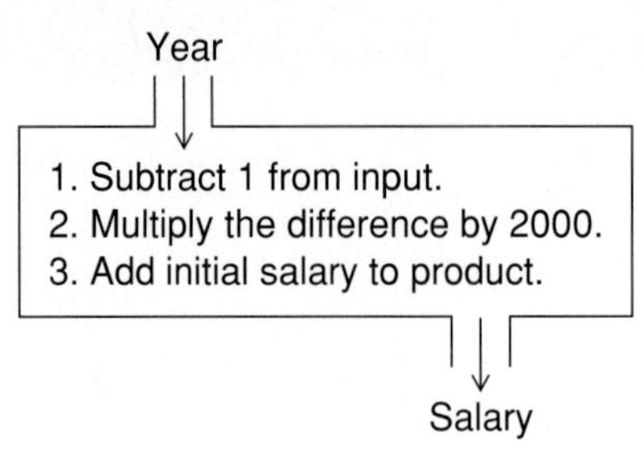

Notice that the function $CD(n) = 35{,}000 + (n-1)2000$ can be simplified.

Distribute the 2000: $CD(n) = 35{,}000 + 2000n - 2000$

Combine like terms: $CD(n) = 2000n + 33{,}000$

Using this formula, the salary in year 35 is found as follows.

$$\begin{aligned} CD(35) &= 2000(35) + 33{,}000 \\ &= 70{,}000 + 33{,}000 \\ &= 103{,}000. \end{aligned}$$

So the salary in year 35 is \$103,000.

Simplifying the function hides some of the structure. Looking at the function $CD(n) = 35{,}000 + (n-1)\,2000$, we notice that the 35,000 is the first element and that 2000 is the common finite difference in successive outputs.

The formula says that the nth element can be found by multiplying the common difference by $(n-1)$ and adding the first element to the product.

Symbolically, if $u(n)$ represents an arithmetic sequence, $u(1)$ is the first element and d is the constant finite difference in successive outputs, then

$$u(n) = u(1) + (n-1)d$$

This is a general function for *any* arithmetic sequence.

Now let's turn our attention to the salaries at Super Dog Burger Bits.

6. Complete Table 3 for the salaries at Super Dog Burger Bits. Record the process (arithmetic computations), rather than the final answer, for the last two columns). The first three have been done for you. Use the last row to write a generalization.

TABLE 3 Generalizing Super Dog Burger Bits Salaries

Year (Input)	Number of 7% Raises	Salary ($) (Output) Process
1	0	25,000
2	1	25,000(1.07)
3	2	25,000(1.07)(1.07) or $25{,}000(1.07)^2$
n		

7. If $SD(n)$ is the salary in year n, write a general function for $SD(n)$. Use the last line of Table 3 as a guide.

8. Write a sentence describing the formula you wrote in words.

9. Use the Super Dog Burger Bits salary function to find the salary in year 25. State what you did.

10. Let $v(n)$ represent a geometric sequence. Let $v(1)$ be the first element and r be the constant finite ratio in successive outputs. Using Super Dog Burger Bits as an example, write a general function for $v(n)$.

Table 4 displays the process required to develop a general rule for the sequence determined by Super Dog Burger Bits salaries.

TABLE 4 Generalizing Super Dog Burger Bits Salaries

Year (Input)	Number of 7% Raises	Salary ($) (Output) Process
1	0	$25,000(1.07)^0$
2	1	$25,000(1.07)^1$
3	2	$25,000(1.07)(1.07)$ or $25,000(1.07)^2$
4	3	$25,000(1.07)^3$
5	4	$25,000(1.07)^4$
6	5	$25,000(1.07)^5$
n	$n-1$	$25,000(1.07)^{n-1}$

To use this approach, we need only know the first element and the common ratio of the geometric sequence. The last row of Table 4 suggests the following algebraic statement for function *SD*.

$$SD(n) = 25,000(1.07)^{n-1}$$

The function machine appears in Figure 3.

FIGURE 3

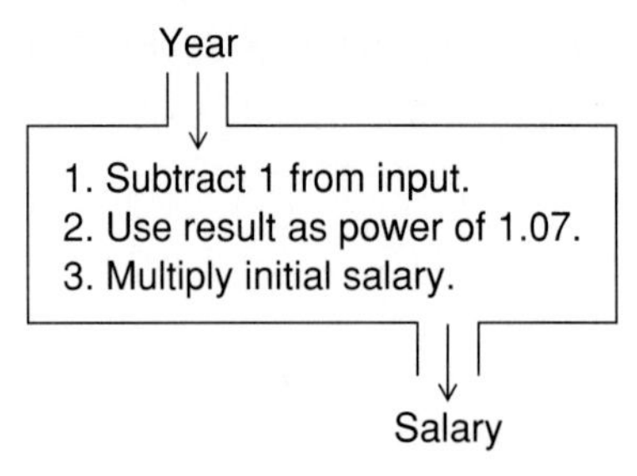

Using the formula $SD(n) = 25,000(1.07)^{n-1}$, you can find the salary in year 25 as follows.

$$\begin{aligned} SD(25) &= 25,000(1.07)^{25-1} \\ &= 25,000(1.07)^{24} \\ &\approx 25,000(5.07237) \\ &\approx 126,809.25 \end{aligned}$$

So the salary in year 25 is approximately \$126,809.25. Note that the symbol $\approx$ means "is approximately equal" to because of rounding.

Looking at the function $SD(n) = 25{,}000(1.07)^{n-1}$, we notice that the 25,000 is the first element and that 1.07 is the common finite ratio in successive outputs.

The formula says that the nth element can be found by raising the common ratio to the power $(n - 1)$ and multiplying the first element by the result.

Symbolically, if $v(n)$ represents a geometric sequence, $v(1)$ is the first element and r is the constant finite ratio in successive outputs, then

$$v(n) = v(1)(r)^{n-1}$$

This is a general function for *any* geometric sequence.

Let's summarize what we have done.

Arithmetic sequences: The function has a constant finite difference between successive outputs (the elements of the sequence). If $u(1)$ is the first element and d is the constant finite difference, then a general formula for the nth element, $u(n)$, is $u(n) = u(1) + (n - 1)d$.

Geometric sequences: The function has a constant finite ratio between successive outputs (the elements of the sequence). If $v(1)$ is the first element and r is the constant finite ratio, then a general formula for the nth element, $v(n)$, is $v(n) = v(1)(r)^{n-1}$.

It is important to analyze these formulas in terms of their use of variables. In both cases, n is the input variable and $u(n)$ or $v(n)$ is the output variable. The letters $u(1)$, $v(1)$, d, and r are constants (***parameters***) determined by the specific sequence under investigation.

The function machines appear in Figure 4.

FIGURE 4

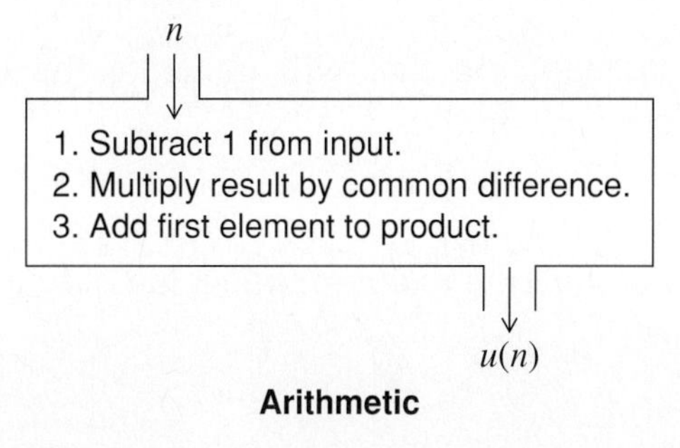

Arithmetic

n

1. Subtract 1 from input.
2. Raise common ratio to (input – 1) power.
3. Multiply by first element.

$v(n)$

Geometric

11. Consider the data in Table 5.

TABLE 5 **Random Data 1**

Input	1	2	3	4	5	6
Output	23	20	17	14	11	8

a. Do the outputs form an arithmetic or geometric sequence? Why?

b. Write an equation that will generate the output from the input. Explain what you did.

c. Use your equation to find the output when the input is 57.

d. Find the input if the output is –85. Explain what you did.

12. Consider the data in Table 6.

TABLE 6 **Random Data 2**

Input	1	2	3	4	5	6
Output	78	70.2	63.18	56.862	51.1758	46.05822

a. Do the outputs form an arithmetic or geometric sequence? Why?

b. Write an equation that will generate the output from the input. Explain what you did.

c. Use your equation to approximate the output when the input is 23.

d. Find the input if the output is approximately 14.45. Explain what you did.

13. a. Create your own arithmetic sequence by writing down the first five elements.

b. Write a general function that generates the elements of the sequence.

c. Use the formula you wrote in part **b** to find the 45th element of your sequence.

14. a. Create your own geometric sequence by writing down the first five elements.

b. Write a general function that generates the elements of the sequence.

c. Use the formula you wrote in part **b** to find the 23rd element of your sequence.

The data

Input	1	2	3	4	5	6
Output	23	20	17	14	11	8

begins at 23 and the successive outputs decrease by 3. Since there is a constant finite difference in outputs, the sequence is arithmetic. Using the formula $u(n) = u(1) + (n-1)d$, where $u(1) = 23$ and $d = -3$, we get

$$\begin{aligned} u(n) &= 23 + (n-1)(-3) \\ &= 23 - 3n + 3 \\ &= -3n + 26. \end{aligned}$$

If the input is 57, then

$$\begin{aligned} u(57) &= -3(57) + 26 \\ &= -171 + 26 \\ &= -145 \end{aligned}$$

So the 57th element of the sequence is –145.

If the output is –85, then

$$-85 = -3n + 26$$

Subtract 26 from both sides: $-111 = -3n$

Divide both sides by –3: $37 = n$

So, the input is 37 if the output is –85.

Input	1	2	3	4	5	6
Output	78	70.2	63.18	56.862	51.1758	46.05822

The data begins at 78 and the successive outputs have a common ratio of $\frac{70.2}{78} = 0.9$. Since there is a constant finite ratio in outputs, the sequence is geometric. Using the formula $v(n) = v(1)(r)^{n-1}$, where $v(1) = 78$ and $r = 0.9$, we get

$$v(n) = 78(0.9)^{n-1}$$

If the input is 23, then

$$\begin{aligned} v(23) &= 78(0.9)^{23-1} \\ &= 78(0.9)^{22} \\ &\approx 78(0.098477) \\ &\approx 7.68. \end{aligned}$$

So the 23rd element of the sequence is approximately 7.68.

If the output is approximately 14.45, then $14.45 = 78(0.9)^{n-1}$ and we must find n. It might be a good idea to search a table for an output of 14.45 (Figure 5).

FIGURE 5

```
Plot1 Plot2 Plot3
\Y1■78(.9)^(X-1)
\Y2=
\Y3=
\Y4=
\Y5=
\Y6=
```

X	Y1
14	19.827
15	17.844
16	16.06
17	14.454
18	13.008
19	11.707
20	10.537

X=17

So, the input is 17 if the output is approximately 14.45.

STUDENT DISCOURSE

Pete: I had a problem with the last two investigations. How do I create a sequence from scratch? There weren't even any numbers given in the question.

Sandy: Which kind of sequence would you like to create first?

Pete: Let's start with arithmetic.

Sandy: Okay. What do you need to know to create an arithmetic sequence?

Pete: Hmmm. It seems that I need to know where to start.

Sandy: That's right—the first element. What else?

Pete: Well, I either need to know the next element or how to get to the next element.

Sandy: True. Let's try one. Pick two numbers at random.

Pete: How about 2 and 5?

Sandy: That's fine. Think of the 2 as your first element and 5 as the second element.

Pete: Okay. That means there is a common difference of 3 if this is arithmetic. Oh, I see, I have everything I need to write the formula. I have the first element and the difference between elements.

Sandy: That's right. We could use the same numbers another way. Suppose 2 is the first element and 5 is the common difference, then what?

Pete: Then there is no problem. I know that $u(n) = 2 + (n-1)5$. Hey, how about that! Okay, what if the sequence is geometric?

Sandy: Then you could still use the 2 as the first element and . . .

Pete: Wait! I know. Use the 5 as the common ratio rather than the common difference. Then the sequence would be 2, 10, 50, 250, etc. multiplying each element by 5 to get the next element. I understand!

1. In the Glossary, define the words in this section that appear in ***italic boldface*** type.

2. Health Care Costs: Health care costs have been rising at a rate of almost 20% annually. Assume that a certain medical test cost $100 in 1981. Consider the sequence defined by the cost of this test in successive years since 1980.

a. Write the first five terms of this sequence.

b. Is the sequence arithmetic or geometric? Why?

c. Identify the first element and either the common difference or the common ratio.

d. Write a function with input the number of years since 1980 and output the cost of the test.

e. Use your function to find the cost of the test in 1995. Describe your answer in a complete sentence.

3. Credit Hour Fees: The fee structure in Table 7 represents the total fees for a college student as determined by the number of credit hours the student takes.

a. Identify the first element and either the common difference or the common ratio for the sequence represented by the total fees.

b. Write a function with input the number of credit hours and output the total fees.

c. Use your function to determine the total fees for 20 credit hours.

TABLE 7 Total Fees as a Function of Credit Hours

Credit Hours	Total Fees ($)
1	57
2	89
3	121
4	153
5	185
6	217
7	249
8	281
9	313
10	345
11	377
12	409

4. **Carla's Bank Account:** Carla opened an account by depositing $1000. She deposits an additional $200 each month. The amount Carla has on deposit each month, not including interest, forms a sequence.

   a. Write the first five terms of this sequence.

   b. Is the sequence arithmetic or geometric? Why?

   c. Identify the first element and either the common difference or the common ratio.

   d. Write a function with input the number of months and output the total amount deposited.

e. Use your function to find how much has been deposited after three years. State your answer in a complete sentence.

5. Write general formulas for the following sequences.

a. The first element is 7 and each element is $\frac{4}{3}$ less than the preceding element.

b. The first element is –3 and each element is 2.1 more than the preceding element

6. Consider the following sequences, which are defined recursively. Write the first five elements of each sequence. Identify each sequence as either arithmetic or geometric. Write the general formula for each sequence.

a. $u(1) = 7$ and $u(n) = u(n-1) - 3$

b. $u(1) = 64$ and $u(n) = \dfrac{u(n-1)}{-2}$

c. $u(1) = 9$ and $u(n) = u(n-1)$

7. Carrie's Job Offer: Carrie has been offered a job with a starting salary of \$31,500. She estimates that she will receive, on the average, a 5% increase each year. Table 8 displays her annual salary for the next 10 years.

a. Identify the first element and either the common difference or the common ratio for the sequence represented by the salary.

b. Write a function with input the year and output the salary.

c. Use your function to determine the salary after 20 years.

TABLE 8 Carrie's Annual Salary by Year

Year	Salary ($)
1	31,500.00
2	33,075.00
3	34,728.75
4	36,465.19
5	38,288.45
6	40,202.87
7	42,213.01
8	44,323.66
9	46,539.85
10	48,866.84

8. **Lottery Doubles:** You have just won a lottery and you will receive money on each of 30 consecutive days. The payoff structure is as follows: 1¢ the first day, 2¢ the second day, 4¢ the third day, 8¢ the fourth day, 16¢ the fifth day, and so on.

   a. Write a general function with input the day number and output the amount paid by the lottery on that day.

   b. Use the function you wrote to determine how much is paid on the 20th day and on the last day.

9. **Triangular Numbers:** Consider the sequence 0, 1, 3, 6, 10, 15, 21, …

   a. Write the next four elements.

   b. Explain how the sequence is generated.

c. Is the sequence arithmetic, geometric, or neither? Why?

10. Fibonacci Sequence: There is a famous sequence named after Fibonacci (look up the name). The rule for generating the sequence is

$$u(1) = 1$$
$$u(2) = 1$$
$$u(n) = u(n-1) + u(n-2), \quad n = 3, 4, 5, 6, \ldots$$

a. Write the first eight elements of the sequence. These are the first eight Fibonacci numbers.

b. Describe how to obtain the elements of the Fibonacci sequence.

c. Is the Fibonacci sequence arithmetic, geometric, or neither? Why?

11. Given $y = -3x + 5$, find the rate of change of y with respect to x as:

a. x increases from –2 to 1

b. y increases from –3 to 4.

12. Given $7x - 3y = 9$, assume that y depends on x.

a. Solve for y.

b. Generate the sequence of y values for $x = 1, 2, 3, \ldots$

c. What kind of sequence is formed by the y values? If arithmetic, what is the common difference? If geometric, what is the common ratio?

d. Find the rate of change of y with respect to x as x increases from 1 to 3. What is the connection between the rate of change and your answer to part **c**?

e. Find the output if the input is 19.

f. Find the input if the output is 11.

13. Find the slope of the line between the points:

a. $(-3, 2)$ and $(-3, -1)$

b. $(4, 0)$ and $(6, -3)$

c.

x	y
−2	−1
0	6

d.

x	1	−3
y	−1.5	−4.7

REFLECTION

The formulas

$$u(n) = u(1) + (n - 1)d$$

$$v(n) = v(1)(r)^{n-1}$$

contain many letters, some used as variables and some used as constants. List all the letters, identify each as either a variable or a constant, and describe what the letter represents. If the letter represents a variable, identify it as an input (independent) or output (dependent) variable.

SECTION 2.4

Domain and Range: Crucial Considerations

Purpose

- Explore the effect of a function's process on domain and range.
- Relate the role of number systems to domain and range considerations.
- Review relationships among number systems.

We have spent time during the first two chapters discussing the domain and range of functional relationships. We'd like to spend a section concentrating solely on the ideas of domain and range. This is important since the understanding and mathematical analysis of a given problem situation often hinges on the domain and range implicit in the situation. We begin by exploring your understanding of domain and range.

1. Describe in your own words the phrase "domain of a function." Provide an example that clarifies your description.

2. Describe in your own words the phrase "range of a function." Provide an example that clarifies your description.

3. You are told the domain of a problem situation is the set of natural numbers. What does this statement mean to you?

4. You are told the domain of a problem situation is the set of real numbers. What does this statement mean to you?

Previously, we have loosely defined the ***domain of a function*** as the set of "legal" inputs to the function. The word "legal" needs clarification, some of which will come in this section. On the other hand, the ***range of a function***

was described as the set of possible outputs of the function. It is important to note that the restrictions inherent in the problem situation and the process defined by the function play crucial roles in the domain and the range of the function. One purpose of this section is to explore these roles.

A natural number domain implies that the only numbers that can be used for input are 1, 2, 3, 4, 5, ... The letter N is usually used to represent this set of numbers. Thus

> **Natural numbers**
> $N = \{1, 2, 3, 4, 5, \ldots\}$

A ***real number*** domain takes any rational or irrational number as input. This includes

- all natural numbers
- zero
- all the opposites of the natural numbers
- all positive and negative fractions, with their decimal equivalents, that have numerators and denominators that are natural numbers or their opposites
- all nonterminating nonrepeating decimals

All the items in this list except the last (nonterminating nonrepeating decimals) are members of the ***rational numbers***. Nonterminating nonrepeating decimals are ***irrational numbers***.

This list may include any number you can think of, though there are some other types of numbers. The letter $\Re$ is usually used to represent the set of real numbers. Thus

> **Real numbers**
> $\Re$ = union of the sets of rational and irrational numbers.

Note that the word *real* in the title of these numbers does not imply that these numbers are any more or less "real" than any other kind of number.

Let's explore these ideas in the context of two problems we have previously studied.

CDs-R-Us Salary Offer: The function that represents the annual salary at CDs-R-Us as a function of the number of years employed is

$$u(n) = 2000n + 33{,}000.$$

5. What is the domain of the salary function? In other words, make a list of the inputs.

6. What is the range of the salary function? In other words, make a list of the outputs.

7. Assume the operations of addition and multiplication using natural numbers are performed on the input. Will a natural number always be the output? Why or why not?

Super Dog Burger Bits Salary Offer: The function that represents the annual salary at Super Dog Burger Bits as a function of the number of years employed is

$$v(n) = 25{,}000(1.07)^{n-1}.$$

8. What does n represent? What is the domain of the salary function? Why?

9. What is the range of the salary function? Why?

10. Assume the operation multiplication using decimal numbers is performed on the input. Will a decimal number be the output? Why or why not?

For both salary offers, the domain is the set of natural numbers because the input represents the number of years worked. We might argue that the domain is a subset of the natural numbers by limiting the larger number that might be used as input. Certainly no one is going to work for either of these companies more than 100 years. This upper limit is somewhat arbitrary. We could write the domain as $\{1, 2, 3, 4, \ldots, N\}$, where N is the maximum number of years a person would work.

While the domains for these two functions are the same, the ranges are quite different. The process for the CDs-R-Us salary function involves multiplying by 2000 and then adding 33,000. Multiplying and adding natural numbers result in natural numbers. The input to this function is a subset of the natural numbers and the output (range) is also a subset of the natural numbers. The specific range is the set $\{35{,}000, 37{,}000, 39{,}000, \ldots, M\}$, where M is the maximum attainable salary.

Now consider the Super Dog Burger Bits salary function. The process requires us to raise the decimal 1.07 to a power equal to the input minus 1 and multiply the result by 25,000. Exponentiations and multiplications involving decimals result in decimal numbers. Often the outputs are not natural numbers. When the outputs involve either terminating or repeating decimals, we have a range that is a subset of the *rational numbers*. We'll formally discuss the rational numbers after we investigate these ideas a bit more.

We see that the types of operations performed affect the range. The operations used may limit the domain. Let's simplify things a little so we can deal with these ideas.

Recall that the four basic operations (addition, subtraction, multiplication, and division) are ***binary operations***. They can be thought of as functions with two inputs producing one output. Figure 1 displays function machines for these operations.

FIGURE 1

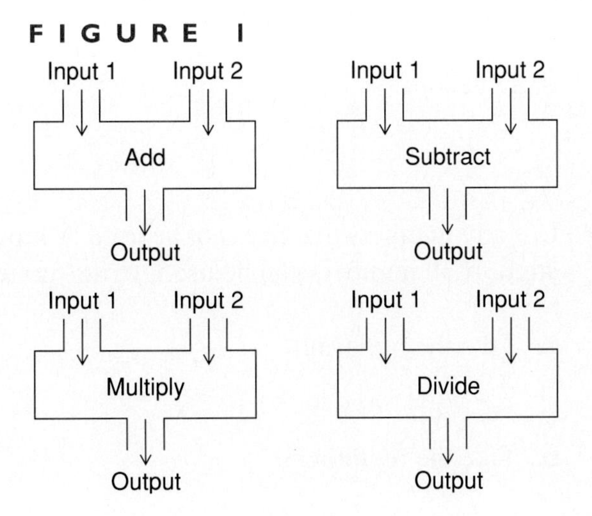

The operations of taking an *opposite*, *reciprocal*, square, and square root are called ***unary operations.*** They can be thought of as functions with a single input producing a single output. Figure 2 displays function machines for these operations.

FIGURE 2

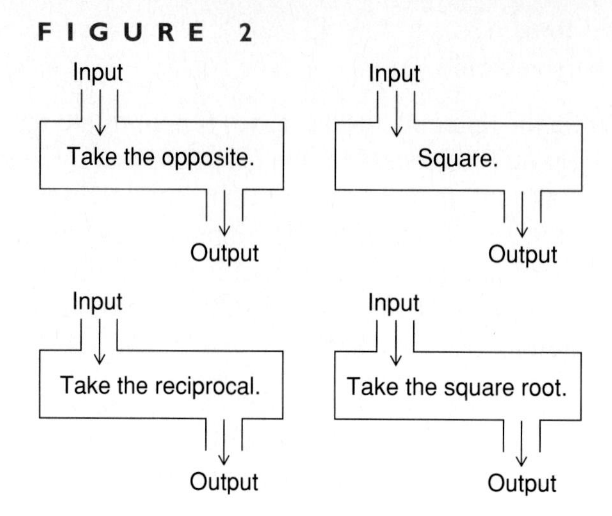

11. List any numbers that may not be used as inputs to the given basic operation. If all numbers can be used, write "no restrictions."

a. Addition:

b. Subtraction:

c. Multiplication:

d. Division:

12. List any numbers that may not be used as inputs to the given basic operation. If all numbers can be used, write "no restrictions."

a. Take the opposite:

b. Take the reciprocal:

c. Square:

d. Take the square root:

First, let's look at the largest collection of numbers available to us: the real numbers. Since any two real numbers can be combined using addition, subtraction, or multiplication, there are no restrictions on the input to these three operations. The domain for these operations is the set of all pairs of real numbers. However, division by zero is undefined. This restricts Input 2 to nonzero real numbers. The domain for the division operation is the set of all pairs of real numbers in which the second number in the pair is not zero.

Now consider the unary operations. Taking the opposite and squaring can be performed on any real number, so the domain of these operations is all real numbers. Taking the reciprocal can only be performed on nonzero numbers. The domain for the reciprocal operation is the set of nonzero real numbers. The square root operation cannot be performed on negative numbers (if we restrict ourselves to the set of real numbers). This limits the domain of the square root operation to *nonnegative* (zero and positive) real numbers.

While we will never be able to remove the restrictions inherent in division by zero, we will be able to remove the restrictions on the domain of the square root operation by defining some new numbers. Something to look forward to!

Functions define a process to be performed on domain elements. The domain limits what can be used as input. The domain and the function's process determine the type of numbers that can be the output. Often the output is different from the type of number that is the input (as we saw with the Super Dog Burger Bits salary function). Let's explore this idea.

13. Given the domain and the process for the function, describe the smallest set that can be used to describe the range of the function.

a. Domain: Whole number pairs Process: Addition

b. Domain: Whole number pairs Process: Subtraction

c. Domain: Integer pairs Process: Multiplication

d. Domain: Integer pairs Process: Division

e. Domain: Integers Process: Squaring

f. Domain: Positive rationals Process: Square rooting

g. Domain: Real numbers Process: Square rooting

Mathematicians have agreed on a hierarchy of numeration systems that allows easier discussion of the domain and range of functions. Some of these numeration systems arose naturally as human beings attempted to deal with numeration and the operations they performed on numbers. Each number system in this hierarchy is a subset of the real number system.

We might begin with the first numbers you learned—the counting, or natural numbers. If we include zero with the natural numbers, we have created the whole numbers. Let's use the letter W to represent the set of whole numbers.

$$W = \{0, 1, 2, 3, 4, \ldots\}$$

If we add two whole numbers, we always get a whole number. This is called ***closure of the whole numbers under addition***. However, if we subtract two whole numbers, we sometimes get a negative number. Negative numbers are not part of the whole numbers. So subtracting two whole numbers does not always result in a whole number. In this case, the whole numbers are not closed under subtraction.

The fact that we sometimes get negative numbers when we subtract whole numbers suggests we create a set of numbers that includes negatives. Such a set is called the ***integers***. The integers are formed by combining the whole numbers and the opposite of the whole numbers into one set. The letter I is often used to represent the integers.

$$I = \{\ldots, -3, -2, -1, 0, 1, 2, 3, \ldots\}$$

So, if the domain is the whole numbers and the process is subtraction, the range is the integers.

Now we consider the integers as the domain. If we multiply two integers, we always get an integer. The integers are closed under multiplication. If we divide two integers, we often get fractions that are not integers. The integers are not closed under division.

The fact that we often get fractions when we divide integers suggests we create a set of numbers that includes fractions. This set is called the ***rational numbers***. The word *rational* comes from the word *ratio*. Common fractions are the ratio of two integers. The symbol used for this set is Q. You might have expected we would use an $\Re$, but $\Re$ is already used for the real numbers. The letter Q is used since the result of a division is called a ***quotient***. The formal set definition for the rationals is

$$Q = \left\{ \frac{a}{b} \middle| \, a \text{ and } b \text{ are integers and } b \neq 0 \right\}.$$

This states that the rational numbers consist of all numbers that are the ratio of two integers as long as the divisor is not zero. The vertical bar in the set brackets represents the phrase "such that."

So, if the domain is the integers and the process is division, then the range is the set of rational numbers.

When we use a domain of the integers and a process of squaring, we end up with a range that is more restricted than the domain. In previous problems we saw that the range often included more numbers than the domain. In the case of squaring, the output will not be negative. If the integers are the domain, then the whole numbers are the range.

Finally we get to the square root function, which offers its own collection of difficulties. While the square roots of positive numbers and zero result in real numbers, the square roots of negative numbers do not. First note that the square root of a positive rational number does not have to be a rational number. Square roots often result in nonterminating, nonrepeating decimals. Such decimals belong to the ***irrational numbers***.

So, if the domain is the positive rational numbers and the process is square rooting, the range is the positive real numbers.

If the domain is the real numbers and the process is square rooting, we must create some new numbers that allow us to define the square root of a negative number. This leads to a number system beyond the real number system called the ***complex numbers***. These numbers are beyond the scope of this section, but you will enjoy learning about them later in the text.

We conclude this section by thinking about the relationship between the number systems in the hierarchy we just discussed.

14. Given two sets, which is a subset of the other? Why?

a. Integers and natural numbers.

b. Whole numbers and real numbers.

c. Rational numbers and real numbers.

15. Draw a diagram that contains the natural numbers, the whole numbers, the integers, the rational numbers, the irrational numbers, and the real numbers. Nest the sets appropriately.

DISCUSSION

The real number system contains all the number systems (except complex) discussed in this section as subsets. The real number system can be divided into two nonoverlapping sets: the rational numbers and the irrational numbers.

The remaining number systems are all subsets of the rational numbers. The integers are rational numbers that reduce to a denominator of 1. The whole numbers are the nonnegative integers. The natural numbers are the nonzero whole numbers. Figure 3 visually displays this hierarchy. The natural numbers, while not displayed, are within the whole numbers.

FIGURE 3

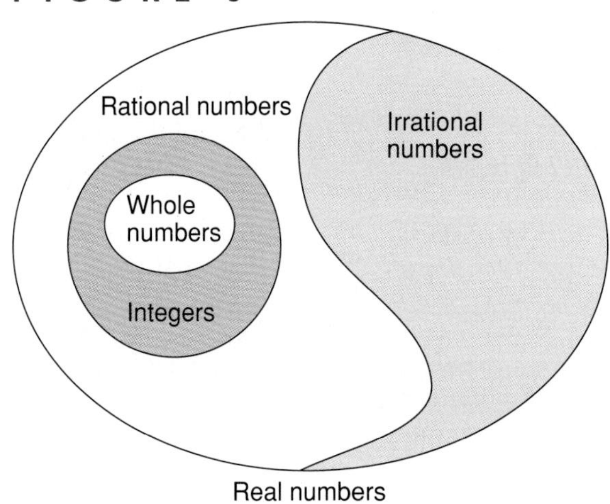

STUDENT DISCOURSE

Pete: I've always had a hard time keeping the number systems straight.

Sandy: Yeah, me too, though Figure 3 did help me think about the relationships between the systems.

Pete: I have a question about the figure.The first thing I notice is that the real numbers seem to include all rational and irrational numbers.

Sandy: I think you're right. Do you know how to distinguish between a rational and irrational number?

Pete: It seems like it's hard to do that. Doesn't it have something to do with decimals?

Sandy: It does. Suppose the decimal form of a number terminates. What kind of number is it?

Pete: That's rational. For example, the decimal for $\frac{1}{4}$ is 0.25. The decimal terminates and $\frac{1}{4}$ is rational.

Sandy: Good. What about something like $\frac{1}{6}$?

Pete: Well, my calculator gives me the decimal 0.1666666667, but I don't think it terminates.

Sandy: That's right. Where did the 7 come from?

Pete: I think the calculator rounded up.

Sandy: Me too. This is an example of a repeating decimal.

Pete: Those are rational too, aren't they?

Sandy: That's right.

Pete: So what's left?

Sandy: What do you think?

Pete: Well, if I want an irrational number, its decimal can't terminate.

Sandy: Correct. And?

Pete: I guess it can't repeat.

Sandy: That's right. Decimals that represent irrational numbers are both nonterminating and nonrepeating. They continue without repeating a set sequence of digits forever.

Pete: Hmmm. I think I get it, but I need to think about this some more. I'll talk to you later.

1. In the Glossary, define the words in this section that appear in ***italic boldface*** type.

2. If the natural numbers are used as input and the operation is taking the opposite, what set represents the range? Justify.

3. If the natural numbers are used as input and the operation is taking the reciprocal, what set represents the range? Justify.

4. If the set of perfect square whole numbers is used as the domain and the operation is square rooting, what set represents the range? Justify.

5. Consider the following list of numbers

$-\sqrt{7}$ $-\frac{5}{8}$ 1 2 51 $\frac{11}{13}$

$0.\overline{2}$ 0 $\sqrt{-7}$ -2 $3\frac{3}{7}$.

List all the:

a. whole numbers

b. integers

c. rational numbers

d. irrational numbers

e. real numbers

f. nonreal numbers

g. even numbers

h. prime numbers

6. Are the rational numbers closed under the operation of taking the reciprocal? Justify.

7. Are the rational numbers closed under the operation of taking the square root? Justify.

8. Are the real numbers closed under the operation of taking the square root? Justify.

9. Selling Costume Jewelry: Remember Erratic Ernie and Steady Sue? The last we looked, they were selling costume jewelry in which their net profit was defined by the function $P(n) = 5n - 100$, where P dollars of profit was realized when n pieces of jewelry were sold. What is the domain and the range of the function $P(n) = 5n - 100$ in the context of selling the jewelry? Explain.

10. Recall the **Virtual Reality Experiences** salary offer. Identify the domain and the range for this problem situation.

11. Your calculator most likely has a function called *INT*. Experiment with various types of input to this function. As a result of the experimentation, state the domain and the range of this function.

REFLECTION

Explain how a function's process can affect the domain of the function. Create an example to illustrate your explanation.

Explain how a function's process can affect the range of a function. Create an example.

Explain how a problem situation can affect both the domain and range of the function that models the situation. Create an example.

SECTION 2.5

Constant and Accelerated Change: Review

Purpose

- Reflect on the key concepts of Chapter 2.

In this section you will reflect upon the mathematics in Chapter 2—on what you've done and how you've done it.

1. State the most important ideas in this chapter. Why did you select each?

2. Identify all the mathematical concepts, processes, and skills you used to investigate the problems in Chapter 2?

There are a number of important ideas that you might have listed, including finite differences, rate of change, finite ratios, sequences, arithmetic sequences, geometric sequences, recursive formulas, common difference, and common ratio.

1. Any two distinct points define a line segment. We can find the slope of that line segment by calculating the rate of change between the two points. Find the slope of the line segment with the following endpoints.

   a. (3, –4) and (13, 6) **b.** (–2, 0) and (–5, –4)

   c. (11, 9) and (–23, 10) **d.** (–3, 7) and (7, 11)

   e. (–7, 6) and (–4, 5)

2. Determine the rate of change for the following functions.

   a. $g(x) = 2x - 3$ **b.** $h(x) = \frac{5}{3}x + 7$

   c. $p(t) = -1.8t - 2.5$ **d.** $k(q) = \frac{2}{5} - \frac{5}{9}q$

3. **Carrie's Job Offer:** Carrie has been offered a job with a starting salary of \$30,000. She estimates that she will receive, on the average, a 5% increase each year. Table 1 displays her annual salary for the next 10 years.

   a. Find the ratio of change in Carrie's annual salary to the change in year for each pair of successive years. Write a sentence interpreting your answer.

   b. What is happening to the rate of change as the years increase?

4. Write recursive definitions for the following sequences.

   a. The first element is –3 and each element is 2.1 more than the preceding element.

TABLE 1 **Carrie's Annual Salary by Year**

Year	Salary ($)
0	30,000.00
1	31,500.00
2	33,075.00
3	34,728.75
4	36,465.19
5	38,288.45
6	40,202.87
7	42,213.01
8	44,323.66
9	46,539.85
10	48,866.84

b. The first element is 0.7 and each element is 4.5 times the preceding element.

5. Write the first five terms of each sequence. Identify if the sequence is arithmetic, geometric, or neither. Why?

a. $u(1) = 2$ and $u(n) = 3u(n-1) - 5$

b. $u(1) = 9$ and $u(n) = u(n-1)$.

6. Selling Costume Jewelry: Erratic Ernie and Steady Sue made and sold costume jewelry. The result was a net profit function $P(n) = 5n - 100$ where P dollars is the net profit when n pieces of jewelry are sold.

a. Write the net profits for n equal to 0 up to 5 inclusive.

b. Do the net profits form a sequence? If yes, what kind? If no, why not?

c. What is the rate of change of net profit per unit change in number of pieces sold?

d. The rate of change equals a very important quantity from sequences. What is the name of that quantity?

7. Given $y = 2x - 7$, find the rate of change of y with respect to x as:

a. x increases from –1 to 3.

b. x increases from 2 to 7.

c. y increases from –3 to 5.

d. y decreases from 5 to –1.

8. Given $-2x + 5y = 9$, assume that y depends on x. Find the rate of change of y with respect to x as x increases from –3 to 6.

9. Find the rate of change of y with respect to x.

a. $2x + y = 5$

b. $0.1x - 4y = 7$

10. Consider the function $f(x) = 7 - 3x$. Find:

a. $f(-8)$

b. $f(h)$

c. $f(x + h)$

d. $f(x + h) - f(x)$

11. Write general formulas for the following sequences.

a. The first element is 0.7 and each element is 4.5 times the preceding element.

b. The first element is $\frac{2}{3}$ and each element is $\frac{1}{2}$ of the preceding element.

12. Find the slope of the line between the points:

a. (–1, 2) and (–3, –5)

b. (3, 0) and (7, –2)

c.

x	y
–3	–1
2	5

d.

x	3	–1
y	–2.5	–4.7

13. Health Care Costs: Health care costs have been rising at a rate of almost 20% annually. Assume that a certain medical test cost $100 in 1981. Consider the sequence defined by the cost of this test in successive years since 1980. What is the domain and the range of the function that defines the cost of the test as a function of the number of years since 1981?

CONCEPT MAP

Construct a Concept Map centered on one of the following:

a. rate of change

b. sequence

c. continuous function

d. relationship

e. slope

f. notation

REFLECTION

Write a paragraph describing the most interesting mathematical idea in this chapter, according to your answer to Investigation 1. How has your thinking changed as a result of studying this idea?

3 Expressing Situations Involving Change Algebraically

SECTION 3.1

Linear Models for Change

Purpose

- Establish connections between arithmetic sequences and linear models.
- Investigate basic linear models and linear relationships.

In Chapter 2, we looked at sequences that had a constant finite difference in successive elements—these were called arithmetic sequences. We also looked at sequences that had a constant finite ratio in successive outputs—these were called geometric sequences.

Each of these cases is modeled by an important class of mathematical function. In the next two sections, we will define these functions and show how they can be used to describe data. In the next chapter, we will see how to use these functions to model other situations involving relationships between data.

Let's look at the salary offer from CDs-R-Us first studied in Section 2.2.

CDs-R-Us: The starting annual salary is \$35,000 and you will receive a fixed raise of \$2000 annually.

We wrote $CD(n) = 2000n + 33{,}000$ here n is the year number and $CD(n)$ is the salary in year n.

1. Enter the function into your calculator.[1] Use the table feature to complete the output column of Table 1. Then complete the rest of the table by computing the finite differences in inputs, the finite differences in outputs, and the rate of change in salary per change in year.

TABLE 1 CDs-R-Us Salaries

$\Delta(n)$	Input Year, n	Output, $CD(n)$	$\Delta(CD)$	Rate of Change, $\frac{\Delta(CD)}{\Delta(n)}$
3	0			
	3			
	7			
	12			
	14			
	20			

2. Consider the function $CD(n) = 2000n + 33{,}000$ first developed in Section 2.3.

a. Where does the coefficient of n, 2000, appear in Table 1?

b. Where does the constant term, 33,000, appear in Table 1?

Given the salary function $CD(n) = 2000n + 33{,}000$, the coefficient of n, 2000, is the rate of change of salaries with respect to year.

Symbolically, the rate of change $\frac{\Delta(CD)}{\Delta(n)} = 2000$.

The constant term, 33,000, is the salary when the year equals 0.

1. The text uses commas in 5- and 6-digit numbers for readability although the calculator does not.

If the rate of change $\frac{\Delta(\text{output})}{\Delta(\text{input})}$ is constant, then the data are modeled by a ***linear function***.

The CDs-R-Us salary data have a constant rate of change. So we can use a linear function to express the relationship between year and salary. In fact, arithmetic sequences are special cases of linear functions.

For the linear data, the algebraic model is $y(x) = ax + b$. Notice that $CD(n) = 2000n + 33{,}000$ fits this form. The variables n and CD have been used in place of x and y. The value of a is 2000 and the value of b is 33,000.

We could write $y(x) = 2000x + 33{,}000$ for the CDs-R-Us salary function, where x is the year number and y is the salary in year x.

There are four letters a, b, x, and y in the model $y(x) = ax + b$.

The letters x and y are the ***variables*** that represent the input (year) and the output (salary), respectively.

The letters a and b represent ***parameters***. These values are constant for a given problem situation. When confronted with a new linear relationship, we must find the values of the constants a and b in order to have the specific algebraic model for the problem.

The parameter a is the constant rate of change, which is usually called ***slope***. Recall that the rate of change (slope) is found by computing

$$\text{Slope} = \frac{\Delta(\text{output})}{\Delta(\text{input})}.$$

This is the ratio of the finite difference in the output to the finite difference in the input. The value of a is 2000 for the CDs-R-Us salaries.

The value of the constant b is equal to the output when the input is zero. The value of b is 33,000 for the CDs-R-Us salaries.

If the slope is known, any data point can be substituted into the model to determine b.

Let's consider another set of data.

3. **Electricity Charges:** Table 2 displays a relationship between the number of kilowatts of electricity used in a month and the total pretax charge.

   a. Complete Table 2. The first row is done for you.

   b. Is the data modeled by a linear function? Why?

TABLE 2 Electricity Charges

$\Delta(k)$	Input Kilowatts Used, k	Output Charge C ($)	$\Delta(C)$	Rate of Change, $\frac{\Delta(C)}{\Delta(k)}$
50	0	27.00	4.5	$\frac{4.5}{50} = 0.09$
	50	31.50		
	110	36.90		
	172	42.48		
	253	49.77		
	475	69.75		

c. Using the general model $y(x) = ax + b$, where x is the number of kilowatts used and y is the pretax charge in dollars, identify the value of a (the slope) and the value of b.

d. Write the specific model for the data using the values of a and b from part **c**.

e. Graph the function on your calculator. Describe the "shape" of the graph.

f. Where does the graph intersect the vertical axis? How could you answer this question by just looking at the equation?

g. Draw a function machine for the function.

h. State where the values of the parameters a and b appear in the function machine.

4. Your friend, Izzy, has some data that is linear. Explain to Izzy how to use the data to write a specific equation for the data.

Table 3 is a completed version of Table 2.

TABLE 3 Electricity Charges

$\Delta(k)$	Kilowatts Used, k	Charge in \$(C)	$\Delta(C)$	Rate of Change, $\frac{\Delta(C)}{\Delta(k)}$
50	0	27.00	4.5	0.09
60	50	31.50	5.4	0.09
62	110	36.90	5.58	0.09
81	172	42.48	7.29	0.09
222	253	49.77	19.98	0.09
	475	69.75		

The rate of change $\frac{\Delta(C)}{\Delta(k)}$ is always 0.09, a constant, so the data are modeled by a linear function $y(x) = ax + b$.

Since a is the slope, $a = 0.09$.

Since b is the output when the input is zero, $b = 27$.

So the specific model for the data is $y(x) = 0.09x + 27$, where x is the number of kilowatts used and y is the pretax charge.

A graph of the model appears in Figure 1.

FIGURE 1

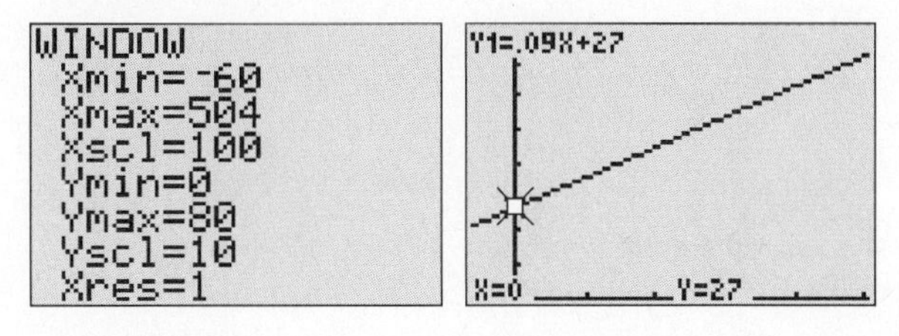

The graph is a straight line, which explains why these functions are called *linear*. Notice that the line intersects the vertical axis (the y-axis) at 27, the value for b.

In general, the ordered pair $(0,b)$, where b is the constant term in $y(x) = ax + b$ is called the ***vertical*, or *y-, intercept***. The vertical intercept is the point where the graph intersects the vertical axis.

Figure 2 displays a function machine.

FIGURE 2

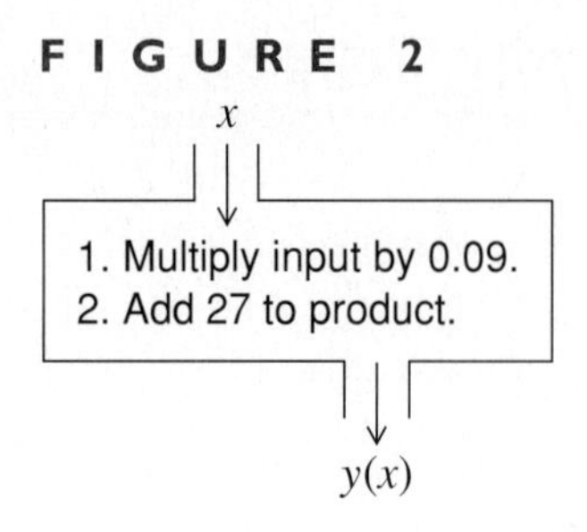

Notice that we multiply the input by the slope a and then add the value of b.

In general, given data that are linear, we need to find values of the parameters a and b in the equation $y(x) = ax + b$ in order to write the specific model. We assign a the value of the slope (change in output over change in input). If the input 0 is given, we use the corresponding output for b.

If the input 0 is not given, we can substitute a data point for x and y to find b.

For example, the slope in the electricity charges problem is 0.09. So we know the equation for the charges is $y(x) = 0.09x + b$.

Let's pick any data point such as (50, 31.5).

Substitute $x = 50$ and $y = 31.5$:	$31.5 = 0.09(50) + b$
Multiply on right side of "=":	$31.5 = 4.5 + b$
Subtract 4.5 from both sides:	$27 = b$

Notice that this is the same value of b that we found before.

Now that we have two examples of linear functions, let's investigate the idea of slope and the corresponding graph a bit more.

5. Suppose the function for CDs-R-Us salaries is $y(x) = 2000x + 40{,}000$ instead of $y(x) = 2000x + 33{,}000$.

a. What has changed and what has remained the same?

b. Graph both functions. Will the graphs of the two lines ever intersect? Why or why not?

6. Suppose the function for electricity charges is $y(x) = 0.09x + 35$ instead of $y(x) = 0.09x + 27$.

   a. What has changed and what has remained the same?

   b. Graph both functions. Will the graphs of the two lines ever intersect? Why or why not?

7. Graph the two functions $y(x) = 2x + 1$ and $y(x) = -\frac{1}{2}x + 1$ in a "square" viewing window. (Ask a buddy or your instructor if you do not know how to do this, or consult the manual.)

   a. What are the slopes of the two lines? How could you calculate one of the slopes given the other slope?

   b. Where do the graphs intersect? Why?

   c. Describe the angle formed by the two graphs at their intersection.

The graphs of the two functions $y(x) = 2000x + 40{,}000$ and $y(x) = 2000x + 33{,}000$ appear in Figure 3.

FIGURE 3

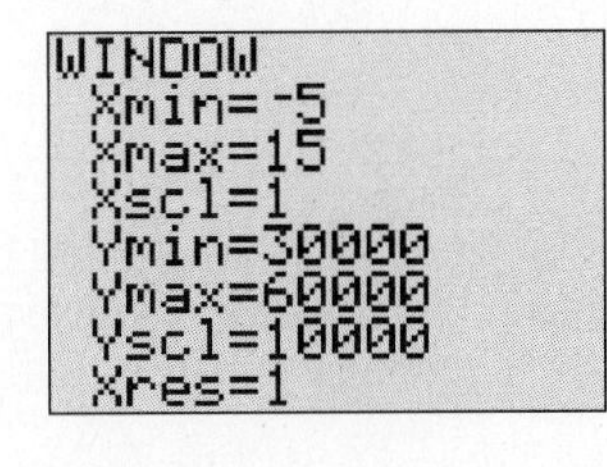

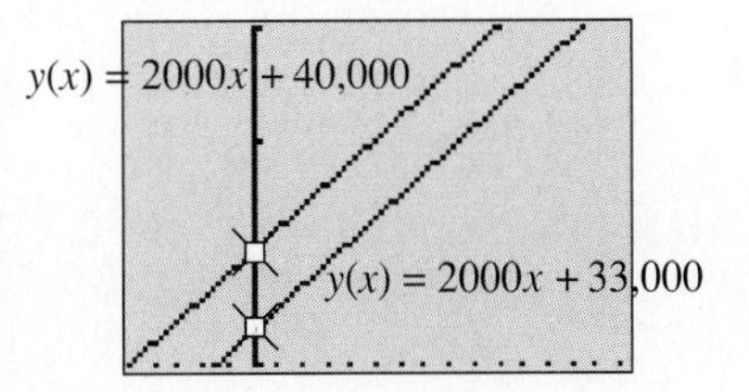

The graphs of the two functions $y(x) = 0.09x + 35$ and $y(x) = 0.09x + 27$ appear in Figure 4.

FIGURE 4

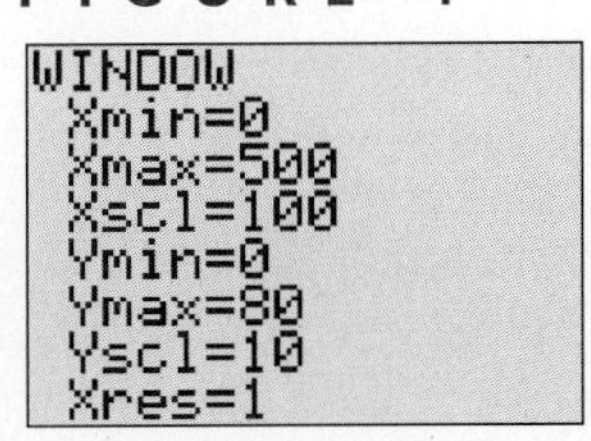

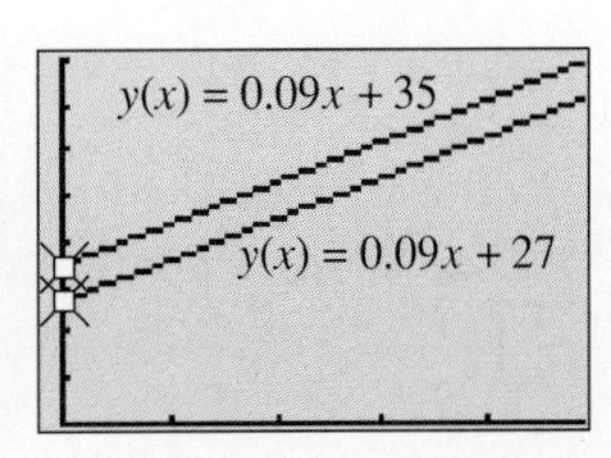

In both cases, the slopes of the two lines are the same, but the vertical intercepts are different. In both cases the two lines will never intersect—they are ***parallel lines***.

In general, if two linear functions have the same slope, but different vertical intercepts, their graphs are parallel lines.

The graphs of the two functions $y(x) = 2x + 1$ and $y(x) = -\frac{1}{2}x + 1$ appear in Figure 5.

FIGURE 5

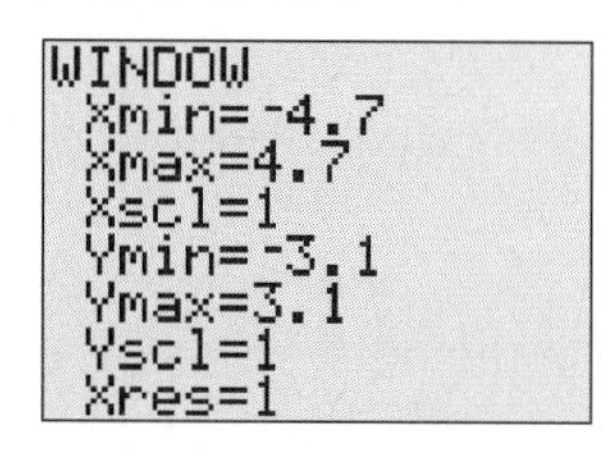

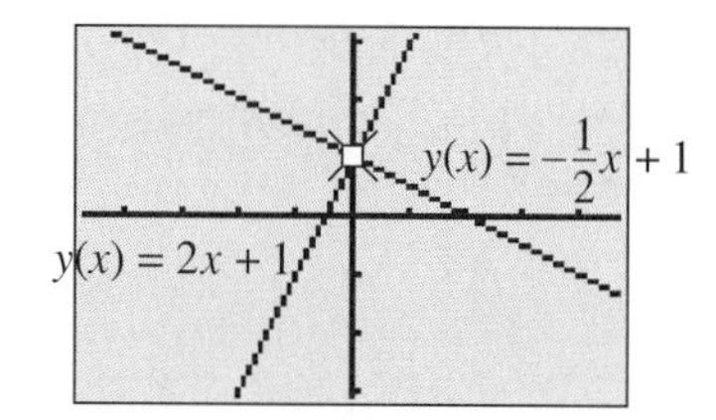

The two lines intersect at their vertical intercept (0, 1). The slopes are 2 and $-\frac{1}{2}$, respectively. The slopes are opposite reciprocals of each other (i.e., the product of the slopes is –1). In this case, the graphs intersect at right, or 90 degree, angles. The two lines are ***perpendicular***.

In general, two lines are perpendicular if their slopes are opposite reciprocals of each other.

Before concluding the section, we look at some other ideas about linear relationships. From the previous investigation, you saw that, given a table, you could compute the slope and then use a point to write the specific linear function. Let's look at creating the equation for the linear function from a graph.

8. Figure 6 displays two points and the graph of a linear function through the two points.

FIGURE 6

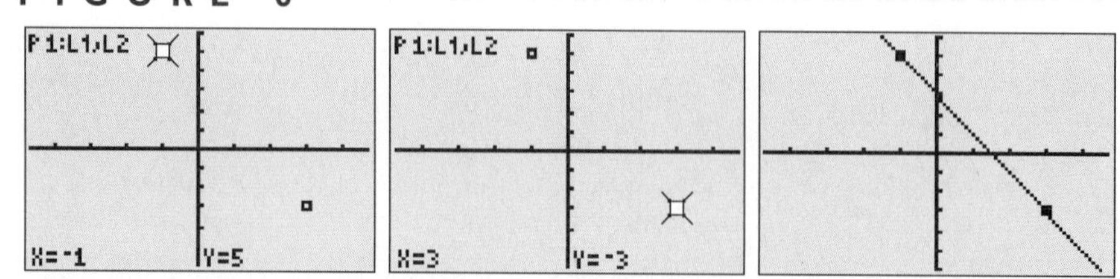

a. Is the information given sufficient to write the equation of the line? Why or why not?

b. Record the two points displayed in Figure 6 as ordered pairs. Enter the two data points in your calculator.

c. Find the slope of the line displayed in Figure 6.

d. Write the linear function whose graph is displayed in Figure 6. Briefly describe what you did. Check your function by graphing the function with the two points on your calculator.

e. Suppose two other points on the line were selected. Would the equation of the function be different? Why or why not?

f. Compare the slope of the function in Figure 1 with the slope of the CDs-R-Us salary function and with the Electricity Charges function. How does the sign of the slope affect the graph of the line?

In order to write the equation of a specific linear function we need at least two pieces of information so that we can determine the values of the two parameters a and b in $y(x) = ax + b$.

There are two ordered pairs, $(-1, 5)$ and $(3, -3)$, displayed in Figure 6. This should be enough information to write the specific equation for the linear function.

Since the parameter a is the slope, we compute the rate of change between the two points as follows:

$$\frac{\Delta\,(\text{output})}{\Delta\,(\text{input})} = \frac{(-3)-(5)}{(3)-(-1)} = \frac{-8}{4} = -2$$

Note that the inputs and outputs must be subtracted in the same order. That is, if we begin the subtraction of outputs with -3, then we must begin the subtraction of inputs with the corresponding input, 3.

We know that $a = -2$ in $y(x) = ax + b$.

So, $y(x) = -2x + b$. We substitute either given point to find b. Suppose we select $(-1, 5)$.

Substitute $x = -1$ and $y = 5$:	$5 = -2(-1) + b$
Multiply on right side of "=":	$5 = 2 + b$
Subtract 2 from both sides:	$3 = b$

The equation of the linear function is $y(x) = -2x + 3$. This can be checked on your calculator.

We would get the same equation regardless of which two points we choose on the line, because the slope is constant and there is only one value of b, the output at the vertical intercept.

The slope of the CDs-R-Us function $y(x) = 2000x + 33{,}000$ is 2000 and the slope of the Electricity Charges function $y(x) = 0.09x + b$ is 0.09. Both slopes are positive. This means the line will rise from left to right, because a ***positive slope*** indicates that, as the input gets larger, so does the output.

The slope of the function whose graph appears in Figure 6 is –2. A negative slope means the line will fall from left to right because the output gets smaller as the input gets larger.

Summarizing, given $y(x) = ax + b$, the ordered pair $(0, b)$ is the vertical intercept of the graph of the line. The parameter a is the slope.

If a is positive, the line rises from left to right (the function is said to be increasing).

If a is negative, the line falls from left to right (the function is said to be decreasing).

If $a = 0$, the output is not changing so the line is horizontal (the function is constant).

The equation $y(x) = ax + b$ is only one of several forms of a linear function. In the next investigation, we look at another form.

9. School Play: Tickets to a school play cost \$10 for adults and \$2 for children. The total receipts were \$988. Let's look at the relationship between the number of adult tickets sold and the number of children's tickets sold. Let x be the number of adult tickets sold and let y be the number of children's tickets sold.

a. Given x adult tickets sold at \$10 apiece, write an expression for the amount of income (dollars) from the sale of adult tickets.

b. Given y children's tickets sold at \$2 apiece, write an expression for the amount of income (dollars) from the sale of children's tickets.

c. Use your answers to parts **a** and **b** to write an equation for the total receipts from the sale of adult and children's tickets. Remember that the total receipts were \$988.

d. Solve the equation you wrote in part **c** for y. This should result in a linear function of the form $y = ax + b$, where x (the input) is the number of adult tickets sold and y (the output) is the number of children's tickets sold.

e. What is the slope of the function you wrote in part **d**? Interpret the slope in terms of the problem situation.

f. What is the vertical intercept of the function you wrote in part **d**? Interpret the vertical intercept in terms of the problem situation.

If x adult tickets are sold at $10 apiece, then the income from adult tickets is $10x$.

If y children's tickets are sold at $2 apiece, then the income from children's tickets is $2y$.

Since the total receipts were $988, we can write $10x + 2y = 988$, which says that the sum of the income from adult tickets and from children's tickets is 988.

This is an example of a linear equation in ***Standard Form***: $Ax + By = C$.

Previously we have worked with linear equations in ***slope-intercept form***: $y = ax + b$.

Linear equations in standard form can be written in slope-intercept form by solving the equation for the output variable—in this case, y. In fact, we usually do this to express the relationship as a function.

Given: $10x + 2y = 988$

Subtract $10x$ from both sides to isolate the y term on one side of "=": $2y = 988 - 10x$

Divide both sides by 2, the coefficient of y: $y = 494 - 5x$

Commute terms on right side of "=": $y = -5x + 494$

The linear function has a slope of –5 and a vertical intercept of $(0, 494)$.

A slope of –5 means that for each adult ticket sold, the number of children's tickets sold decreases by 5.

The vertical intercept $(0, 494)$ indicates that, if no adult tickets are sold, there would be 494 children's tickets sold.

STUDENT DISCOURSE

Pete: Sometimes I get confused by all the letters. For example, $y(x) = ax + b$ has four letters in it. Some are supposed to stand for input and output and some are not. How do I know?

Sandy: I know what you mean—they confused me also. You need to keep in mind that the equation $y(x) = ax + b$ is a very general one and that the data help us create a specific equation.

Pete: I understand that.

Sandy: How can you identify the variables used for input and output?

Pete: The left side tells me that. Since I see $y(x)$, that's function notation, so x is the input and y is the output.

Sandy: Good. What letters are left?

Pete: That's easy—a and b.

Sandy: These letters get their values from the data. In other words, to write the specific function, you have to replace the a and the b with numbers.

Pete: So the a and the b will not be in the specific function I write?

Sandy: Correct. You'll end up with an equation like $y(x) = 2x - 5$. See, there's no a and b in there.

Pete: Would a be 2 and b be 5 in that example?

Sandy: Almost—watch your sign on b.

Pete: Oh, you mean b would be –5?

Sandy: Right.

Pete: Let me see if I have it. We begin with a general model like $y(x) = ax + b$ and the data helps us find a specific function by finding a and b.

Sandy: Yes, and this section gave you some procedures to follow to find a and b. That's what modeling data with functions is all about.

Pete: One other question—at the end of the section, there were two different forms of a linear equation $y = ax + b$ and $Ax + By = C$. Are the a and the A the same? Are the b and the B the same?

Sandy: Good question. All the letters you mention are parameters, but they represent different numbers. So a is usually different from A. That's why one is capitalized and the other is not. The same goes for b and B.

Pete: That's what I figured, but I wasn't sure. Thanks.

1. In the Glossary, define the words in this section that appear in ***italic boldface*** type.

2. **Credit Hour Fees:** The fee structure listed in Table 4 represents the total fees for a college student as determined by the number of credit hours the student takes.

TABLE 4 Total Fees as a Function of Credit Hours

Credit Hours	Total Fees ($)
1	57
3	121
6	217
7	249
9	313
11	377
12	409

a. Total fees are a function of credit hours. Is this function linear or not? Why?

If the data is not linear, you do not have to do parts **b** and **c**.

b. Write a function with input the number of credit hours and output the total fees. Describe what you did.

c. What is the slope of your function? What does the slope mean in this problem situation?

d. Use your function to determine the total fees for 20 credit hours.

3. Carrie's Job Offer: Carrie has been offered a job with a starting salary of $31,500. She estimates that she will receive, on the average, a 5% increase each year. Table 5 displays her annual salary for the next 10 years.

TABLE 5 Carrie's Annual Salary by Year

Year	Salary ($)
1	31,500.00
2	33,075.00
3	34,728.75
4	36,465.19
5	38,288.45
6	40,202.87
7	42,213.01
8	44,323.66
9	46,539.85
10	48,866.84

a. Salary is a function of year. Is this function linear or not? Why?

If the data is not linear, you do not have to do parts **b** and **c**.

b. Write a function with input the year and output the salary. Describe what you did.

c. Use your function to determine the salary after 20 years. Describe what you did.

4. Health Care Costs: Health care costs have been rising at a rate of almost 20% annually. Assume that a certain medical test cost $100 in

1981. Consider the sequence defined by the cost of this test in successive years since 1980.

a. The cost of the medical test is a function of the years since 1980. Is this function linear or not? Why?

If the data is not linear, you do not have to do parts **b** and **c**.

b. Write a function with input the number of years since 1980 and output the cost of the test. Describe to a friend how you created this function.

c. Use your function to find the cost of the test in 1995. Describe your answer in a complete sentence.

5. Carla's Bank Account: Carla opened an account by depositing $1000. She deposits an additional $200 each month.

a. The amount Carla has on deposit, not including interest, is a function of the month. Is this function linear or not? Why?

If the data is not linear, you do not have to do parts **b** and **c**.

b. Write a function with input the number of months and output the total amount that has been deposited. How did you create this function?

c. Use your function to find how much has been deposited after three years. Describe your answer in a complete sentence.

6. Consider the data in Table 6.

TABLE 6 **Random Data I**

Input	1	5	9	13	17	20
Output	23	20	17	14	11	8

a. Is this function linear or not? Why?

If the data is not linear, you do not have to do parts **b–d**.

b. Write an equation that will generate the output from the input. Explain what you did.

c. Use your equation to find the output when the input is 57.

d. Find the input if the output is –85. Explain what you did.

7. Consider the data in Table 7.

TABLE 7

Input	1	2	3	4	5	6
Output	78	70.2	63.18	56.862	51.1758	46.05822

a. Is this function linear or not? Why?

If the data is not linear, you do not have to do parts **b–d**.

b. Write an equation that will generate the output from the input. Explain what you did.

c. Use your equation to approximate the output when the input is 23.

d. Find the input if the output is approximately 14.45. Explain what you did.

8. Triangular Numbers: Consider the sequence 1, 3, 6, 10, 15, 21,...

a. The sequence element is a function of its position. Is this function linear or not? Why?

b. Can you write a specific function with input position and output the sequence element? Why or why not?

9. Given the following linear equations in standard form, write them in the form $y = ax + b$ and identify the slope and vertical intercept.

a. $5x + y = 2$

b. $x + 7y = 4$

c. $4x + 9y = -3$

d. $-5x - 7y = 1$

e. $11x - 3y = 0$

f. $-4x + 2y = -14$

10. A line contains the points $(-3, 4)$ and $(5, 7)$.

a. Write the linear function for the line.

b. Write the equation of a line parallel to the given line through $(2, -3)$.

c. Write the equation of the line perpendicular to the given line through $(2, -3)$.

11. Write the equation of the line

a. parallel to $5x + 4y = -2$ through the point $(6, -1)$.

b. perpendicular to $5x + 4y = -2$ through the point $(6, -1)$.

12. Which of the following graphs could be the graph of $y = 40 - 0.2x$? Justify your choice.

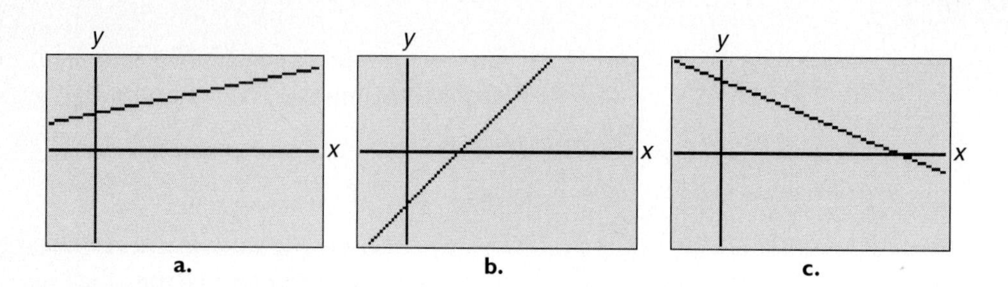

13. A linear function satisfies $f(2) = 7$ and $f(-3) = 14$. Find parameters a and b and write the linear function $f(x) = ax + b$.

14. If $y = ax + b$, what can be said about a and b for each graph below.

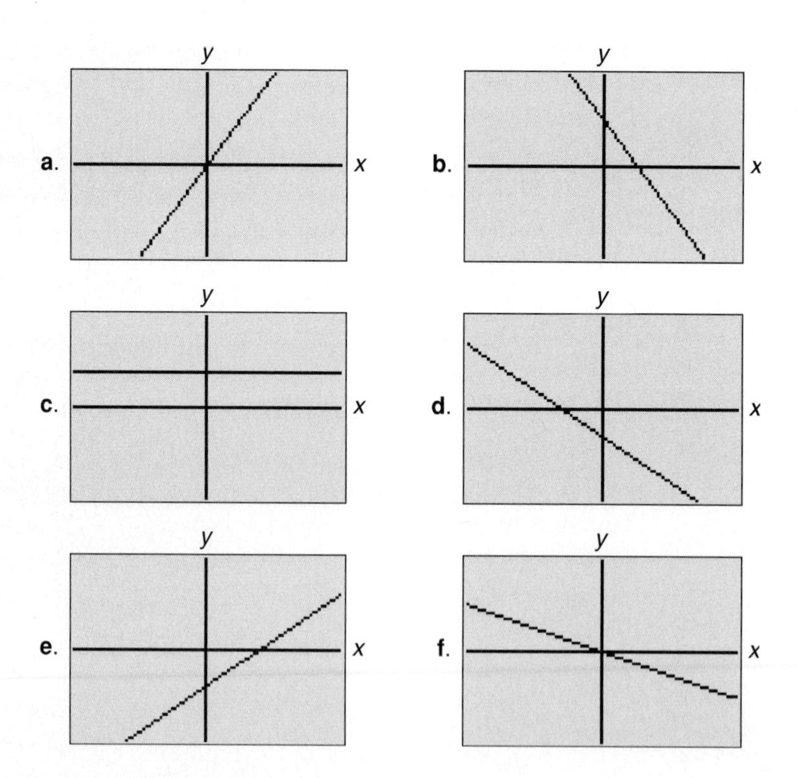

REFLECTION

Write a paragraph describing what you know about linear functions. Be sure to address numeric (table), symbolic (equation), and geometric (graph) aspects of linear functions. Include statements in your paragraph like:

I used to think . . .

Now I think . . .

because . . .

SECTION 3.2

Exponential Models for Change

Purpose

- Establish connections between geometric sequences and exponential models.
- Compare and contrast basic linear and exponential models.

In the last section we focused on linear functions. Linear functions have a constant rate of change. In this section, we consider functions in which the finite ratios of successive outputs are constant. This allows us to introduce a new class of functions: exponential functions. To do so, we look back at the Super Dog Burger Bits salary data.

The Super Dog Burger Bits salary data is not linear. The rates of change in salaries per year are increasing rather than remaining constant. However, the finite ratios between successive salaries are constant.

We wrote $SD(n) = 25{,}000(1.07)^{n-1}$, where n is the year number and $SD(n)$ is the salary in year n. The 1.07 is called the ***base*** of the exponent.

This function can be written as $SD(n) = 23{,}364.48598(1.07)^n$. Let's see if this is correct.

1. Enter the function $SD(n) = 25{,}000(1.07)^{n-1}$ and the function $SD(n) = 23364.48598\,(1.07)^n$ into your calculator. Use the table feature to complete the second column and third columns of Table 1. Complete the rest of the table by computing the finite ratios in successive outputs.

2. How do the salaries in columns 2 and 3 compare?

TABLE 1 Super Dog Burger Bits Salaries

Year, n	Salary Using $25000(1.07)^{n-1}$	Salary Using $23{,}364.48598(1.07)^n$	Finite Ratios of Salaries, ®S
0			
1			
2			
3			
4			
5			
6			

3. Consider the salary function $SD(n) = 23364.48598\,(1.07)^n$.

a. Where does the base, 1.07, appear in Table 1?

b. Where does the constant factor, 23364.48598, in the function $SD(n) = 23364.48598\,(1.07)^n$ appear in Table 1?

DISCUSSION

We first note from Table 1 that the two salary functions $SD(n) = 25000\,(1.07)^{n-1}$ and $SD(n) = 23364.48598\,(1.07)^n$ have the same outputs.

We will focus on the function $SD(n) = 23364.48598\,(1.07)^n$ for the remainder of this discussion.

The base 1.07 is equal to the finite ratios in salaries (the fourth column of Table 1).

The constant factor, 23364.48598, is the output (salary) when the input is zero (year zero).

If the finite ratios of successive outputs are constant, then the data are modeled by an ***exponential function***. Note that ***successive outputs*** means that the corresponding inputs are increasing by 1 (i.e. 1, 2, 3, 4, . . .).

The Super Dog Burger Bits salary data have a constant finite ratio in consecutive outputs. Therefore we can use an exponential function to express the relationship between year and salary. In fact, geometric sequences are special cases of exponential functions.

For exponential data, an algebraic model is $y(x) = a(b)^x$. The function $y(x)$ is called the "exponential function" with **base** b. Notice that $SD(n) = 23364.48598(1.07)^n$ fits this form. The variables n and SD have been used in place of x and y. The value of b is 1.07 and the value of a is 23364.48598.

We could write $y(x) = 23364.48598(1.07)^x$ for the Super Dog Burger Bits salary function, where x is the year number and y is the salary in year n.

There are four letters a, b, x, and y in the model $y(x) = a(b)^x$.

The letters x and y are the variables that represent the input (year) and the output (salary), respectively. When we developed the linear model, the input variable was multiplied by the constant slope. For exponential functions, the base (finite ratio) is constant and the exponent acts as the input variable. One of the reasons this is called an *exponential model* is the fact that the input variable is an *exponent*.

The letters a and b represent ***parameters***. These values are constant for a given problem situation. When confronted with a new exponential relationship, we must find the values of the constants a and b in order to have the specific algebraic model for the problem.

The value of constant b is the constant finite ratio in successive outputs. Since we want successive outputs, the finite differences in inputs are all 1.

The value of constant a is the output when the input is zero.

Let's consider another set of data.

4. Television Depreciation: Terra just bought a new television. The value of the television depreciates over time. Table 2 displays the relationship between the number of months Terra has owned the television and how much the television is still worth. The numbers in the second column have been rounded to the nearest cent.

a. Complete Table 2. The first row is done for you.

b. Is the data modeled by an exponential function? Why?

TABLE 2 Terra's Television

Month, n	Television's Worth, W ($)	Finite Ratios of Televisions Worth, ®W
0	599.00	$\frac{569.05}{599} = 0.95$
1	569.05	
2	540.60	
3	513.57	
4	487.89	
5	463.49	
6	440.32	

c. Using the general model $y(x) = a(b)^x$, where x is the number of months Terra has owned the television and y is the worth of the television, identify the value of b (the base) and the value of a.

d. Write the specific model for the data using the values of a and b from part **c**.

e. Graph the function on your calculator. Describe the "shape" of the graph.

f. Where does the graph intersect the vertical axis? How could you answer this question by just looking at the equation?

g. Draw a function machine for the function.

h. Identify where the values of the parameters a and b appear in the function machine.

5. Your friend, Izzy, has some data that is exponential. Explain to Izzy how to use the data to write a specific equation for the data.

The third column of Table 2 contains the constant 0.95. The fact that the finite ratios in successive outputs is constant demonstrates that the data is exponential.

We model the data with the exponential function $y(x) = a(b)^x$.

Since b is the finite ratio in successive outputs (the base), $b = 0.95$.

Since a is the output when the input is zero, $a = 599$.

So the specific model for the data is $y(x) = 599(0.95)^x$, where x is the number of months since Terra bought the television and y is the value of the set ($).

A graph of the model appears in Figure 1.

FIGURE 1

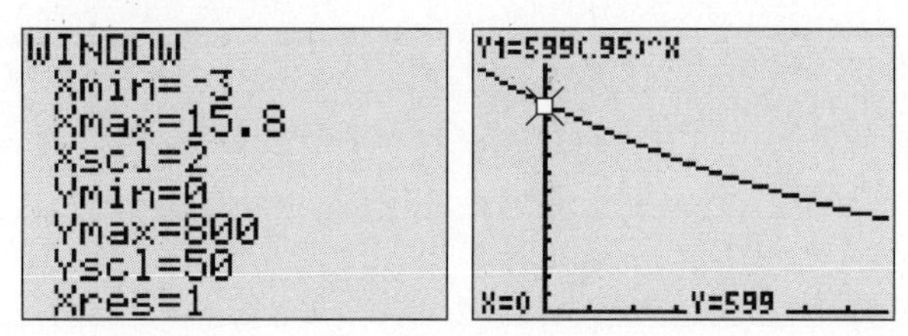

Notice that the graph is *not* a straight line and that the curve intersects the vertical axis (the y-axis) at 599, the value of a.

In general, the ordered pair $(0, a)$, where a is the constant factor in $y(x) = a(b)^x$, is called the vertical, or y, intercept.

Notice, also, that as the input increases, the output decreases. This occurs when the base b (in this case, 0.95) is less than 1, but larger than 0.

By comparison, the base of the Super Dog Burger Bits salary function is 1.07, which is larger than 1. So, as the inputs increase, the outputs of the Super Dog Burger Bits salary function also increase.

Figure 2 displays a function machine.

FIGURE 2

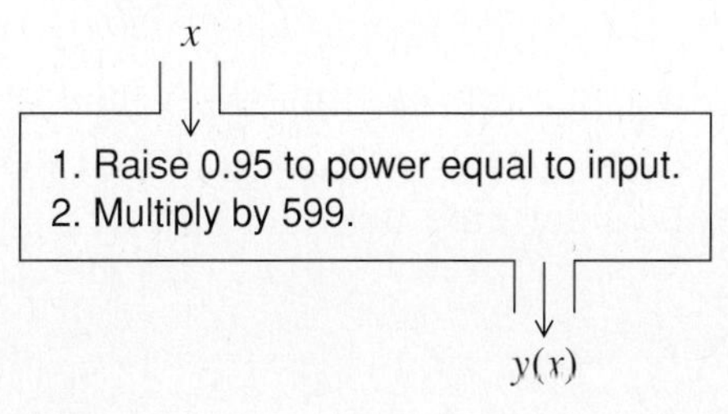

In general, given data that is exponential, we need to find the values of the parameters a and b in the equation $y(x) = a(b)^x$ in order to write the specific model. We assign b the value of the constant ratio of change in consecutive outputs. If the input 0 is given, we use the corresponding output for a.

Before concluding the section, let's use the salary functions for CDs-R-Us and Super Dog Burger Bits to answer some interesting questions. Remember from Section 3.1 that the linear function for the CDs-R-Us salary data is $y(x) = 2000x + 33{,}000$, where x is the year number and y is the salary in year x.

6. Use your linear function for CDs-R-Us and your exponential function for Super Dog Burger Bits to complete Table 3.

TABLE 3 Comparative Salaries: Years 7–15

Year	CDs-R-Us Salary	Super Dog Burger Bits Salary
7		
8		
9		
10		
11		
12		
13		
14		
15		

7. Use parametric mode to graph both salary functions. (Ask your instructor or a friend how to use parametric mode, or consult your manual.) This involves defining the functions as follows (Figure 3).

FIGURE 3

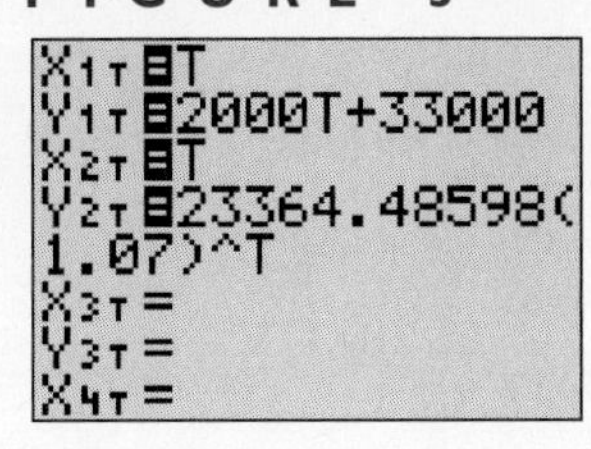

Be sure to set an appropriate viewing window based on your tables of data. Estimate the year when the salaries would be approximately equal. How can you use the graph to answer this question?

8. Suppose that you stay at the job for 30 years.

a. What will be your salary for the 30th year at CDs-R-Us? Defend your answer.

b. What will be your salary for the 30th year at Super Dog Burger Bits? Defend your answer.

9. Your goal is to make $100,000 annually.

a. Use a table or graph to determine when this will happen at CDs-R-Us.

b. Use a table or graph to determine when this will happen at Super Dog Burger Bits.

When two variables, x and y, are defined in terms of a third variable t, the resulting equations are called ***parametric equations***. This is accomplished by letting the input variable equal t and letting the output variable equal the mathematical expression that defines it, but in the variable t, rather than the variable x.

Figure 4 displays a table of comparative salaries and the corresponding graphs, all done in parametric mode.

The reason for using parametric mode relates to the domain of these functions. For both problems, the domain of the problem set is the set of natural numbers 1, 2, 3, 4,.... Graphing in standard function mode will not easily

FIGURE 4

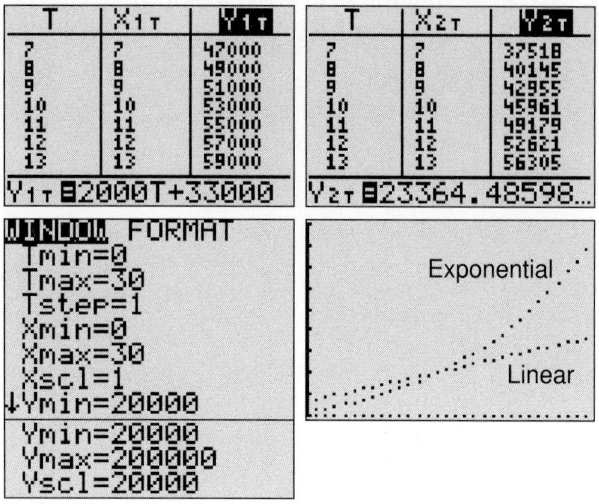

T	X_{1T}	Y_{1T}
7	7	47000
8	8	49000
9	9	51000
10	10	53000
11	11	55000
12	12	57000
13	13	59000

Y_{1T}=2000T+33000

T	X_{2T}	Y_{2T}
7	7	37518
8	8	40145
9	9	42955
10	10	45961
11	11	49179
12	12	52621
13	13	56305

Y_{2T}=23364.48598...

WINDOW FORMAT
Tmin=0
Tmax=30
Tstep=1
Xmin=0
Xmax=30
Xscl=1
↓Ymin=20000
Ymin=20000
Ymax=200000
Yscl=20000

Exponential
Linear

allow us to limit inputs to natural numbers. However, in parametric mode, we can control the values of the input by the way we set the values of t.

We now have three techniques available to answer questions about the models. To determine the salary in year 30, we can

a. move down the table until the input equals 30,

b. substitute the input of 30 into the defining functions, thus evaluating each function when the input is 30, or

c. trace on the graph until the input reaches 30.

To determine when the annual salary would reach $100,000, we can

a. use the table to determine when the output is approximately 100000.

b. use the graph to trace to the point when the salary is approximately 100000 (that is, $100,000).

It is more complex to use the model to find the input when the output is given. This involves ***solving an equation*** rather than ***evaluating an expression***. The two equations that represent the case where the annual salary reaches $100,000 are as follows (shown without commas as on the calculator.)

$$100000 = 2000x + 33000$$

and

$$100000 = 23364.48598(1.07)^x$$

We will investigate solving these algebraically later in the text.

Finally, we can determine the year when the salaries are approximately equal by

a. scanning the table until the two outputs are approximately equal, or

b. locating the point of intersection of the two graphs.

Again, to do this task using the defining functions, we would have to solve a relatively complex equation.

Pete: I have a question from way back in the beginning of the section.

Sandy: Go ahead.

Pete: Well, I saw on the calculator that $SD(n) = 23364.48598(1.07)^n$ and $SD(n) = 25000(1.07)^{n-1}$ produce the same outputs. So they must be the same function, but I don't know why.

Sandy: Yeah, that's a tough one. You have to remember some exponent rules. Let me see if I can help. Divide 25000 by 1.07.

Pete: That's about 23364.48598, which is the value of a in $SD(n) = 23364.48598(1.07)^n$.

Sandy: Okay. So you know that $SD(n) = 23364.48598(1.07)^n$ is the same as $SD(n) = \frac{25000}{1.07}(1.07)^n$.

Pete: I'm with you so far.

Sandy: Now write the last equation as $SD(n) = 25000\frac{(1.07)^n}{1.07}$.

Pete: Okay, I'm starting to see some similarity. So $\frac{(1.07)^n}{1.07}$ must be the same as $(1.07)^{n-1}$. But why?

Sandy: You need to recall a rule of exponents. What is $\frac{x^7}{x^3}$?

Pete: I think it is x^4. Basically, three factors of x divide out.

Sandy: That's correct. But how could you answer that just by looking at the exponents?

Pete: Oh, yeah. I think I remember now. I subtract the exponents, $7-3$, to get 4. And I keep the base of x. In fact, don't I have to have the same base in both numerator and denominator?

Sandy: That's right. Now think about $\frac{(1.07)^n}{1.07}$ again.

Pete: Well, the base is 1.07 in both numerator and denominator. So I guess I can subtract exponents.

Sandy: What do you get?

Pete: Hmmm. The exponent in the denominator is 1, I guess.

Sandy: That's right.

Pete: Oh, so I get an exponent of $n-1$. That's it! $\frac{(1.07)^n}{1.07}$ is equal to $(1.07)^{n-1}$. I see why they are the same function. Thanks. That was tough.

Sandy: You are welcome. I had a tough time with that one myself.

1. In the Glossary, define the words in this section that appear in ***italic boldface*** type.

2. Carrie's Job Offer: Carrie has been offered a job with a starting salary of \$31,500. She estimates that she will receive, on the average, a 5% increase each year. Table 4 displays her annual salary for the next 10 years.

a. Salary is a function of year. Is this function linear, exponential, or neither? Why?

b. Write a function with input the year and output the salary. Describe what you did.

c. Use your function to determine the salary after 20 years. Describe what you did.

TABLE 4 **Carrie's Annual Salary by Year**

Year	Salary ($)
1	31,500.00
2	33,075.00
3	34,728.75
4	36,465.19
5	38,288.45
6	40,202.87
7	42,213.01
8	44,323.66
9	46,539.85
10	48,866.84

3. **Health Care Costs:** Health care costs have been rising at a rate of almost 20% annually. Assume that a certain medical test cost $100 in 1981. Consider the sequence defined by the cost of this test in successive years since 1980.

   a. The cost of the medical test is a function of the years since 1980. Is this function linear, exponential, or neither? Why?

   b. Write a function with input the number of years since 1980 and output the cost of the test. Describe to a friend how you created this function.

   c. Use your function to find the cost of the test in 1995. Explain your answer in a complete sentence.

4. Consider the data in Table 5.

TABLE 5 Random Data 1

Input	2	4	6	8	10	12
Output	42	35	28	21	14	7

a. Is this data linear, exponential, or neither? Why?

b. Write an equation that will generate the output from the input. Explain what you did.

c. Use your equation to find the output when the input is 57.

d. Find the input if the output is –85. Explain what you did.

5. Consider the data in Table 6.

TABLE 6 Random Data 2

Input	1	2	3	4	5	6
Output	78	70.2	63.18	56.862	51.1758	46.05822

a. Is this data linear, exponential, or neither? Why?

b. Write an equation that will generate the output from the input. Explain what you did.

c. Use your equation to approximate the output when the input is 23.

d. Find the input if the output is approximately 14.45. Explain what you did.

6. Lottery Doubles: Recall the lottery problem: You have just won a lottery in which you will receive money on each of 30 consecutive days. The payoff structure is as follows: 1¢ the first day, 2¢ the second day, 4¢ the third day, 8¢ the fourth day, 16¢ the fifth day, and so on.

a. The amount received on a given day is a function of the day. Is this function linear, exponential, or neither? Why?

b. Write a general function with input the day number and output the amount paid by the lottery on that day.

c. Use the function you wrote to determine how much is paid on the last day.

7. Lottery Doubles: Suppose you are paid 2¢ on the first day instead of 1¢.

a. Write a general function with input the day number and output the amount paid by the lottery on that day.

b. Use the function you wrote to determine how much is paid on the last day.

c. How much more do you receive on the last day under this plan than under the original plan?

8. Triangular Numbers: Consider the sequence 1, 3, 6, 10, 15, 21, . . .

a. The sequence element is a function of its position. Is this function linear, exponential, or neither? Why?

b. Can you write a general function with input position and output the sequence element? Why or why not?

9. Given the following linear equations in standard form, write them in the form $y = ax + b$ and identify the slope and vertical intercept. Graph each function.

a. $-4x + 3y = 7$

b. $x - 7y = -2$

10. Consider the linear function f in which $f(2) = -5$ and $f(-6) = 7$

a. Use the information given to write two ordered pairs on the line.

b. Write the linear function for the line.

c. Write the equation of a line parallel to the given line through (0, 1).

d. Write the equation of the line perpendicular to the given line through (0, 1).

11. Write the equation of the line

a. parallel to $3x - 2y = -5$ through the point (–4, 2).

b. perpendicular to $3x - 2y = -5$ through the point (–4, 2).

12. Write an example of an exponential function in which

a. the function is increasing and the vertical intercept is (0, 8).

b. the function is decreasing and the vertical intercept is (0, 17).

13. An exponential function contains the points (0, 3) and (1, 6).

a. Identify the base of the exponential function.

b. Write the specific equation of the exponential function.

14. An exponential function contains the points (1, 4.2) and (2, 2.52).

a. Identify the base of the exponential function.

b. Write the specific equation of the exponential function.

15. Consider the graph of the linear function.

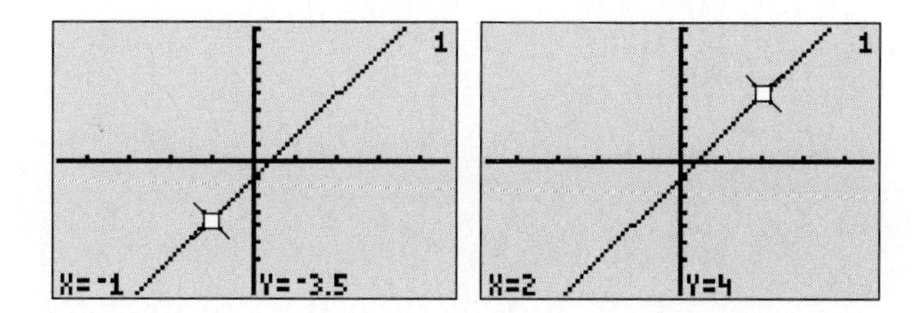

a. Find the slope of the line.

b. Write the specific equation of the linear function.

16. Consider the graph of the exponential function.

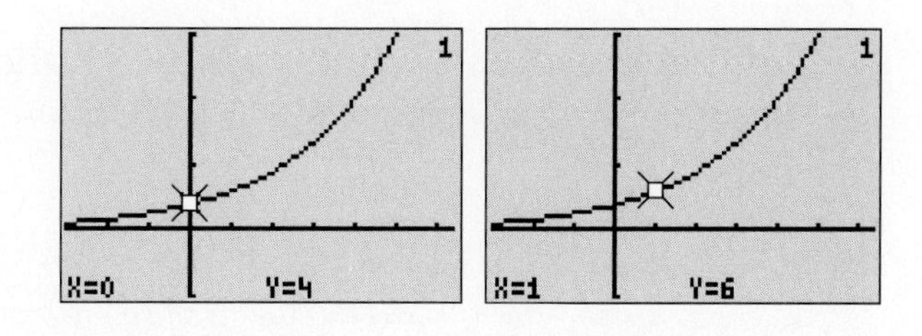

a. Find the base of the exponential function.

b. Write the specific equation of the exponential function.

REFLECTION

Write a paragraph comparing and contrasting linear functions and exponential functions.

SECTION 3.3

Building Quadratic Models

Purpose

- Introduce quadratic models.
- Develop techniques for recognizing the value of the parameters in quadratic models.

We have just learned how to create specific linear and exponential models from data by finding the values of the parameters for the general models.

We use a linear model when there is a constant rate of change in output with respect to input. We use an exponential model when there is a constant finite ratio in successive outputs.

What if neither of these situations occurs? Sometimes, the first finite differences in consecutive outputs are *not* constant, but the second finite differences in outputs are constant. In order to model such data, we need to study ***quadratic models***, which have the form $y(x) = ax^2 + bx + c$. Notice that x is the input variable and y is the output variable and that there are three parameters a, b, and c. Our task, once we know the data are quadratic, will be to find the values for a, b, and c.

Let's look at three specific problem situations that have a quadratic as a model. From these examples, we will explore how the model is created from the data.

Virtual Reality Experiences Revenue: You probably recognize, as a consumer, that the demand for a product often increases as the price decreases. A ***demand equation*** expresses the relationship between price and the number of items demanded.

In particular, the demand for a Virtual Reality Experiences headset is given by the equation

$$x = \frac{1}{2}(10{,}000 - p),$$

where x is the number of headsets sold and p is the price per headset. Since we want the input to be the number of headsets sold, we solve this equation for p in terms of x to get

$$p = 10{,}000 - 2x.$$

This equation has as its input x (number of headsets sold) and expresses the price per headset as a function of x.

Companies use a ***revenue equation*** to calculate the revenue if they sell a certain number of units. This equation has the form:

$$R = (\text{number of units sold})\,(\text{price per unit}).$$

Let's investigate the revenue equation for Virtual Reality Experiences headsets.

1. Given that x headsets are sold and p is the price per headset, write the revenue equation using only variables x and p.

2. Replace the variable p in your answer to Investigation 1 with the equivalent expression involving x. (*Hint*: See the definition of p as a function of x in the problem introduction.)

3. Multiply out the indicated product in your answer to Investigation 2.

4. A general quadratic model has form $y(x) = ax^2 + bx + c$. Your answer to Investigation 3 is a specific quadratic model. For this specific model, identify the values of the constants a, b, and c.

5. Use your answer to Investigation 3 to complete Table 1.

TABLE 1 Virtual Reality Experiences Headset Revenue

Number of Headsets Sold	Revenue, R (\$)	First Finite Difference, ΔR	Second Finite Difference, $\Delta(\Delta R)$
0			
1			
2			
3			
4			
5			
6			

6. Describe what you notice about the first and second finite differences in consecutive revenues.

The revenue equation for the Virtual Reality Experiences headsets can be written in any of the following three ways.

$$R = (x)(p)$$

Substitute $10{,}000 - 2x$ for p: $\quad R = (x)(10{,}000 - 2x)$

Multiply and commute: $\quad R = -2x^2 + 10{,}000x$

We have our first example of an actual quadratic model.

In general, quadratic models have the form

$$y(x) = ax^2 + bx + c.$$

This model has five letters a, b, c, x, and y.

The letters x and y are the input and output variables, respectively.

The letters a, b, and c are parameters. They are constants that, once they are known, give us a specific quadratic model. To create a quadratic model from data, we must determine the values of a, b, and c.

In the specific model $R = -2x^2 + 10{,}000x$, we see that

$$a = -2, \quad b = 10{,}000, \quad \text{and} \quad c = 0.$$

While the first finite differences in revenue are not constant, the second finite difference in revenue is a constant of –4. The fact that this second finite difference is constant indicates that the model to be quadratic.

Let's look at a second problem situation that has a quadratic as a model.

Fencing In the Dog: We will use 200 meters of fence to create a rectangular dog pen for our golden retriever, Katie. (Katie eats Super Dog Burger Bits, by the way.) Eventually, we want to consider the relationship between the length of the sides of the pen and the area of the pen. For now though, let's look at the relationship between the length of the sides of the pen when the perimeter is known.

7. Figure 1 displays five different pens with the length given. Write the width of each in the space provided. Remember we have only 200 m of fence and we will use it all.

FIGURE 1

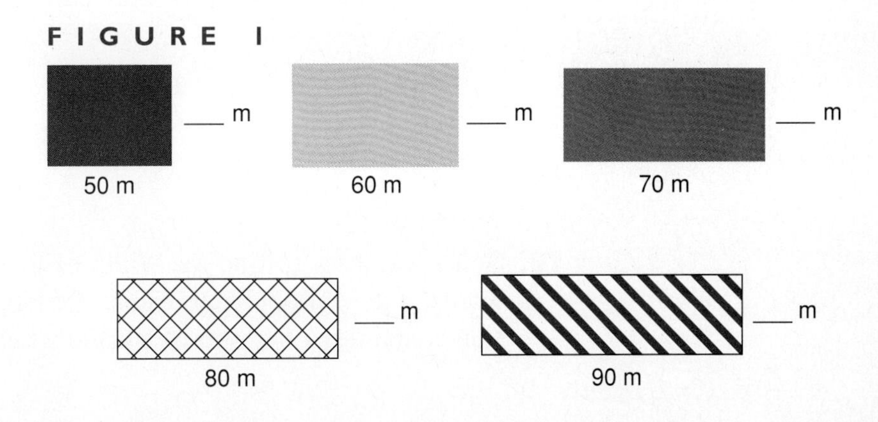

8. Describe the process you used to calculate the width from the length.

9. Record the results of Figure 1 in Table 2. In the third column, record the process for calculating the width rather than the answer to the width. Included in the table are several other length–width combinations.

TABLE 2 **Length and Width of Dog Pen**

Length of Dog Pen (m)	Width of Dog Pen (m)	Process for Finding Width
50		
55		
60		
65		
70		
75		
80		
85		
90		

10. Generalize how to find the width of the pen given the length. In other words, if the variable l represents the length of the pen, write an expression containing l that represents the width.

Now we consider the area of the pen.

11. a. Write an area formula for a rectangle in terms of the letters l (for length) and w (for width).

b. Write an area formula for the dog pen that uses only the letter l (for length). Use the result of Investigation 10 to help you.

c. If you haven't done so, multiply out your answer to part **b**.

12. The area function you wrote should be quadratic. Identify the values of the parameters a, b, and c.

13. Use the area function to complete Table 3.

TABLE 3 Area of Dog Pen

Length, l (m)	Area, A (sq. m.)	Finite Difference in Area ΔA	Second Finite Difference in Area $\Delta(\Delta A)$
0			
1			
2			
3			
4			
5			
6			
7			

14. What do you notice about the first and second finite differences in consecutive areas?

You may have noticed by completing Figure 1 or Table 2 that the sum of the length and the width is 100. A portion of Table 2 is replicated as Table 4 to display the process of finding the width from the length.

For each length l, the process to find the width is $100 - l$.

TABLE 4 Length and Width of Dog Pen

Length of Dog Pen (m)	Width of Dog Pen (m)	Process for Finding Width
50	50	100 – 50
55	45	100 – 55
60	40	100 – 60
65	35	100 – 65
70	30	100 –70

With that in mind, we can create the function for area of a rectangle.

Area of a rectangle is the product of the length and width:

$$A = l\,w$$

But w is equal to $100 - l$: $A = l(100 - l)$

Multiply out and commute: $A = -l^2 + 100l$

This is a quadratic model with $a = -1$, $b = 100$, and $c = 0$.

For this model the first finite differences vary, but the second finite difference is a constant of –2.

Before we summarize, let's look at one more problem situation commonly modeled by a quadratic function.

First, a bit of background. If an earthly object is projected straight up into the air, then the height h (in feet) of the object above the Earth is given by a quadratic function of the time t (in seconds) elapsed since the object was shot or thrown into the air. The general model is

$$h(t) = -16t^2 + v_0 t + h_0,$$

where v_0 is the initial velocity (in feet per second) and h_0 is the initial height (in feet) above the Earth.

Note that we have given the formula $h(t) = -16t^2 + v_0 t + h_0$ without derivation. The derivation can be done in a later math course or in a physics course.

For now, we make use of the formula without knowing all the background about where it comes from.

Let's look at a particular instance where this formula is useful.

Jordan's Power Toss: Michael Jordan throws a baseball vertically into the air with an initial velocity of 125 feet per second. He releases the ball 8 feet above the ground. After reaching some maximum height, the ball stops and turns downward. Let h feet represent the height of the ball after t seconds.

15. Write the specific algebraic model for the power toss.

16. Write the values for a, b, and c for the specific quadratic model of the power toss.

17. Complete Table 5.

TABLE 5 Jordan's Power Toss

Time t Since Launch	Height, h	First Finite Differences in Height, Δh	Second Finite Differences in Height, $\Delta(\Delta h)$
0			
1			
2			

18. What do you notice about the first and second finite differences in consecutive heights?

The specific model for the power toss is

$$h(t) = -16t^2 + 125t + 8$$

In the original general form, there were two parameters v_0 and h_0. Both of these contain subscripts to differentiate them from a variable. The subscript zero is used to indicate the value at the start of the motion; in other words, the value of the parameters when $t = 0$.

For the power toss, $v_0 = 125$ and $h_0 = 8$.

Once we have $h(t) = -16t^2 + 125t + 8$, we can identify the values of the parameters in this specific quadratic model.

For this model, $a = -16$, $b = 125$, and $c = 8$.

Again, we see that, while the first finite differences vary, the second finite differences in output are the constant –32.

We want to finish this section by figuring out how to determine the values for the parameters a and c if we just have the data.

19. For each of the problems investigated so far, complete Table 6 by identifying the value of a, the value of the second finite difference in successive outputs, and the value of c.

TABLE 6 Organizing Information on Quadratics

Function	Value of a	Second Finite Difference in Output	Value of c
$R = -2x^2 + 10{,}000x$			
$A = -l^2 + 100l$			
$h(t) = -16t^2 + 125t + 8$			

20. Given a set of quadratic data, use the results of Table 6 to write a conjecture explaining:

a. the relationship between a in $y(x) = ax^2 + bx + c$ and the second finite difference in consecutive outputs.

b. how you might determine the value of c in $y(x) = ax^2 + bx + c$ using an input–output table. (*Hint*: Think of c as an output and notice the value of the input.)

21. Test your conjectures from Investigation 20 for the given quadratic functions by completing Table 7. If they are not correct, modify your conjectures.

TABLE 7 Testing Conjectures on Quadratics

Function	Value of a	Second Finite Difference in Output	Value of c	Input if c is the Output
$y(x) = 6x^2 - 5x - 3$				
$y(x) = 5x^2 + 83x + 92$				
$y(x) = 0.3x^2 - 9x - 15$				

By looking at the tables for each function, we see that the value of c in the quadratic model is always the value of the output when the input is zero. That is, we find the value of c by looking for an input of 0 and assigning c the value of the corresponding output.

Table 8 suggests that the value of a appears to be one-half the value of the second finite difference in the outputs.

This suggests that the second finite difference in outputs of

$y(x) = 6x^2 - 5x - 3$ is 12, which is $2(6)$.

$y(x) = 5x^2 + 83x + 92$ is 10, which is $2(5)$.

$y(x) = 0.3x^2 - 9x - 15$ is 0.6, which is $2(0.3)$.

TABLE 8 Process for Computing *a* from Second Finite Difference

Function	Value of a	Second Finite Difference in Output	Process to Find a from the Second Finite Difference
$R = -2x^2 + 10{,}000x$	-2	-4	$\frac{1}{2}(-4)$
$A = -l^2 + 100l$	-1	-2	$\frac{1}{2}(-2)$
$h(t) = -16t^2 + 125t + 8$	-16	-32	$\frac{1}{2}(-32)$

Why? Check it out and see if we are correct.

Let's practice one problem before ending the section.

22. Consider the data in Table 9.

TABLE 9 Some Nice Data

Input	Output	Δ(output)	Δ(Δ(output))
−3	8		
−2	−2		
−1	−6		
0	−4		
1	4		
2	18		

a. Find the values of the parameters a and c in the general model $y(x) = ax^2 + bx + c$. Describe what you did.

b. Use the values of a and c to write the quadratic model with just the parameter b. The letters that remain in the equation should be the variables x and y along with parameter b.

c. Write one ordered pair that is listed in the table. Do not use $(0, -4)$.

d. Substitute the ordered pair from part **c** into the model in part **b**. Solve the resulting equation for b.

e. Use your answers to parts **b** and **d** to write the specific quadratic model that represents the data.

f. Use a table on your calculator to check your answer to part **e**. If it doesn't check, correct your function.

23. Explain to your friend Izzy how to create a specific quadratic function from a table of quadratic data.

Looking at the fourth column of Table 9, we note that $\Delta(\Delta(\text{output})) = 6$. The value of a in $y(x) = ax^2 + bx + c$ is one-half of 6, or 3.

If the input is 0, then the output is –4. So $c = -4$ in $y(x) = ax^2 + bx + c$.

So far, the specific model for the data is $y(x) = 3x^2 + bx - 4$.

Now we use a data point to find the value of b. Let's say we use $(1, 4)$.

Substitute $x = 1$ and $y = 4$:	$4 = 3(1)^2 + b(1) - 4$
Simplify right side of "=":	$4 = 3 + b - 4$
	$4 = b - 1$
Add 1 to both sides:	$5 = b$

Now we can write the specific model for the data: $y(x) = 3x^2 + 5x - 4$

We can check that the model is correct by looking at the table on the calculator and comparing to the data in Table 9 (Figure 2).

FIGURE 2

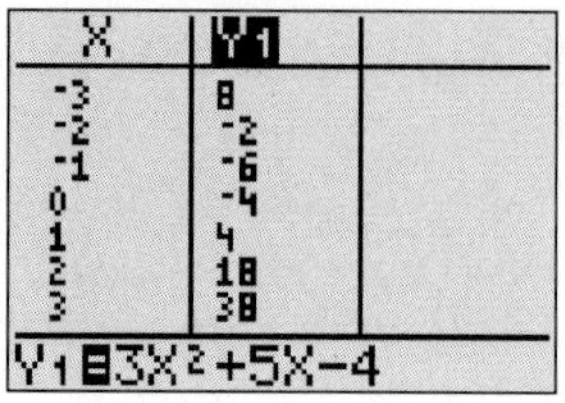

X	Y1
-3	8
-2	-2
-1	-6
0	-4
1	4
2	18
3	38

Y1=3X²+5X-4

Summarizing, if data has a constant second finite difference in successive outputs, it is modeled by a quadratic function of the form

$$y(x) = ax^2 + bx + c.$$

The values of the parameters are found as follows.

- The parameter a is one-half of the constant second finite difference in successive outputs.
- The parameter c is the output when the input is 0.
- Write the model $y(x) = ax^2 + bx + c$ with the specific values substituted for a and c.
- Choose a data point other than the point $(0, c)$.
- Substitute the data point (x and y) in the model.
- Solve for b.
- Write the specific model.
- Check your result by matching the data with a table on your calculator.

The ideas encountered in this section will allow us to create a quadratic model related to the salary data in the next section. I'll bet you can't wait!

STUDENT DISCOURSE

Pete: I think I understand the material in this section, but I do have one question.

Sandy: What's that?

Pete: When I had to find b in the model, I read that I shouldn't use the point $(0, -4)$. Why?

Sandy: At that point, the model was $y(x) = 3x^2 + bx - 4$. What happens if you substitute 0 for x and -4 for y?

Pete: Then I'd have $-4 = 3(0)^2 + b(0) - 4$. Wait a minute. I just get $-4 = -4$. That doesn't help me find b.

Sandy: That's right. If you substitute 0 for x, the terms with x are zero and you are just left with c.

Pete: I guess I see the problem. I can use any other point from the data, though. Is that right?

Sandy: That's my understanding.

1. In the Glossary, define the words in this section that appear in ***italic boldface*** type.

2. Consider the quadratic data in Table 10.

TABLE 10 Some More Nice Data

Input	Output		
−2	17.6		
−1	6.3		
0	1.2		
1	2.3		
2	9.6		
3	23.1		

a. Find the values of the parameter a and c in the general model $y(x) = ax^2 + bx + c$. Describe what you did.

b. Use the values of a and c to write the quadratic model with just the parameter b. The letters that remain in the equation should be the variables x and y along with parameter b.

c. Write one ordered pair that is listed in the table. Do not use (0, 1.2).

d. Substitute the ordered pair from part **c** into the model in part **b**. Solve the resulting equation for b.

e. Use your answers to parts **b** and **d** to write the specific quadratic model that represents the data.

f. Use a table on your calculator to check your answer to part **e**. If it doesn't check, correct your function.

3. The demand equation for the headset made by **Virtual Reality Experiences** has changed to $p = 6000 - 3x$.

a. Write the new revenue function.

b. Identify the values of the parameters a, b, and c in the new revenue function.

c. What is the second finite difference in successive outputs of the new revenue function? Defend your answer.

d. What is the slope of the demand function? Interpret the slope in terms of the problem situation. Defend your answer.

e. What is the slope of the original demand function? Interpret the slope in terms of the problem situation. Defend your answer.

4. Dog Pen Adjustment: Our dog, Katie, destroyed 50 meters of fencing. Instead of having 200 meters of fencing for the dog pen, we now have only 150 meters.

a. If l represents the length of the dog pen, write a function that expresses the width w as a function of the length l.

b. What kind of function is the width function written in part **a**?

c. What is the slope of the function written in part **a**? Interpret your answer in terms of the problem situation.

d. Express the area as a function of the length l.

e. What is the second finite difference in successive outputs of the new area function? Defend your answer.

5. Rocket Firing: We are going to fire the rocket from the top of an apartment building. The top of the apartment building is 30 feet off the ground. The rocket is fired with an initial velocity of 150 feet per second.

a. Write the equation of the specific model for this rocket firing. That is, express the height of the rocket as a function of time since firing. (*Hint*: Refer to Jordan's Power Toss.)

b. What is the second finite difference in successive outputs. Justify.

c. Create an input–output table for the rocket firing that shows the height in 1-second intervals from 0 to 10 seconds.

d. Extend that table in part **c** to display the first and second differences in height.

e. In what 1-second time interval does the rocket's height change fastest?

f. In what 1-second time interval does the rocket's height change slowest?

g. What does the varying rate of change in output with respect to input represent in this problem situation?

h. About how long will it take before the rocket hits the ground? Why?

6. Refer to Exploration 5. Explain in terms of the rocket what it means when the rate of change in height with respect to time is

a. positive.

b. negative.

c. near zero.

d. very large in magnitude.

7. This section introduced three different problem situations (**Virtual Reality Experiences Revenue**, **Fencing in the Dog**, **Jordan's Power Toss**). Explain why the models developed for these problems are neither linear nor exponential.

8. Given the following quadratic functions, identify the second finite difference in outputs by using only the value of a.

a. $y(x) = 0.2x^2 + x + 2$

b. $y(x) = x^2 - 11x - 7$

c. $y(x) = -4.7x^2 + 8.9x - 0.33$

d. Briefly describe what you did to solve this problem.

9. Consider the data given in Table 11.

TABLE 11 **Random Data I**

Input	−2	−1	0	1	2	3
Output	1.28	3.2	8	20	50	125

a. Identify the data as linear, quadratic, or exponential. Give a reason for your answer.

b. Construct the specific algebraic model for the data.

c. Find the output if the input is –5. Explain what you did.

d. Find the input if the output is 4882.8125. Explain what you did.

10. Consider the data given in Table 12.

TABLE 12 Random Data 2

Input	–8	–5	2	4	11	19
Output	–18.5	–10.4	8.5	13.9	32.8	54.4

a. Identify the data as linear, quadratic, or exponential. Give a reason for your answer.

b. Construct the specific algebraic model for the data.

c. Find the output if the input is 9. Explain what you did.

d. Find the input if the output is –32. Explain what you did.

11. Write the equation of a line parallel to $y = \frac{2}{3}x - 15$ that has a vertical intercept of $(0, -2)$.

12. Write the equation of a line perpendicular to $y = \left(-\frac{5}{8}\right)x - 7$ that has a vertical intercept of $(0, 1)$.

13. Write the equation of an exponential function with base 1.2 that contains the point $(2, 7)$.

REFLECTION

Create a set of data and a problem context. Describe how you can recognize if the data is modeled by a

a. linear function.

b. exponential function.

c. quadratic function.

SECTION 3.4

Quadratic Models for Change

Purpose

- Investigate the behavior of sums of arithmetic sequences.
- Establish a connection between sums of arithmetic sequences and quadratic models.
- Use algebraic models to answer questions.
- Compare and contrast basic linear, quadratic, and exponential models.

In the last section, you saw how to identify if data is quadratic and, if it is, how to develop the algebraic model (equation) for the data. Previously you saw that data in which the rate of change is constant could be modeled by a linear function $y(x) = ax + b$, where the letter a represents the rate of change of the output with respect to the input and b is the value of the output when the input is zero. This resulted in a method for using linear functions to model arithmetic sequences.

You also learned to model data in which the finite ratios of consecutive outputs is constant. This required an exponential function $y(x) = a(b)^x$, where the letter b represents the constant ratio of change and the letter a represents the output when the input is zero. This provides a method for using exponential functions to model geometric sequences.

Now you will apply what you studied in Section 3.3 to a problem related to the CDs-R-Us salaries. Then you will learn to create algebraic models for several sets of data.

Recall the scenario involving annual salary and increases in salary at CDs-R-Us.

CDs-R-Us: The starting annual salary is $35,000 and you will receive a fixed raise of $2000 annually.

We determined that the salaries at CDs-R-Us were modeled by the linear function $y(x) = 2000x + 33,000$, where x is the year number and y is the salary in year x. Now we want to look at the total salary earned over x years; in other words, the accumulated salary.

1. Complete Table 1. To complete the third column, sum the annual salaries up to and including that year's salary.

TABLE 1 Accumulated Salaries

Year	Salary ($)	Accumulated Salary ($)
1	35,000	35,000
2	37,000	35,000 + 37,000 = 72,000
3		
4		
5		
6		

2. The accumulated salaries form a sequence. Write down the first six elements of this sequence.

Consider the data formed by the year number as input and the accumulated salary as output. Let's try and create an algebraic model for this data.

3. Complete the first five columns of Table 2.

TABLE 2 Analyzing Accumulated Salary Sequence

Δ(Year)	Year	Accumulated Salary *A* ($)	Finite Differences in *A*, Δ*A*	Finite Ratios in *A*, ®*A*	Second Finite Differences in *A*, Δ(Δ*A*)
1	1	35,000			
	2	72,000			
	3				
	4				
	5				
	6				

4. Is the data linear, quadratic, or exponential? Why?

DISCUSSION

The accumulated salary data appear in Table 3. Note that the second finite difference in consecutive outputs is constant.

TABLE 3 Analyzing Accumulated Salary Sequence

Year	Accumulated Salary *A* ($)	Finite Differences in *A*, Δ*A*	Finite Ratios in *A* ®*A*	Second Finite Differences in *A* Δ(Δ*A*)
1	35,000	37,000	2.057142857	2000
2	72,000	39,000	1.541666667	2000
3	111,000	41,000	1.369369369	2000
4	152,000	43,000	1.282894737	2000
5	195,000	45,000	1.230769231	
6	240,000			

The first finite differences in accumulated salary are *not* constant, so the related function is *not* linear.

The finite ratios in accumulated salary are *not* constant, so the related function is *not* exponential.

The second finite differences in accumulated salary *are* constant—this implies a defining function that is ***quadratic***.

The algebraic model is $y(x) = ax^2 + bx + c$.

There are three parameters, or constants, in this equation: a, b, and c. The variable x represents the input (year) and the variable y represents the output (accumulated salary).

If we want the specific algebraic model for the data, we must determine the values of the parameters a, b, and c.

Recall the following facts from Section 3.3.

- If the finite difference in the input is 1 (the change in successive years is 1), then the value of a is one-half of the second finite difference.
- The value of c is the accumulated salary after zero years.
- Once we know the values of a and c, we can use substitution to find the value of b.

Let's find the algebraic model.

5. For the data in Table 3, find the value of the parameter a. Write the general model $y(x) = ax^2 + bx + c$, but replace a with the value you just computed.

6. Find the value of parameter c. Write the general model $y(x) = ax^2 + bx + c$, but replace a and c with the values you have computed.

7. Find parameter b by selecting a data point from Table 3, substituting the point into the function, and solving for b.

8. Use your calculator to display a table for your function. Compare to Table 3. Is your function correct? How do you know?

9. State the domain of this function. Justify.

10. Graph the function on your calculator. Sketch your screen and briefly describe what you see.

Once we have a general algebraic model, we can ask questions about the data.

11. Use the general model for the accumulated salaries to determine the total salary you have earned after 10 years and after 20 years. Describe to your friend, Izzy, how you did these problems.

12. Use the table or the graph to determine how many years it will take to accumulate $500,000 and $1 million. Explain to your friend, Izzy, how you did these problems.

For the ***general model*** $y(x) = ax^2 + bx + c$, we find a by computing one-half of the second finite difference in accumulated salaries. Thus,

$$a = \frac{1}{2}(\Delta(\Delta A)) = \frac{1}{2}(2000) = 1000$$

The value of c equals the output when the input is zero. This represents the accumulated salary after zero years. Since we haven't worked yet, our accumulated salary must be zero. Thus

$$c = 0$$

The model so far is

$$y(x) = 1000x^2 + bx + 0$$

or just

$$y(x) = 1000x^2 + bx.$$

Now we pick a data point. How about (1, 35,000)?
So $x = 1$ and $y = 35{,}000$.

Substitute into the model: $35{,}000 = 1000(1)^2 + b(1)$

Simplify: $35{,}000 = 1000 + b$

We need to isolate b. Subtract 1000 from both sides:

$$35{,}000 - 1000 = 1000 + b - 1000$$

Simplify: $34{,}000 = b$

Using this value for b, we finally have a specific model for the quadratic function that determines accumulated salaries (output) from the year (input). Here it is!

$$y(x) = 1000x^2 + 34{,}000x$$

The domain of this function is the set of natural numbers 1, 2, 3, 4, ... since we can only input the number of years we have worked at CDs-R-Us. Figure 1 displays a table and a graph of this model in both function mode and parametric mode.

FIGURE 1

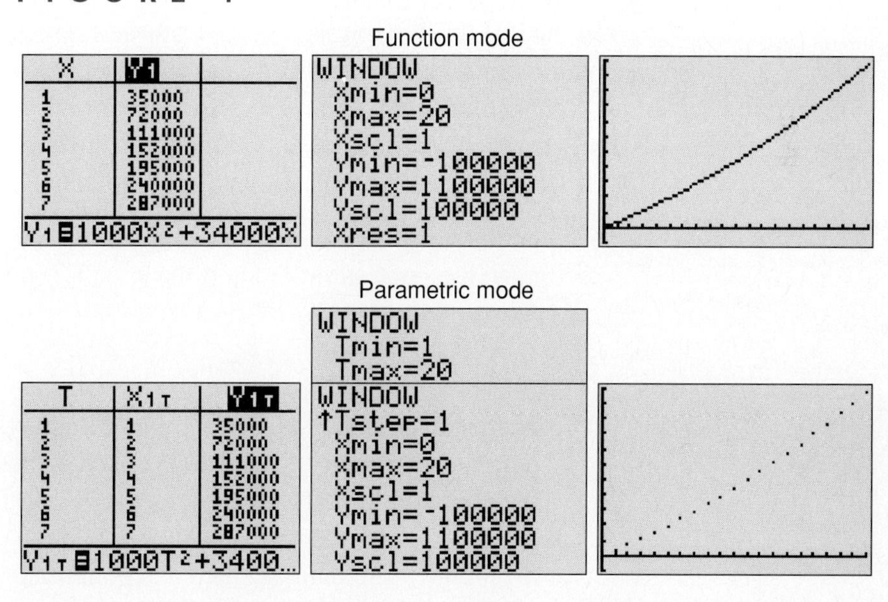

Since the domain is the set of natural numbers, it is appropriate to use parametric mode in which the input only takes on natural number values. The graph must be disconnected to indicate data points at only natural number inputs.

We could find the accumulated salary after 20 years by using 20 as the input to the function, as follows.

$$
\begin{aligned}
y(20) &= 1000(20)^2 + 34{,}000(20) \\
&= 1000(400) + 680{,}000 \\
&= 400{,}000 + 680{,}000 \\
&= 1{,}080{,}000
\end{aligned}
$$

Thus we will have accumulated $1,080,000 after 20 years.

Figure 2 displays both a tabular and graphical way to determine when the accumulated salary reaches $1 million.

FIGURE 2

T	X_{1T}	Y_{1T}
13	13	611000
14	14	672000
15	15	735000
16	16	800000
17	17	867000
18	18	936000
19	19	1.01E6

Y_{1T}=1007000

T=19
X=19 Y=1007000

Note the 1.01E6 displayed in the table. This is an approximation of the result 1,007,000 shown at the bottom of the screen. 1.01E6 is ***scientific notation*** and means 1.01 times 10^6.

In the last three sections, we have studied how to recognize linear, quadratic, and exponential models from data. After determining the general model, we found the specific parameters for each model in order to create a specific algebraic model for the data. Once we had the specific model we could use the model to answer questions about the data.

We conclude the chapter by analyzing three sets of data. Let's see if you can recognize the general model and the specific model.

13. Natural Gas Charges: Table 4 displays a relationship between the number of therms of natural gas used to heat a building in a month and the charge in dollars for the gas.

a. Identify the data as linear, exponential, quadratic, or none of these. Give reasons for your answer.

b. Find the specific algebraic model. Show all work. Check your answer by displaying a table on your calculator and comparing it to the given table.

TABLE 4 Charges for Natural Gas

Therms of Natural Gas	Charge in Dollars ($)
100	69
200	131
300	193
400	255
500	317
600	379

c. What is the charge if you use 352 therms of natural gas? Explain how you found the answer.

d. The charge is \$311.41. How many therms of natural gas were used? Explain how you found the answer.

The **Natural Gas Charges** data are modeled by a linear function of the form $C(t) = at + b$, where the charges C depend on the therms t of natural gas used. This is true because the change in t is constant ($\Delta t = 100$) and the change in charges C is constant ($\Delta C = 62$). Thus the rate of change in the output with respect to the input is given by

$$a = \frac{\Delta C}{\Delta t} = \frac{62}{100} = 0.62.$$

Since this value is constant, the function that models the data is linear. Thus the value of the parameter a is 0.62. Recall that this is also known as the slope.

To find the value of parameter b, we choose a point from Table 4 and substitute the values into the model $C(t) = 0.62t + b$. Let's use $(100, 69)$.

Let $t = 100$ and $C = 69$	$69 = 0.62(100) + b$
Multiply on right side of "=":	$69 = 62 + b$
Subtract 62 from both sides:	$7 = b$

The model for the data is $C(t) = 0.62t + 7$.

To determine the charge if we use 352 therms, we substitute 352 for the input variable t and compute the value of C.

To determine the number of therms used if the charge is \$311.41, we substitute \$311.41 for the output. This leaves an equation to solve.

Substitute for output:	$311.41 = 0.62t + 7$
Subtract 7 from both sides:	$311.41 - 7 = 0.62t + 7 - 7$
Simplify:	$304.41 = 0.62t$
Divide both sides by 0.62:	$\frac{304.41}{0.62} = \frac{0.62t}{0.62}$
Simplify:	$491 \approx t$

Approximately 491 therms were used if the charge was \$311.41.

14. Super Sam's Rock Fling: Super Sam, a truly amazing individual, picks up a rock and throws it as hard as he can. Table 5 displays the relationship between the rock's horizontal distance (in feet) from Sam and the rock's height.

TABLE 5 Super Sam's Rock Fling

Horizontal Distance (feet)	Height (feet)
0	5.0
1	9.9
2	14.6
3	19.1
4	23.4
5	27.5

a. Identify the data as linear, exponential, quadratic, or none of these. Give reasons for your answer.

b. Find the specific algebraic model. Show all work. Check your answer by displaying a table on your calculator and comparing it to the given table.

c. What is the height of the rock when it is 20 feet horizontally from Sam? Explain how you found the answer.

d. How far is the rock from Sam when the rock hits the ground? Explain how you found the answer.

Super Sam's Rock Fling data is modeled by a quadratic function of the form $h(s) = as^2 + bs + c$, where the height h is a function of the horizontal distance s of the rock from Sam. This is true since the first finite differences in consecutive outputs are varying, but the second finite differences in consecutive outputs are all -0.2. Note that the change in the inputs is a constant of 1.

Now let's find the parameters a, b, and c.

a is one-half of the second finite difference in the outputs.

$$a = \frac{1}{2}(-0.2) = -0.1$$

c is the output when the input is zero. From Table 5, $c = 5$.

The model so far is $h(s) = -0.1s^2 + bs + 5$.

We just need to find b. To do so, we substitute an ordered pair from Table 5. Let's use $(1, 9.9)$

Substitute $s = 1$ and $h = 9.9$:	$9.9 = -0.1(1)^2 + b(1) + 5$
Simplify:	$9.9 = -0.1 + b + 5$
Add constants:	$9.9 = 4.9 + b$
Subtract 4.9 from both sides:	$5 = b$

We now have the specific model for **Super Sam's Rock Fling**.

$$h(s) = -0.1s^2 + 5s + 5.$$

To find the height when the rock is 20 feet horizontally from Sam, we substitute 20 for the input s.

$$h(20) = -0.1(20)^2 + 5(20) + 5 = 65$$

The rock is 65 feet high after 20 seconds.

To determine how far the rock is from Super Sam when it hits the ground, we note that the height is zero when the rock hits the ground. This means we must find the input s when the output h is zero. We have a rather messy equation to solve.

$$0 = -0.1s^2 + 5s + 5$$

Let's estimate the answer using a table or a graph. We will see how to solve the equation algebraically later in the text. Figure 3 displays some approximate solutions.

FIGURE 3

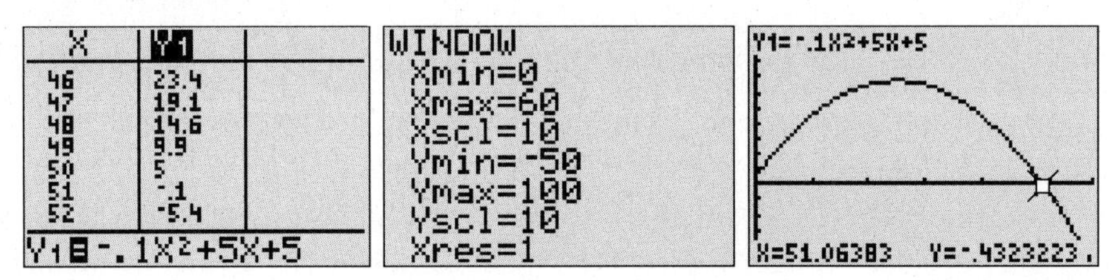

The information in Figure 3 suggests the rock hits the ground just under 51 feet from Sam.

15. Drug in Bloodstream: Five milliliters of a drug is injected into the bloodstream. After one hour 60% of the original amount remains in the bloodstream. For each succeeding hour, 60% of the amount present in the previous hour remains in the bloodstream.

Table 6 displays the amount of drug present at the end of each of the first 6 hours.

a. Identify the data as linear, exponential, quadratic, or none of these. Give reasons for your answer.

b. Find the specific algebraic model. Show all work. Check your answer by displaying a table on your calculator and comparing it to the given table.

c. How many milliliters of the drug remain in the bloodstream after 10 hours? Explain how you found the answer.

TABLE 6 Drug Present in Bloodstream

Elapsed Time (hours)	Amount of Drug in Bloodstream (ml)
0	5
1	3
2	1.8
3	1.08
4	0.648
5	0.3888
6	0.23328

d. After how many hours will the amount of drug in the bloodstream be less than 0.01 ml? Explain how you found the answer.

The **Drug in Bloodstream** data is an example of ***exponential decay,*** which occurs when the ratios between consecutive outputs are constant but less than 1. The general model is

$$A(t) = a(b)^t,$$

where a is the value of P when $t = 0$ and b is the decay rate that equals the constant finite ratio in output of 0.6. So, $b = 0.6$.

From Table 6, the output is 5 when the input is 0. So, $a = 5$.

The specific exponential model is $A(t) = 5(0.6)^t$.

To find the amount of the drug in the bloodstream after 10 hours, substitute 10 for t in the function $A(t) = 5(0.6)^t$. So

$$A(10) = 5(0.6)^{10} \approx 0.03$$

After 10 hours, only 0.03 ml of the drug remains in the bloodstream.

From the previous answer we note that it will take longer than 10 hours for there to be less than 0.01 ml of the drug remaining in the bloodstream. We can find a more accurate answer however. We need to find the input to the function for which the output is 0.01. Since the function outputs are decreasing, any larger input will produce an output that is less than 0.01. Use either a table or a graph to answer this question.

Figure 4 displays a table and a graph with the corresponding viewing window.

FIGURE 4

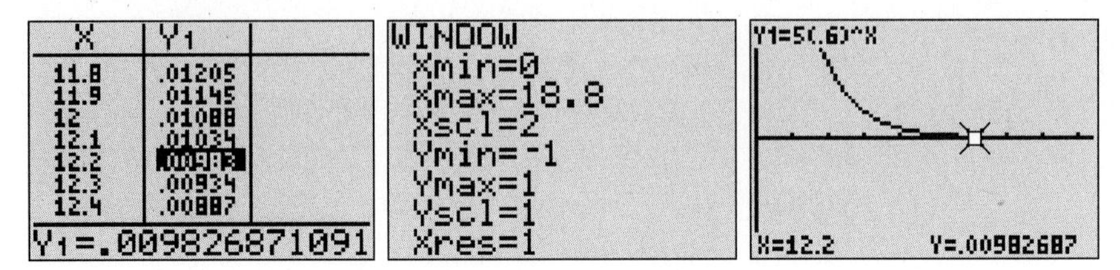

Thus, there will be about 0.01 ml in the bloodstream approximately 12.2 hours after the drug enters the bloodstream. This problem could also be solved numerically by substituting values for t until the output is approximately 0.01. Thus, after more than 12.2 hours there will be less than 0.01 ml of the drug in the bloodstream.

Now it's time to create some data and challenge a classmate to find the algebraic model.

16. Write a linear function, a quadratic function, and an exponential function. Don't show them to anyone. They are your secret.

17. On a separate sheet of paper, use the functions you created to make an input–output table for each function. Include five data points in each table. Make the finite differences in the input one unit. Include the input value of 0 in your tables.

18. Give your tables to a classmate. Challenge the classmate to discover your functions.

The previous investigation was designed as a challenge to the class. Since all the students made up their own data, answers will vary. That completes our initial study of linear, quadratic, and exponential models. We will see how to apply these to other types of data in Chapter 4. We also will see how to better manipulate these models.

Pete: Does every set of data have to be linear, exponential, or quadratic?

Sandy: What do you think?

Pete: I don't think the world is that organized.

Sandy: You're right. In fact, in Chapter 4, you will analyze some real-world data that can only be approximated by one of these models.

Pete: But are there other kinds of models?

Sandy: Sure. In fact, you'll learn about some of them in Chapter 4 too.

Pete: But does data necessarily fit any of these models?

Sandy: No. Any model of real-world data is, at best, an approximation of what is really happening.

Pete: Why's that?

Sandy: The models you have been working on have only two variables—an input and an output.

Pete: So . . .

Sandy: Most situations in the real-world have many variables. To create models, the situation has to be simplified by holding many of the variables constant. For example, you might consider the relationship between the math you had in high school and your final grade in this class. Will the math you had in high school determine your grade in this class?

Pete: It has an effect, but how much I study and attend class has a lot to do with it too.

Sandy: Exactly, the math you had in high school is only one input to the function that will determine your course grade.

Pete: Hmmm, I see what you mean. Well, I will say I never thought about trying to write equations that attempted to simulate actual data. I actually enjoy it.

Sandy: I did too.

1. In the Glossary, define the words in this section that appear in ***italic boldface*** type.

2. Water Heater Costs: The data in Table 7 display the relationship between the annual energy cost of a water heater and the cost of the water heater over its life cycle.

TABLE 7 Cost of Water Heater over Its Life Cycle

Annual Energy Cost ($)	Life-cycle Cost ($)
70	1060
71	1073
72	1086
73	1099
74	1112
75	1125

a. Are the data modeled by a linear, quadratic, or exponential function? Why?

b. Find an algebraic model for the water heater data.

c. What is the domain for the function written in part **b**?

d. Use the model created in part **b** to find the life cycle cost if the annual energy cost is $90.

3. Recall the **Campus Vehicle Congestion** data. We have changed the input so that it represents the number of hours since 8 A.M. rather than the time of day (Table 8).

a. Determine if the data are linear, quadratic, or exponential. Justify your answer.

b. Create an algebraic model for the data.

TABLE 8 **Car Leaving Campus by Hours Since 8 A.M.**

Hours Since 8 A.M.	Number of Cars Leaving Campus
0	159
1	341
2	467
3	537
4	551
5	509
6	411
7	257

c. What is the domain of the algebraic model? Why?

4. Consider the **Triangular Number Sequence**, 1, 3, 6, 10, 15, 21, ...

a. Numerically express the sequence as a function by creating an input–output table.

b. Is the algebraic model for the data linear, quadratic, or exponential? Why?

c. Create the algebraic model of the function that models the data.

d. What is the domain of this function? Why?

e. Find the 100th element of the sequence. Explain what you did.

5. For each of the problems investigated in this section, state the domain and justify your answer.

a. Natural Gas Charges

b. Super Sam's Rock Fling

c. Drug in Bloodstream

6. For each set of data, create the specific algebraic model.

a.

Input	7	8	9	10	11	12
Output	289	378	479	592	717	854

b.

Input	4	5	6	7	8
Output	89.1	267.3	801.9	2405.7	7217.1

c.

Input	−3	−1	1	3	5	7
Output	10.1	7.5	4.9	2.3	−0.3	−2.9

7. For each graph, record the data in a table and create the specific algebraic model.

a.

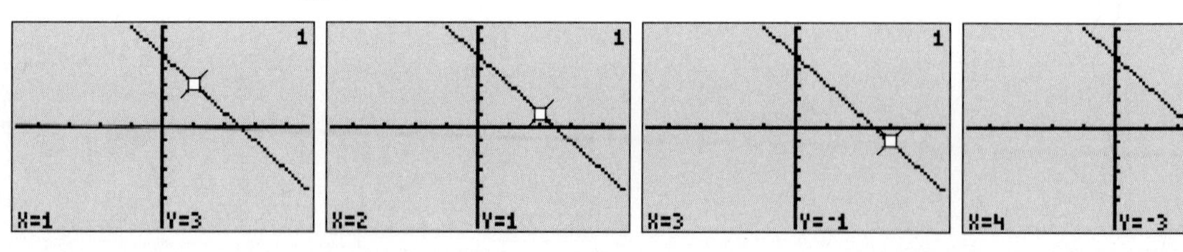

b.

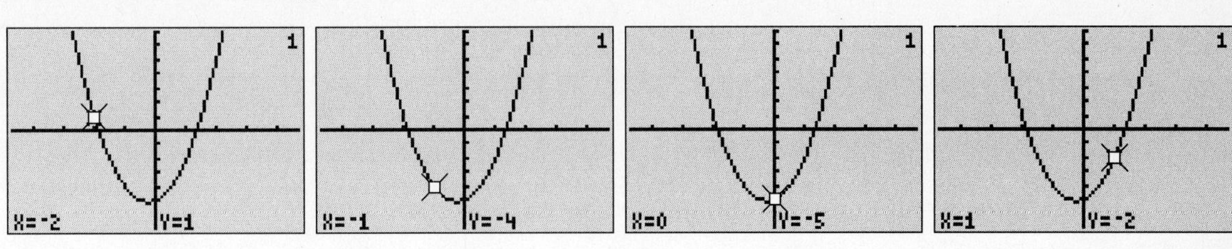

c.

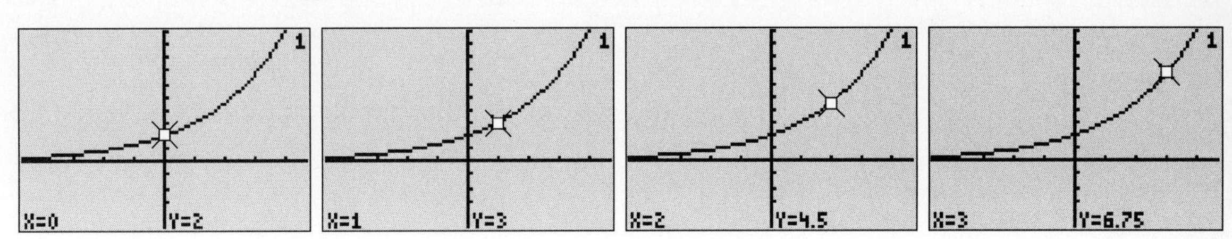

8. In this section, you saw that the sequence of accumulated salaries of the arithmetic sequence (CDs-R-Us salaries) is modeled by a quadratic function. There is an alternative way to compute such sums. Consider the sequence formed by the natural numbers $1, 2, 3, 4, \ldots$. This sequence is arithmetic with a common difference of 1.

Now consider the sequence formed by summing the elements. In particular, let's look at

$$1 + 2 + 3 + 4 + 5 + 6 + 7 + 8 + 9 + 10$$

Let's find a fast way to calculate this sum.

a. Compute the sum with your calculator so we can check it later.

b. Add the pairs of numbers connected by the dotted lines. Record your answers on the dotted lines in Figure 6.

FIGURE 6

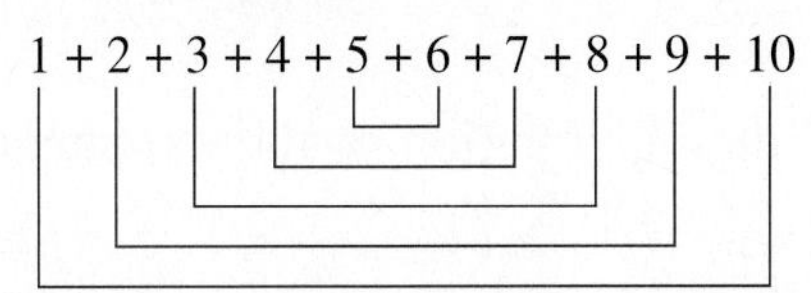

c. What is the sum of each pair of numbers from Figure 6? How could you use the first number and the last number to obtain this answer?

d. How many pairs of numbers are there? How could you determine this number if you knew the number of addends in the sum?

e. Find the sum of the first 10 natural numbers using your answers to parts **b** and **c**.

f. Find the sum of the first 20 natural numbers using the same technique.

g. Find the sum of the first 100 natural numbers using the same technique.

h. Find the sum of the first 1000 natural numbers using the same technique.

9. Write an algebraic expression for the sum $1 + 2 + 3 + 4 + \cdots + n$. As an aid, refer to the approach used in Exploration 8. Check your formula using your answers from Exploration 8.

10. Using a similar approach, compute the total salary earned for 20 years if you take the job at CDs-R-Us.

11. Write the equation of the linear function that contains the points $(4, -2)$ and $(-5, 17)$.

12. Write the equation of the exponential function that contains the points $(2, 5)$ and $(3, 6)$.

13. Write the equation of a quadratic function that contains the points $(0, -7)$ and $(3, -8)$ if the second finite difference in consecutive outputs is -4.

14. The following numbers are written in scientific notation. Write their decimal value.

a. 1.37 E 8

b. 2.463 E–6

c. –2.3 E 14

d. 1.001 E–13

15. Consider the quadratic model $h(t) = -16t^2 + 60t + 27$ for the height h (feet) of an object t seconds after the object has been projected straight up into the air.

a. Find $h(4)$ and interpret its meaning.

b. What was the original height? Why?

c. Estimate the maximum height. Describe what you did.

d. Estimate when the object will hit the ground. Describe what you did.

REFLECTION

Write a paragraph comparing and contrasting linear functions, quadratic functions, and exponential functions. Be sure to list the critical features of each type of function. Discuss numeric, algebraic, and graphical representations.

SECTION 3.5

Expressing Situations Involving Change Algebraically: Review

Purpose

- Reflect on the key concepts of Chapter 3.

In this section you will reflect upon the mathematics in Chapter 3—on what you've done and how you've done it.

1. State the most important ideas in this chapter. Why did you select each?

2. Identify all the mathematical concepts, processes, and skills you used to investigate the problems in Chapter 3.

There are a number of important ideas that you might have listed including linear functions, slope, vertical intercepts, parallel lines, perpendicular lines, standard form of a line, slope-intercept form of a line, exponential functions, base, quadratic functions, and parameters.

1. Write an example of:

   a. a linear function with a positive slope.

   b. a linear function with a vertical intercept at $(0, -5)$

   c. a linear function with a slope of 0.

   d. an exponential function with a vertical intercept of $(0, 63)$.

   e. an exponential function in which the output decreases as the input increases.

f. a quadratic function with a vertical intercept of $(0, -10)$.

g. a quadratic function with a second finite difference in consecutive outputs of –9.

h. two linear functions that are parallel.

i. two linear functions that are perpendicular.

j. two linear functions that are perpendicular and have the same vertical intercept.

2. Sue's Bank Account: Sue opened an account by depositing $5500. She deposits an additional $450 each month.

a. The amount Sue has on deposit, not including interest, is a function of the month. Is this function linear, exponential, or quadratic? Why?

b. Write a function with input the number of months and output the total amount that has been deposited. How did you create this function?

c. How much has been deposited after two and one-half years? Describe your answer in a complete sentence.

d. How long will it take until she has deposited at least $20,000? Defend your answer.

3. More Health Care Costs: Health care costs have been rising at a rate of almost 15% annually in one area of the country. Assume that a certain medical test cost $30 in 1986. Consider the sequence defined by the cost of this test in successive years since 1985.

a. The cost of the medical test is a function of the years since 1985. Is this function linear, exponential, or quadratic? Why?

b. Write a function with input the number of years since 1985 and output the cost of the test. Describe to a friend how you created this function.

c. Use your function to find the cost of the test in 1998. Describe your answer in a complete sentence.

4. Consider the data in Table 1.

TABLE 1 Random Data 1

Input	–2	1	4	7	10	13
Output	42	40	38	36	34	32

a. Is this function linear, exponential, or quadratic?

b. Write an equation that will generate the output from the input. Explain what you did.

c. Use your equation to find the output when the input is 42.

d. Find the input if the output is –20. Explain what you did.

5. Consider the data in Table 2.

TABLE 2 Random Data 2

Input	1	2	3	4	5	6
Output	40	20	10	5	2.5	1.25

a. Is this function linear, exponential, or quadratic?

b. Write an equation that will generate the output from the input. Explain what you did.

c. Use your equation to approximate the output when the input is 23.

d. Find the input if the output is approximately 0.01953125. Explain what you did.

6. Consider the data in Table 3.

TABLE 3 Random Data 2

Input	0	1	2	3	4	5
Output	1	2	9	22	41	66

a. Is this function linear, exponential, or quadratic?

b. Write an equation that will generate the output from the input. Explain what you did.

c. Use your equation to approximate the output when the input is 23.

d. Find the input if the output is approximately 209. Explain what you did.

7. Given the following linear equations in standard form, write them in the form $y = ax + b$ and identify the slope and vertical intercept.

a. $x + y = 11$

b. $3x - 10y = -14$

c. $4x + y = 0$

d. $-7x + 3y = 19$

8. A line contains the points $(0, 5)$ and $(4, -7)$.

a. Write the linear function for the line.

b. Write the equation of a line parallel to the given line through $(11, 2)$.

c. Write the equation of the line perpendicular to the given line through $(11, 2)$.

9. Write the equation of the line

a. parallel to $7x - 2y = -23$ through the point $(2, 31)$.

b. perpendicular to $7x - 2y = -23$ through the point $(2, 3)$.

10. Given the graph, write the equation of the function.

a.

WINDOW
Xmin=-4.7
Xmax=4.7
Xscl=1
Ymin=-4
Ymax=14
Yscl=2
Xres=1

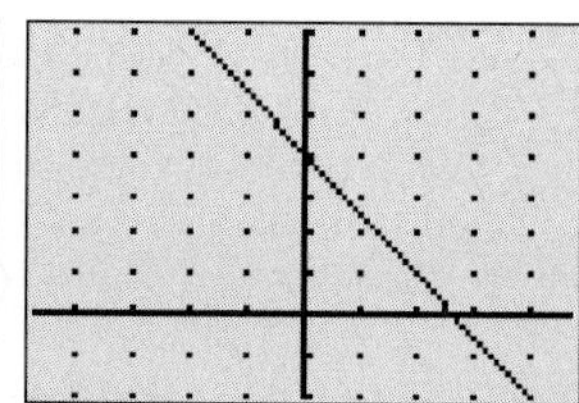

b.

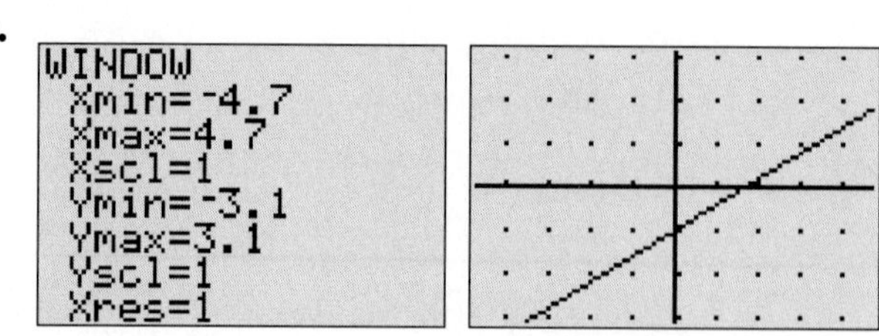

11. Write an example of an exponential function in which

a. the function is increasing and the vertical intercept is $(0, 47)$.

b. the function is decreasing and the vertical intercept is $(0, 23)$.

12. An exponential function contains the points $(0, 9)$ and $(1, 10.8)$.

a. Identify the base of the exponential function.

b. Write the specific equation of the exponential function.

13. An exponential function contains the points $(2, 9.6)$ and $(3, 7.68)$.

a. Identify the base of the exponential function.

b. Write the specific equation of the exponential function.

14. Consider the graph of the exponential function.

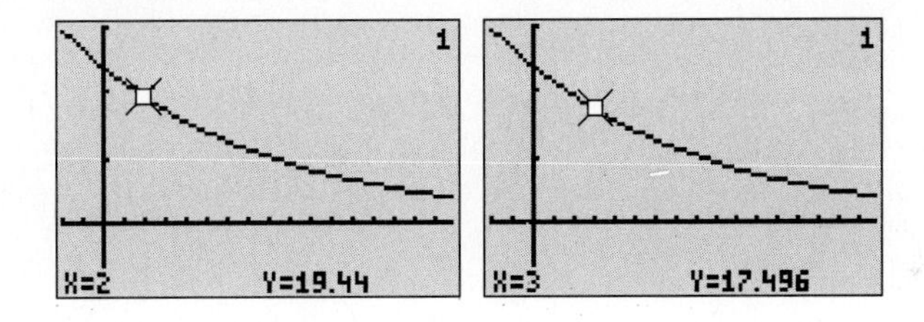

a. Find the base of the exponential function.

b. Write the specific equation of the exponential function.

c. Find the output if the input is 9. What did you do?

15. Rocket Firing 2: We are going to fire the rocket from the top of an apartment building. The top of the apartment building is 35 feet off the ground. The rocket is fired with an initial velocity of 140 feet per second.

a. Write the specific model for this rocket firing. That is, express the height h of the rocket as a function of time t since firing. (*Hint*: Refer to Jordan's Power Toss.)

b. What is the second finite difference in successive outputs? Justify.

c. Find $h(6)$ and interpret in terms of the problem situation.

d. How high will the rocket be after 2 seconds? What did you do to answer the question?

e. How long will it take the rocket to hit the ground? What did you do to answer the question?

16. Write the equation of an exponential function with base 0.7 that contains the point $(3, 49)$.

17. Write the equation of a quadratic function that contains the points $(0, 4)$ and $(7, 52)$ if the second finite difference in consecutive outputs is 6.

18. Find the y-intercepts of the following. Briefly describe what you did in each case.

a. $2x - 7y = 12$

b. $y = 3x^2 + 7x - 5$

c. $y = (x + 3)^2 + 7$

d. $y = \frac{2}{3}(0.96)^x$

CONCEPT MAP

Construct a Concept Map centered on one of the following:

a. linear function

b. exponential function

c. quadratic function

REFLECTION

Select one of the important ideas you listed in Investigation 1. Write a paragraph to Izzy explaining your understanding of this idea. How has your thinking changed as a result of studying this idea?

4 Data: Creating Models and Answering Questions

SECTION 4.1

Exploring the Meaning of Mathematical Actions

Purpose

- Explore the meaning of the words "evaluate," "simplify," "solve," "equation," "inequality," and "function."

You have heard the words "evaluate," "simplify," "solve," "equation," "inequality," and "function" many times before. It is likely that you have not thought much about what these words really mean, but misunderstanding them can lead to errors in your work. A precise understanding of these words is crucial as we continue our work in this text. As we create mathematical models to describe data from the world around us, we will concentrate on asking questions about these models using this terminology. We begin this chapter by exploring your understanding.

1. The salary offer from CDs-R-Us is modeled by the function $y(x) = 2000x + 33{,}000$, where x is the number of years worked and y is the annual salary ($).

a. What is the salary in year 22? What did you do?

b. In what year will the salary reach $55,000? What did you do?

2. What is different about the two problems you solved in Investigation 1?

3. What is meant by each of the following words?

a. simplify

b. evaluate

c. solve

To determine the salary at CDs-R-Us in year 22, we ***substitute*** the given value, 22, for the input variable in the function $y(x) = 2000x + 33{,}000$.

$$\begin{aligned} y(22) &= 2000(22) + 33{,}000 \\ &= 44{,}000 + 33{,}000 \\ &= 77{,}000 \end{aligned}$$

This is an example of both evaluating and simplifying.

We ***evaluate*** the function to find the output of a function when the input is given by substituting the input value, 22, for x. This is called ***evaluating the function*** at the given input. The first line in the preceding computation demonstrates this action.

Once the substitution is complete, we simplify using the various mathematical properties (commutative, associative, distributive, identity, etc.) and operations to obtain the final answer (the output).

This process of applying mathematical properties and operations to an expression to obtain the "simplest" result is called ***simplifying***. What is the "simplest form" is arguable sometimes, so the word "simplifying" is less than precise. The final two steps in the computation of $y(22)$ would be called simplifying.

To find the year when the salary will reach $55,000 requires a much different process. We are given the output ($55,000) and asked to find the corresponding input as demonstrated below. This process is referred to as ***solving an equation***.

Original function: $y(x) = 2000x + 33{,}000$

Substitute 55,000 for output: $55{,}000 = 2000x + 33{,}000$

Subtract 33,000 from both sides:
$$55{,}000 - 33{,}000 = 2000x + 33{,}000 - 33{,}000$$

Simplify: $22{,}000 = 2000x$

Divide both sides by 2000: $\dfrac{22{,}000}{2000} = \dfrac{2000x}{2000}$

Simplify: $11 = x$

Therefore, the salary will reach $55,000 in the 11th year worked.

If we have to find the input when given the output, we create a new equation by substituting the $55,000 for the output variable y(x) (second step above). The process of finding the actual input is called ***solving***.

We solve a linear equation algebraically by performing "legal" mathematical operations so that a variable is isolated on one side of the "=" symbol. Once the operations are completed, the process of simplification occurs. We solve an equation when the input for a function must be found given the output.

Many equations can be solved numerically or graphically. It is often neither possible nor desirable to solve an equation algebraically. We'll investigate this idea later.

The word "equation" has been mentioned several times. An ***equation*** is a mathematical statement with an equal sign. What can you learn about equations?

4. Explain the difference, from the point of view of English grammar, between "delicious Super Dog Burger Bits" and "My yellow lab's favorite meal is delicious Super Dog Burger Bits."

5. What is the difference between a mathematical expression and a mathematical statement?

6. Identify each of the following as an equation, an inequality, or an expression. Give supporting reasons for your answers.

   a. $3x - 2$

   b. $3x - 2 = 5$

   c. $7 \le 3x - 2$

   d. $f(x) = 3x - 2$

7. Complete the sentence: An equation is... Write an equation different from any used previously.

8. Complete the sentence: An inequality is... Write an inequality different from any used previously.

Mathematical expressions are equivalent to phrases in English. "delicious Super Dog Burger Bits" is a phrase—an incomplete thought. This expression is an incomplete thought because it does not contain a verb. Reading $3x - 2$ out loud, we might say "three times x minus two." This is an incomplete thought—it is not linked to another expression with a mathematical verb.

The comment, "My yellow lab's favorite meal is delicious Super Dog Burger Bits" is an example of an English sentence—a complete thought. $3x - 2 = 5$, $7 \le 3x - 2$, and $f(x) = 3x - 2$ are all examples of ***mathematical statements***, equivalent to English sentences. $3x - 2 = 5$ and $f(x) = 3x - 2$ are equations that contain the mathematical verb "=". $7 \le 3x - 2$ is an inequality whose verb is "≤." Both equations and inequalities are mathematical statements.

We have an equation if the verb is "equals." An ***equation*** is a statement that two expressions have the same value.

We have an inequality if the verb "is" is followed by any of the modifiers "less than" (<), "less than or equal to" (≤), " greater than" (>)," greater than or equal to" (≥), or "not equal to" (≠). An ***inequality*** is a statement that one expression is less than, less than or equal to, greater than, or greater than or equal to a second expression.

Equations contain the symbol "=." It is vital we are clear about the meaning of this symbol. Let's explore.

9. What is the "=" symbol telling you if you follow the expression $2 + 3(-5)$ with the "equal" symbol?

10. What is the "=" symbol telling you in the statement $2x - 3 = 5 - 11x$?

11. What is the "=" symbol telling you in the statement $y = 2x - 3$?

12. What is the "=" symbol telling you in the statement $f(x) = 2x - 3$?

13. Is $2x - 3 = 5 = 2x = 8 = x = 4$ correct? Why or why not?

14. What is the "=" symbol telling you in the statement $x + 5 = 5 + x$?

15. What is the "=" symbol telling you in the statement $2 + 3 = 5 - 1 = 4(2) = 8$?

16. How are functions and equations different? How are they the same?

In every equation the expression on the left represents the same thing as the expression on the right. However, past experience has led to three common interpretations of the "=" symbol which may ignore the basic meaning.

The "=" symbol may express

- a ***relationship*** between two or more variables. The "=" does not represent an action but rather states a relationship between two varying quantities. This is the use we have seen when defining functions. The relation may represent a covariation between the variables (a change in x causes a change in y), or perhaps we have named the expression y or $f(x)$. For example, in Investigation 11, $y = 2x - 3$ expresses a functional relationship between the variables x and y. The statement $f(x) = 2x - 3$ expresses the same functional relationship as $y = 2x - 3$.

In $f(x) = 2x - 3$, we are using function notation, whereas we are using two-variable notation in $y = 2x - 3$.

- an ***equality*** between two functions. This often results in an equation containing only one variable. The "=" does not represent an action but rather states a conditional relationship between two expressions. For example, in Investigation 10, $2x - 3 = 5 - 11x$ states the equality of two functions defined by the processes $2x - 3$ and $5 - 11x$. Clearly these two expressions are generally not the same. What is implied by this notation is that we are looking for an input value (or values) of x so that when the value replaces x, we have the same outputs on each side of the equation. Such an equation can be solved for the variable giving the value of the input that results in equal outputs. Sometimes the equation is true for every value of the variable as in Investigation 14. $x + 5 = 5 + x$ is called an ***identity***.

- an ***action*** to be performed. This is similar to the most common use of "=" in basic arithmetic. The "=" in arithmetic generally tells you to perform an operation. The action usually involves simplifying. For example, in Investigation 9, $2 + 3(-5)$ followed by "=" represents an expression to be simplified using order of operations. This is not an equation. It is waiting for us to do something. It becomes an equation when we complete the thought as

$$2 + 3 \cdot (-5) = -13.$$

Keep in mind that "=" is used between two quantities that are the same. Thus $2x - 3 = 5 = 2x = 8 = x = 4$ indicates that the expressions $2x - 3$, 5, $2x$, 8, x, and 4 are all equal to one another. This makes no sense for several reasons.

Among the reasons are the following:

$2x - 3 = 5$ and $5 = 2x$ contradict each other, and

$8 = x$ and $x = 4$ contradict each other.

If we are simplifying or evaluating, we use the equal sign (=) to express equality between steps in the simplification. This is the "action" use of the "=". When we are solving an equation, we write equivalent forms of the equation one below the other. Do not connect equations with "=". The problem in Investigation 13 is written correctly as follows.

$$2x - 3 = 5$$
$$2x - 3 + 3 = 5 + 3$$
$$2x = 8$$
$$\frac{2x}{2} = \frac{8}{2}$$
$$x = 4.$$

Each of these steps represents an equation that is equivalent to the others since a input value of 4 makes each equation true.

Likewise, $2 + 3 = 5 - 1 = 4(2) = 8$ is notation to give your math teacher gray hair. This is a lazy representation of a series of computations and clearly violates the convention that what is on the left represents the same number as what is on the right. $5 = 4 = 8$ deserves a big YUK!!

An equation expresses equality of two expressions. The algebraic representation of a function is often written as an equation. Not all equations represent functions. If the equation contains only one variable, such as $2x - 3 = 5 - 11x$, it does not represent a function. Rather, it represents the equality of the outputs of two functions. Equations that contain two variables always represent a relation. They only represent a function if there is just one output for any input.

STUDENT DISCOURSE

Pete: I always thought that "solving" meant finding the answer to a math problem, regardless of the type of problem.

Sandy: So did I until I took this course. Everything I did in math, I called solving. But I learned that I was using the word "solve" incorrectly.

Pete: In the same connection, I had no idea where equations came from or what they meant. I just knew I had to solve them.

Sandy: So how has your thinking changed?

Pete: Now I see that I can interpret an equation like $3x - 7 = 12$. I can think that there is a function $y(x) = 3x - 7$ and I want to find the input if the output $y(x)$ is 12.

Sandy: Very good. How would you find the input?

Pete: That's what solving means. Find the input when I know the output. For the equation I wrote, that isn't too hard. I add 7 to both sides to get the x term alone and then divide both sides by 3.

Sandy: That's great. One other question. At some point, you probably solved equations like $5x - 11 = 2x + 5$. How could you interpret this equation in terms of functions?

Pete: Hmmm, let's see. I see two functions $y(x) = 5x - 11$ and $y(x) = 2x + 5$.

Sandy: What does the "=" mean?

Pete: It says that the processes of the two functions must produce the same output. Oh, okay. So I have to find the value of x so that the outputs are the same.

Sandy: And what does x stand for?

Pete: The input. Great. That helps me see equations in a new light.

1. In the Glossary, list and define the words in this section that appear in ***italic boldface*** type.

2. What are the "simplify" steps in the solution to

$$55{,}000 = 2000t + 33{,}000$$

given on page 231? What are the "equation-solving" steps?

3. **Super Dog Burger Bits:** The Super Dog Burger Bits annual salaries were described by the function $S(t) = 23364.48598(1.07)^t$.

   a. What mathematical process is being performed if you are asked to find the salary in year 12? Why?

   b. What mathematical process is being performed if you are asked to find the year when the salary will reach $100,000? Why?

   c. Write a mathematical statement that could be used to find the years when the annual salary will exceed $60,000. Is this statement an equation or an inequality? Why?

4. Create an example of a mathematical expression and a mathematical statement. Explain the difference between your two answers.

5. Create examples of the following.

   a. Simplifying an expression

   b. Evaluating a function

c. Solving an equation

6. Create an example in which the "=" is used to represent an action.

7. Create an example in which the "=" expresses the equality of two functions. Solve the equation you wrote using the appropriate notation.

8. Create an example of an equation and an example of an inequality. Solve both of your examples.

9. Can all functions be represented by equations? Provide evidence, such as examples, to support your answer.

10. Aside from equations, what other methods are used to represent functions? Provide an example of each.

11. Write an example of:

    a. a linear function with a negative slope.

    b. a linear function with a vertical intercept at $(0, 3)$.

    c. a linear function with a slope of 8.

    d. an exponential function with a vertical intercept of $(0, 12)$.

    e. an exponential function in which the output increases as the input increases.

    f. a quadratic function with a vertical intercept of $(0, 5)$.

g. a quadratic function with a second finite difference in consecutive outputs of 6.

h. two linear functions that are parallel.

i. two linear functions that are perpendicular.

j. two linear functions that are perpendicular and that have different vertical intercepts.

12. Consider the data in Table 1.

TABLE 1 Random Data 1

Input	−3	−1	1	3	5	7
Output	−5	−2	1	4	7	10

a. This table arranges the inputs and outputs horizontally. Rearrange the table so that it is vertical. Which form is easier for you to use?

b. Is this function linear, exponential, or quadratic?

c. Write an equation that will generate the output from the input. Explain what you did.

d. Use your equation to find the output when the input is 37.

e. Find the input if the output is −87. Explain what you did.

13. Consider the function $y(x) = \frac{7}{3}x - 1$.

a. What happens to the output if the input increases by 3? What did you do to answer the question?

b. What change in the input causes the output to increase by 7? What did you do to answer the question?

c. What happens to the output if the input decreases by 12? What did you do to answer the question?

d. What change in the input causes the output to increase by 14? What did you do to answer the question?

REFLECTION

Write a paragraph comparing and contrasting the words "simplify," "evaluate," and "solve."

Write a paragraph comparing and contrasting the words "expression," "equation," "inequality," and "function."

SECTION 4.2

What Does It Mean to Solve?

Purpose

- Distinguish between evaluating and solving.
- Investigate how equations arise.
- Investigate methods for solving equations.

In the preceding section, we investigated the meaning of several mathematical words. The word "solve" is the focus of this section. You probably are familiar with "solving" linear equations in one variable, such as $3x + 11 = 23$. By subtracting 11 from both sides and then dividing both sides by 3, you can "solve" the equation for x. This is the simplest form of solving. Many types of equations are either difficult or impossible to solve by manipulating the symbolic form, as was done with $3x + 11 = 23$. In such cases, we resort to using either tables or graphs to "solve." Sometimes the solutions may only be approximations, but they may be the best we can do. To investigate this idea, we look at three complex functions that describe the sales of records, CDs, and cassette tapes over a 15-year period.

Audio Media Sales: Since 1980, the number of compact discs (CDs) sold in the United States has drastically increased, while the number of long-playing albums (LPs) has greatly decreased. During the same time period the number of cassettes first increased, then decreased.

Using data-analysis techniques, the following functions approximate the relationship between the year and the sales of each product.

The function $LP(x) = -0.01488x^4 + 0.737x^3 - 12.9x^2 + 60.9x + 257$ expresses the relationship between the number of LPs sold (in millions) in the United States and the time x years since 1975 ($x = 0$ in 1975).

The function $CD(x) = 2.626x^2 - 22.55x + 12.25$ expresses the relationship between the number of CDs sold (in millions) in the United States and the time x years since 1984. This expression cannot be used for years before 1984 since CDs were not sold prior to 1984. $CD(x) = 0$ expresses the relationship between the number of CDs sold in the years from 1975 to 1983. ($x = 0$ in 1975). $CD(x)$ is called a piecewise function since we have two expressions. The value of the input x determines which expression we use.

The function $Cass(x) = -0.38x^3 + 8.67x^2 - 15.6x + 16.7$ expresses the relationship between the number of cassettes sold (in millions) in the United States and the time x years since 1975 ($x = 0$ in 1975).

1. a. Find the number of LPs sold in 1975. Explain what you did.

b. Find the number of LPs sold in 1970. Explain what you did.

2. a. Find the number of CDs sold in 1985. Explain what you did.

b. Find the number of CDs sold in 1980. Why is the first expression in the model not valid for 1980?

3. Find the number of cassettes sold in 1975. Explain what you did.

4. Find the number of LPs, the number of CDs, and the number of cassettes sold in 1990. Explain what you did.

5. In each of the previous investigations, are you evaluating or solving? Justify your answers.

In the previous investigation, you were given an input value and asked to find the corresponding output value. This requires evaluating the appropriate function and simplifying the result.

To find the number of LPs and the number of cassettes sold in 1975, evaluate the *LP* and *Cass* functions by substituting zero for the input variable x in the respective functions.

$$\begin{aligned} LP(0) &= -0.01488(0)^4 + 0.737(0)^3 - 12.9(0)^2 + 60.9(0) + 257 \\ &= 257 \end{aligned}$$

$$\begin{aligned} Cass(0) &= -0.38(0)^3 + 8.67(0)^2 - 15.6(0) + 16.7 \\ &= 16.7. \end{aligned}$$

So 257 million LPs and 16.7 millions cassettes were sold in 1975.

To find the number of LPs sold in 1970, evaluate by substituting –5 for the input variable in the *LP* function.

$$\begin{aligned} LP(-5) &= -0.01488(-5)^4 + 0.737(-5)^3 - 12.9(-5)^2 + 60.9(-5) + 257 \\ &= -471.425. \end{aligned}$$

Notice that the output (the number of LPs sold) is negative, which does not make sense because we used a value for input that is outside the domain of the problem situation. The year 1975 was given as the arbitrary zero for these functions. The functions are likely to be inaccurate before this time.

To find the number of CDs sold in 1985, evaluate by substituting 10 for the input variable x in the CD function.

$$\begin{aligned} CD(10) &= 2.626(10)^2 - 22.55(10) + 12.25 \\ &= 262.6 - 225.5 + 12.25 \\ &= 49.35. \end{aligned}$$

So 49.35 million CDs were sold in 1985.

If the *CD* function is evaluated at 1980 ($x = 5$), the result is

$$\begin{aligned} CD(5) &= 2.626(5)^2 - 22.55(5) + 12.25 \\ &= -34.85. \end{aligned}$$

Again the output (the number of CDs sold) is negative. This indicates that 5 is not in the domain of this problem situation. This suggests why years before 1984 could not be used.

For 1990, substitute 15 for the input variable into each of the functions.

$$LP(15) = -0.01488(15)^4 + 0.737(15)^3 - 12.9(15)^2 + 60.9(15) + 257$$
$$= 2.075.$$

$$CD(15) = 2.626(15)^2 - 22.55(15) + 12.25$$
$$= 264.85.$$

$$Cass(15) = -0.38(15)^3 + 8.67(15)^2 - 15.6(15) + 16.7$$
$$= 450.95.$$

The last three computations indicate that 2.075 million LPs, 264.85 million CDs, and 450.95 million cassettes were sold in the United States in 1990.

Each of these investigations is an example of ***evaluating a function.*** When the directions say ***evaluate***, the value of the input variable is given and you must find the output. The word ***simplify*** is used to describe the process of using order of operations to compute the output once the value of the input has been substituted for the variable.

Reverse the problem. Suppose the output is known, but the input is not. Use both tables and graphs as you explore this idea in the next investigations.

6. For each recording medium, approximate the year when the number of items sold was 100 million. Describe your procedure in each case.

a. LPs

b. CDs

c. Cassettes

7. In what year does the algebraic model suggest that no LPs will be sold? Describe your procedure.

8. In what year did the sales of LPs match the sales of CDs? Describe your procedure.

9. In what year did the sales of CDs equal the sales of cassettes? Describe your procedure.

DISCUSSION

The just-completed investigations provide us with information about output and requires us to find the input. This is an example of solving. Each investigation gives rise to an equation in which the value of x must be found. This is a much harder proposition than evaluating. To evaluate, we must substitute and simplify. To solve, we must find a way to determine the value of the input variable so that a certain output condition is met. This is accomplished for "simple" functions by reversing the function machine process. However, this is not practical for the functions explored in this section.

Solving often can be accomplished by investigating a table of output values or by investigating a graph of the functions.

Let's look closely at each of the investigations. Begin by creating an algebraic statement that reflects the question. Then use a table or a graph to assist in answering the question. Note that it is easy to check results. Since we are determining input for a specific output, we can evaluate the function at our answer to see if it produces the desired output. What fun!

Investigation 6 states that the output is 100. Figure 1 displays both a table solution and a graphical approximate solution for each function.

FIGURE 1

X	Y1
11.4	115.36
11.5	111.96
11.6	108.57
11.7	105.2
11.8	101.85
11.9	98.509
12	95.184

X=11.8

WINDOW
Xmin=0
Xmax=15
Xscl=1
Ymin=-100
Ymax=500
Yscl=50
Xres=1

Y1=-.01488X^4+.737X^3-12_
X=11.85 Y=100.17667

a. LP sales reach 100 million.

X	Y2
11.4	96.455
11.5	100.21
11.6	104.02
11.7	107.89
11.8	111.8
11.9	115.77
12	119.79

X=11.5

WINDOW
Xmin=0
Xmax=15
Xscl=1
Ymin=-100
Ymax=500
Yscl=50
Xres=1

Y2=2.626X²-22.55X+12.25
X=11.5 Y=100.2135

b. CD sales reach 100 million.

X	Y3
4.5	87.44
4.6	91.41
4.7	95.448
4.8	99.552
4.9	103.72
5	107.95
5.1	112.24

X=4.9

WINDOW
Xmin=0
Xmax=15
Xscl=1
Ymin=-100
Ymax=500
Yscl=50
Xres=1

Y3=-.38X^3+8.67X²-15.6X+_
X=4.9 Y=103.72008

c. Cassette sales reach 100 million.

Reading from either the table or the graph, LP sales reached 100 million approximately 11.85 years after 1975, which is the year 1986. CD sales reached 100 million 11.5 years after 1975, which is the year 1986. While both the sales of LPs and CDs were 100 million in 1986, the graphs suggest that the number of LPs sold was declining while the number of CDs sold was increasing. Cassette sales passed 100 million approximately 4.9 years after 1975, which is near the end of the year 1979.

By answering Investigation 6, we have used both a numerical (table) and a graphical (graph) approach to solve the following equations.

$$100 = -0.01488x^4 + 0.737x^3 - 12.9x^2 + 60.9x + 257$$

$$100 = 2.626x^2 - 22.55x + 12.25$$

$$100 = -0.38x^3 + 8.67x^2 - 15.6x + 16.7.$$

To determine the year when no LPs will be sold, use the table or graph to approximate when the output of the LP function is zero (Figure 2).

FIGURE 2

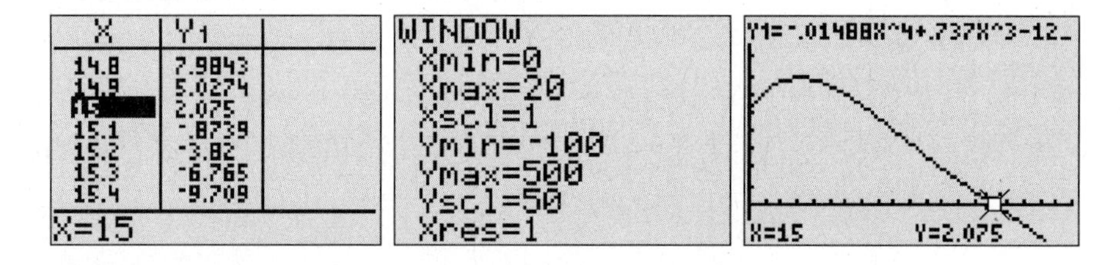

Both table and graph predict no LP sales 15 years after 1975. This would be 1990. We know from experience that this is inaccurate. The equation that was solved in this case is

$$0 = -0.01488x^4 + 0.737x^3 - 12.9x^2 + 60.9x + 257$$

In Investigation 8 we want to find the input for which the outputs of the *LP* function and the *CD* function are equal; that is, $LP(x) = CD(x)$. This is equivalent to solving the equation

$$-0.01488x^4 + 0.737x^3 - 12.9x^2 + 60.9x + 257 = 2.626x^2 - 22.55x + 12.25$$

Numerical and graphical solutions appear in Figure 3.

FIGURE 3

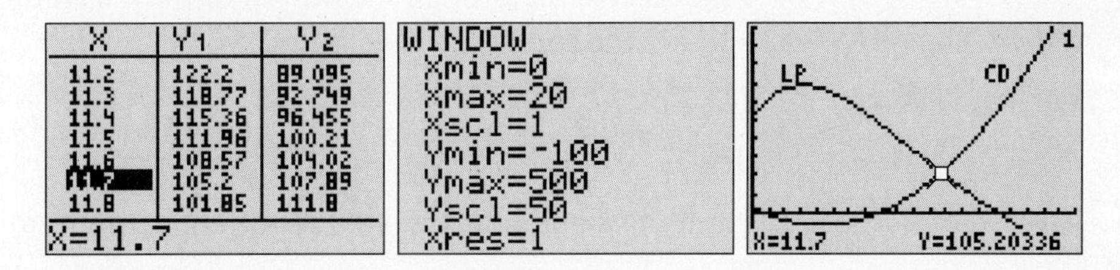

Eleven years after 1975 is 1986. The sales of LPs and CDs were approximately the same in 1986.

In Investigation 9 we want to find the input for which the outputs of the *CD* function and the *Cass* function are equal, that is, where $CD(x) = Cass(x)$. This is equivalent to solving the following equation.

$$2.626x^2 - 22.55x + 12.25 = -0.38x^3 + 8.67x^2 - 15.6x + 16.7$$

Numerical and graphical solutions appear in Figure 4.

FIGURE 4

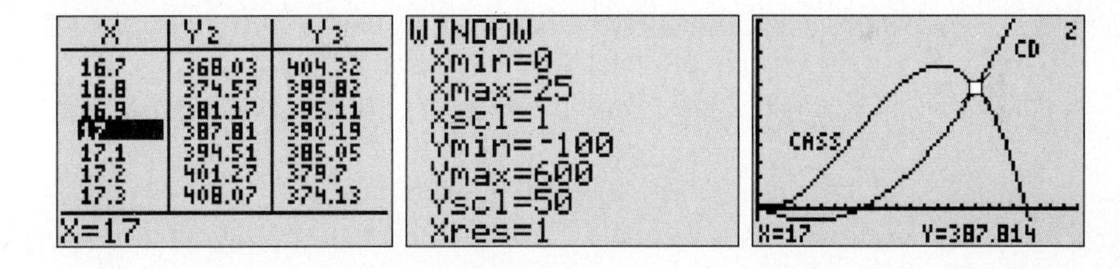

Seventeen years after 1975 is 1992. The sales of CDs and cassettes were approximately the same in 1992.

10. For what years did the sale of LPs exceed the sale of cassettes? Defend your answer.

11. For what years did the sale of cassettes exceed the sale of CDs? Defend your answer.

Investigations 10 and 11 involve an ***inequality*** instead of an equation. In Investigation 10 we want to know the years when LP sales exceeded cassette sales. This means that we want to know the years when the output of the *LP* function is greater than the output of the *Cass* function. This is equivalent to solving the following inequality.

$$-0.01488x^4 + 0.737x^3 - 12.9x^2 + 60.9x + 257 > -0.38x^3 + 8.67x^2 - 15.6x + 16.7$$

Numerical and graphical solutions appear in Figure 5.

FIGURE 5

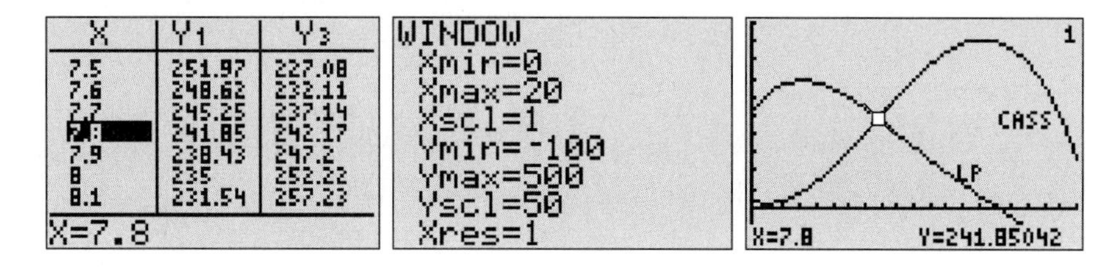

The output of Y_1 (the *LP* function) becomes less than the output of Y_2 (the *Cass* function) 8 years after 1975. Thus the sales of LPs exceeded the sales of cassettes until 1983. The graph shows us that the LP output is larger than the Cass until the two curves intersect at about 7.8 years.

Investigation 11 asks for the years when cassette sales exceeded CD sales. This means that we want the years when the output of the *Cass* function is greater than the output of the *CD* function. This is equivalent to solving the inequality

$$-0.38x^3 + 8.67x^2 - 15.6x + 16.7 > 2.626x^2 - 22.55x + 12.25.$$

Numerical and graphical solutions appear in Figure 6.

FIGURE 6

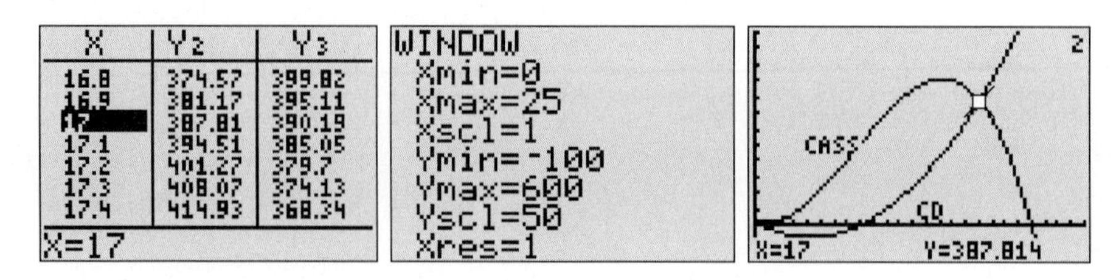

In Investigation 9, we saw that the sales of CDs and cassettes were approximately equal in 1992. Thus the sale of cassettes exceeded the sales of CDs for 17 years after 1975, or until 1992. The table in Figure 6 shows that the output of Y_3 (the *Cass* function) is larger than the output of Y_2 (the *CD* function) for input values up to 17. The graph in Figure 6 shows that the *Cass* function is above the *CD* function from $x = 0$ through $x = 17$.

In summary, we continued to investigate the relationship between the words "evaluate," "simplify," and "solve" in mathematics.

- Investigations 1–4 asked for the output given a value for input. We answer these by evaluating the function at the given input. The process of evaluating involves first substituting the given input for the input variable. Simplifying is required to find the output.

- Investigations 6–11 requested the input given specific restrictions on output. This required the creation of an equation (Investigations 6–9) or an inequality (Investigations 10–11). These were solved numerically and graphically. We did not solve them algebraically because of the difficulty involved. These numerical and graphical techniques are very powerful for approximating the real number solutions to most algebraic equations and inequalities.

STUDENT DISCOURSE

Pete: I always thought you had to solve by getting a variable, like x, alone on one side of the "=" symbol.

Sandy: So did I, but my instructor last semester explained that graphing calculators provide other ways to "solve."

Pete: Yeah, I can see that. For example, solving by graphing wouldn't be too practical if it was difficult to draw and manipulate the graph. With the calculator, it isn't too hard to investigate the graph.

Sandy: That's right. The same is true with tables. Using tables to solve would be very tedious if we couldn't easily create a table of values in which we can change the inputs and increments or amount of change.

Pete: So how do I know what technique to use?

Sandy: Well, for simple linear equations like $5x + 1 = 2$, it's probably quickest to solve without the table or the graph.

Pete: I can see that—I could solve that equation faster than I could enter it into the calculator. I guess I could check my answer on the calculator.

Sandy: That's right. On the other hand, you may not be able, at this point, to solve any other types of equations without the table or the graph.

Pete: You mean equations that may be quadratic or exponential, for example.

Sandy: Yes, and the third and fourth degree equations in this section. I mean the ones with an exponent of 3 or 4 in the equation.

Pete: Will I learn to solve any other types of equations without using the table or graph?

Sandy: Yes. Later in the book, you learn how to solve quadratic and exponential equations without the calculator.

Pete: Are they hard to do?

Sandy: Well, I found that they took me a little longer than the linear ones.

Pete: What's the advantage of knowing how to solve those without the calculator?

Sandy: One advantage is that you get exact answers. The table or graph sometimes only give you approximate answers.

Pete: Thanks for the information. It's interesting to see how the various methods of "solving" all fit together.

1. In the Glossary, list and define the words in this section that appear in ***italic boldface*** type.

2. Decide whether the given problem is one of evaluation or one of solving.

   a. If $N = 13b$, find N if $b = 6$.

   b. If $N = 13b$, find b if $N = 936$.

   c. If $C = 4z^3$, find C if $z = 7$.

   d. If $Q = 5m^2$, find Q if $m = 12$.

   e. If $Q = 5m^2$, find m if $Q = 245$.

   f. If $A = 3 + 2b^2$, find A if $b = 13$.

   g. If $A = 3 + 2b^2$, find b if $A = 75$.

h. If $L = 3r^2 - 2r + 7$, find L if $r = 6$.

3. Write an equation or inequality for the given situation.

a. The receipts R for a concert are a function of the number of tickets t sold. If each ticket costs \$19, the function is $R(t) = 19t$. Determine the number of tickets that must be sold so that the receipts exceed \$50,000.

b. The accumulated salary function at CDs-R-Us is $y(x) = 1000x^2 + 34{,}000x$ where x is the years on the job and y is the total salary earned. For how many years will the accumulated salary be less than \$200,000?

4. a. Create a function that reflects an application. Clearly identify the input and the output variable.

b. Use your function to write a problem that requires evaluation to answer.

c. Use your function to write a problem that requires solving to answer.

5. Write an equation or inequality for the given situation. These questions refer to the functions given at the beginning of this section.

a. In what year did CD sales equal approximately 100 million?

b. In what years did cassette sales exceed 400 million?

c. In what year will CD sales be double the sales of cassettes?

6. Discuss the use of zero as a measure of time in this section.

7. Investigation 8 was solved by setting the expression representing the *LP* function equal to the expression representing the *CD* function. Another approach would be to state that the difference between the two expressions must be zero.

a. Write the equation representing the fact that the difference is zero.

b. Use a table or a graph to solve the equation. Briefly explain what you did.

8. Investigation 8 was solved by setting the expression representing the *CD* function equal to the expression representing the *Cass* function. Another approach would be to state that the difference between the two expressions must be zero.

a. Write the equation representing the fact that the difference is zero.

b. Use a table or a graph to solve the equation. Briefly explain what you did.

9. In your Chaos class, your performance is evaluated based on five exams. Assume that your scores on the first four exams were 79, 87, 81, and 90 (out of 100). Both the mean and the median are functions of the score on the last exam.

a. Write an equation that represents the *mean* function.

b. Find the mean if you score 95 on the last exam. Did you "evaluate" or "solve" to answer this question?

c. Find the score you need on the last exam to have a mean of 80. Did you "evaluate" or "solve" to answer this question?

d. Find the median if you score 95 on the last exam. Did you "evaluate" or "solve" to answer this question?

e. Find the score required on the last exam to have a median of 82. Did you "evaluate" or "solve" to answer this question?

10. Consider the functions $f(x) = x^2 + x - 4$ and $g(x) = -3x + 1$.

a. When will the output $g(x)$ be equal to –8? Explain what you did.

b. When will the output $f(x)$ be equal to 5? Explain what you did.

c. When will the outputs of the two functions be the same? Explain what you did.

d. When will the outputs of $f(x)$ be greater than the outputs of $g(x)$? Explain what you did.

11. Given the following linear equations in standard form, write them in the form $y = ax + b$ and identify the slope and vertical intercept.

a. $2x - 3y = -91$

b. $5x + 11y = 14$

c. $3x + 4y = 12$

d. $4x - 3y = 0$

12. A line contains the points $(5, -1)$ and $(-2, -11)$.

a. Write the linear function for the line.

b. Write the equation of a line parallel to the given line through (7, –8).

c. Write the equation of the line perpendicular to the given line through (7, –8).

13. Write the equation of the following lines.

a. The line parallel to $7x - 2y = 10$ through the point (1, 3)

b. The line perpendicular to $7x - 2y = 10$ through the point (1, 3)

REFLECTION

Write a paragraph to a friend explaining the difference between the word "evaluate" and the word "solve" in mathematics. Include a discussion of various techniques that could be used to "solve" a mathematical statement.

SECTION 4.3

Models and Their Representations

Purpose

- Summarize the key features of polynomial and exponential models.
- Compare numeric, symbolic, and graphic representations of polynomial and exponential models.

Chapters 2 and 3 focused on the development of *linear functions*, *exponential functions*, and *quadratic functions*. Sections 4.1 and 4.2 discussed various mathematical actions that can be performed using these types of functions. In this section, we investigate the similarities and differences between these function types. Understanding the characteristics of each function is vital when attempting to select a particular kind of function to serve as a model for data.

Once we have investigated the key characteristics of each function type, we will use this information to select and create an appropriate *model* for real-world data. By real-world data, we mean data collected in an actual setting which can be approximated by a model. Up to this point, we have used data that exactly fits a linear, quadratic, or exponential function. One important use of mathematics is its ability to numerically describe real-world phenomena. Once we have laid the groundwork in this section, we will see how this might be done in the remaining sections of this chapter.

1. You read the statement: "There is a linear relationship between the input and output." What does this mean to you?

2. Given a table of data, how do you use the table to recognize if the data are linear?

3. The general symbolic form of a linear function is $y(x) = ax + b$, where x is the input and y is the output.

a. What are the parameters in the general form? What does each parameter represent?

b. What is the least number of points you need to determine the equation of a specific linear function? How does this compare to the number of parameters in the general symbolic form?

4. Graph the functions $y = 3x - 2$, $y = -3x - 2$, and $y = -2$ on your calculator using a standard viewing window.

a. What are the similarities and differences in the graphs of the functions?

b. What is the slope of each function?

c. What effect does the slope have on the graph of the function?

5. Consider the general linear function. State or draw the key characteristics of the function for each of the following representations. (*Hint:* Refer to your answers to Investigations 2–4.)

a. Numeric (table)

b. Symbolic (equation)

c. Geometric (graph)

The word "relationship" appears often in mathematical discussions. A relationships between two things just means that the value of one thing will determine the value of the second thing.

So what does it mean to say a relationship is "linear"? We can answer this several ways.

- *Numerically,* a linear relationship means that the change between any two outputs divided by the change between the corresponding inputs always produces the same number.

For example, consider the data in Table 1.

TABLE 1 Linear Data 1

Input	−8	6	3	−2
Output	23	−5	1	11

Choosing (−8, 23) and (3, 1), we get a slope of $\dfrac{23-1}{-8-3} = \dfrac{22}{-11} = -2.$

Choosing (6, −5) and (−2, 11), we get a slope of $\dfrac{-5-11}{6-(-2)} = \dfrac{-16}{8} = -2.$

Choosing (−8, 23) and (−2, 11), we get a slope of $\dfrac{23-11}{-8-(-2)} = \dfrac{12}{-6} = -2.$

Since the data is linear, we get the same slope, −2, regardless of which pair of points we choose.

- *Symbolically,* a linear relationship means that the data must satisfy some equation of the form $y = ax + b$ for some specific values of a and b. The value of a equals the slope of the function and is computed by finding the rate of change between two points. If the input is 0, the output is b.

- *Graphically,* a linear relationship means that the data, when graphed, lie in a straight line. The line intersects the vertical axis at $(0, b)$. The slope determines the "slant" of the line.

For example, the functions $y = 3x - 2$, $y = -3x - 2$, and $y = -2$ all have a value of –2 for parameter b. This means all three lines intersect the vertical axis at $(0, -2)$. The slopes of the functions are 3, –3, and 0, respectively.

A slope of 3 means that the line goes *up* three units for every one unit to the right.

A slope of –3 means that the line goes *down* three units for every one unit to the right.

A slope of 0 means that the line is horizontal since a zero slope means that the output does not change.

Figure 1 displays the graphs of these three functions. Each is displayed in a standard viewing window.

FIGURE 1

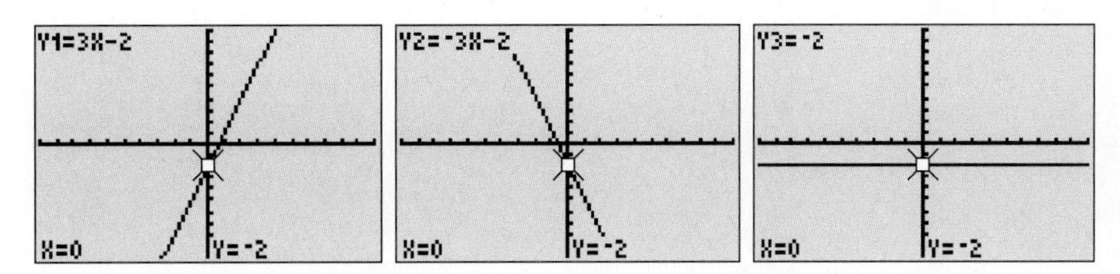

Finally, exactly two points will determine a specific line. Thus, we need, minimally, two points to determine a linear function. Notice that the number of points needed is the same as the number of parameters in the general linear equation.

Next we summarize some ideas about exponential functions.

6. You read the statement: "There is an exponential relationship between the input and output." What does this mean to you?

7. Given a table of data, how do you use the table to recognize if the data are exponential?

8. The general symbolic form of a exponential function is $y(x) = a(b)^x$, where x is the input and y is the output.

a. What are the parameters in the general form? What does each parameter represent?

b. What is the least number of points you need to determine the equation of a specific exponential function? How does this compare to the number of parameters in the general symbolic form?

9. Graph the functions $y = 5(1.2)^x$ and $y = 5(0.8)^x$ on your calculator using Xmin = –10, Xmax = 10, Ymin = 0, and Ymax = 20.

a. What are the similarities and differences in the graphs of the functions?

b. What is the base of each function?

c. What effect does the base have on the graph of the function?

10. Consider the general exponential function. State or draw the key characteristics of the function for each of the following representations. (*Hint:* Refer to your answers to Investigations 7–9.)

a. Numeric (table)

b. Symbolic (equation)

c. Geometric (graph)

What does it mean to say a relationship is "exponential"? Let's look at each representation to answer this question.

◆ *Numerically*, an exponential relationship means that the ratio of successive outputs always produces the same number.

For example, consider the data in Table 2.

TABLE 2 Exponential Data I

Input	–1	0	1	2
Output	1.25	2	3.2	5.12

Choosing (–1, 1.25) and (0, 2), we get a ratio of outputs of $\frac{2}{1.25} = 1.6$.

Choosing (0, 2) and (1, 3.2), we get a ratio of outputs of $\frac{3.2}{2} = 1.6$.

Choosing (1, 3.2) and (2, 5.12), we get a ratio of outputs of $\frac{5.12}{3.2} = 1.6$.

Since the data is exponential, we get the same ratio of outputs, 1.6, regardless of which pair of successive points we choose.

- *Symbolically*, an exponential relationship means that the data must satisfy some equation of the form $y = a(b)^x$ for some specific values of a and b. The value of b equals the ratio of outputs (the base). It is computed by dividing two successive outputs. If the input is 0, the output is a.

- *Graphically*, an exponential relationship means that the data, when graphed, lie on a special curve we look at below. The curve intersects the vertical axis at $(0, a)$. The base b determines whether the curve rises from left to right or falls from left to right. The size of the base also determines how quickly the curve rises or falls.

For example, the functions $y = 5(1.2)^x$ and $y = 5(0.8)^x$ have a value of 5 for parameter a. This means both curves intersect the vertical axis at $(0, 5)$. The bases of the functions are 1.2 and 0.8, respectively.

A base larger than 1 means that the output gets larger moving from left to right on the graph.

A positive base smaller than 1 means that the output gets smaller moving from left to right on the graph.

Figure 2 displays the graphs of these two functions.

FIGURE 2

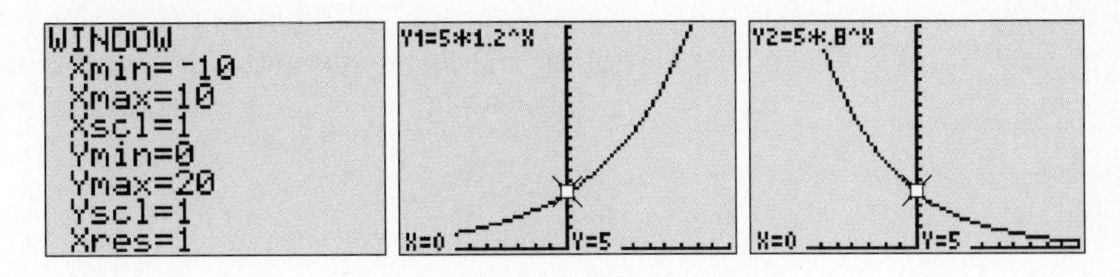

Finally, exactly two points will determine a specific exponential function. Essentially, we need two points to determine the base and, thus, the value of parameter b. Once the base is known, one point can be substituted in for (x, y) and we can solve for the parameter a. Notice that the number of points needed is the same as the number of parameters in the general exponential equation.

Next we focus on the quadratic function. Actually, both the linear and quadratic function are members of the general class of polynomial functions. A ***polynomial function*** is one in which the input variable is raised to some whole number power. The largest exponent on the input variable is called the ***degree of the polynomial.***

For example, the linear function $y = 5x - 7$ has degree 1 since the largest exponent on the variable x is 1.

The quadratic function $y = 5x^2 - 7x + 16$ has degree 2 since the largest exponent on the variable x is 2.

Exponential functions are *not* polynomial functions. Instead of the input variable being raised to a whole number power, the input variable is the exponent in the exponential function.

11. You read the statement: "There is a quadratic relationship between the input and output." What does this mean to you?

12. Given a table of data, how do you use the table to recognize if the data are quadratic?

13. The general symbolic form of a quadratic function is $y(x) = ax^2 + bx + c$, where x is the input and y is the output.

a. What are the parameters in the general form? What does each parameter represent?

b. What is the least number of points you need to determine the equation of a specific quadratic function? How does this compare to the number of parameters in the general symbolic form?

14. Graph the quadratic functions $y(x) = 2x^2 - 3x - 4$ and $y(x) = -2x^2 - 3x - 4$ in a standard viewing window. What are the similarities and differences in the graphs of the functions? What difference in the equations is causing the difference in the graphs?

15. Consider the general quadratic function. State or draw the key characteristics of the function for each of the following representations. (*Hint:* Refer to your answers to Investigations 12–14.)

a. Numeric (table)

b. Symbolic (equation)

c. Geometric (graph)

So what does it mean if we say a relationship is "quadratic"? We can answer this several ways.

- *Numerically*, a quadratic relationship means that the second difference in successive outputs always produces the same number.

For example, consider the data in Table 3.

TABLE 3 Linear Data I

Input	−1	0	1	2
Output	6	−3	4	27
Δ(Output)	−9	7	23	
Δ(Δ(Output))	16	16		

Note that Δ(Δ(Output)) = 16, a constant.

Since the data is quadratic, we get the same second difference in consecutive outputs, 16, regardless of which three consecutive points we choose.

- *Symbolically*, a quadratic relationship means that the data must satisfy some equation of the form $y = ax^2 + bx + c$ for some specific values of a, b, and c. The value of a equals one-half of Δ(Δ(Output)). If the input is 0, the output is c. The value of b is found after a and c are found, by substituting a data point for (x, y) and solving for b.

- *Graphically*, a quadratic relationship means that the data, when graphed, form a curve called a ***parabola***. The parabola intersects the vertical axis at $(0, c)$. The sign of a determines if the parabola opens upward or downward.

For example, the quadratic $y(x) = 2x^2 - 3x - 4$ and $y(x) = -2x^2 - 3x - 4$ both intersect the vertical axis at $(0, -4)$. The graph of $y(x) = 2x^2 - 3x - 4$ opens upward since a, 2, is positive. The graph of $y(x) = -2x^2 - 3x - 4$ opens downward since a, -2, is negative (Figure 3 using a standard viewing window).

FIGURE 3

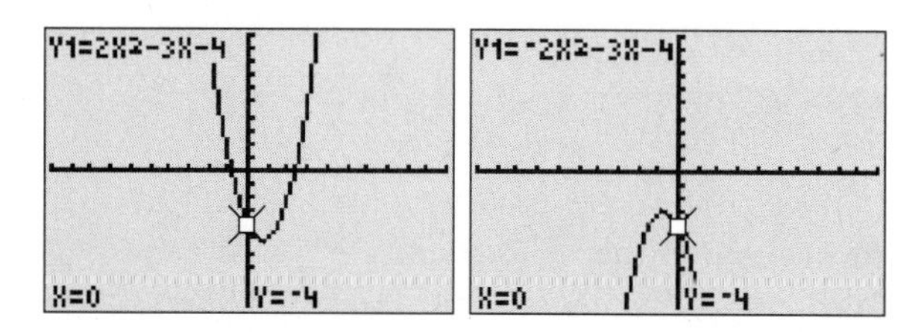

Finally, exactly three points will determine a specific quadratic function. Essentially, we need three points to determine the second finite difference in outputs and, thus, the value of parameter a. Notice that the number of points needed is the same as the number of parameters in the general quadratic function.

As stated before, both linear and quadratic functions belong to the class of polynomial functions. Linear functions have degree 1 and quadratic functions have degree 2. Before concluding this section let's briefly investigate polynomial functions of degrees 3 and 4.

16. Consider the cubic (third-degree) functions

$$y(x) = -x^3 - 2x^2 + 3x + 4 \quad \text{and} \quad y(x) = x^3 + 2x^2 - 3x - 4.$$

a. Graph both functions in a standard viewing window. Describe what is similar and what is different about the two graphs. How do you account for the differences?

b. Complete Table 4 for the function $y(x) = -x^3 - 2x^2 + 3x + 4$. Compute successive finite differences in output until the finite differences become constant.

TABLE 4 **Cubic Function Numeric Analysis**

x	−3	−2	−1	0	1	2	3	4	5
y									
Δ(y)									
Δ(Δ(y))									
Δ(Δ(Δ(y)))									

c. How does the degree of the polynomial compare to the number of finite differences in outputs needed to get a constant?

17. Consider the quartic (fourth-degree) functions

$$y(x) = x^4 - 2x^3 - 2x^2 + 4x + 1 \quad \text{and}$$

$$y(x) = -x^4 + 2x^3 + 2x^2 - 4x - 1.$$

a. Graph both functions in a standard viewing window. Describe what is similar and what is different about the two graphs. How do you account for the differences?

b. Complete Table 5 for the function $y(x) = x^4 - 2x^3 - 2x^2 + 4x + 1$. Compute successive finite differences in output until the finite differences become constant.

TABLE 5 **Quartic Function Numeric Analysis**

x	−3	−2	−1	0	1	2	3	4	5
y									
Δ(y)									
Δ(Δ(y))									
Δ(Δ(Δ(y)))									
Δ(Δ(Δ(Δ(y))))									

c. How does the degree of the polynomial compare to the number of finite differences in outputs needed to get a constant?

The functions

$$y(x) = -x^3 - 2x^2 + 3x + 4 \quad \text{and} \quad y(x) = x^3 + 2x^2 - 3x - 4$$

are called ***cubic***, or ***third-degree***, functions.

The graphs appear in Figure 4.

FIGURE 4

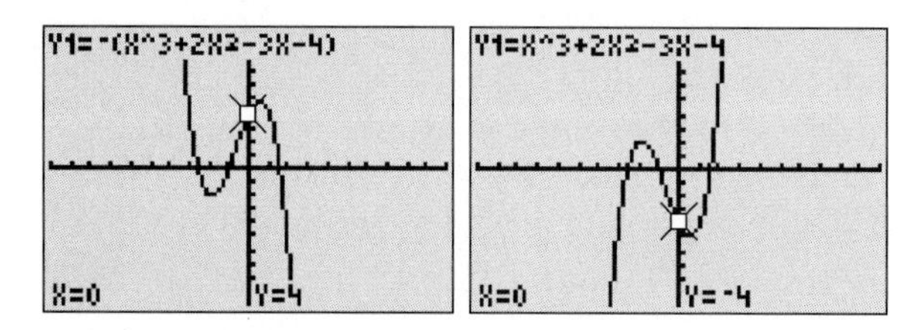

Most cubic functions have graphs similar to those in Figure 4. The main difference between the graphs of cubic functions is the vertical distance between the turns, which can be zero as in $y = x^3$.

The third finite differences in successive outputs are all 6. If the third finite differences in successive outputs are constant (and the first and second finite differences in successive outputs vary), then the function has degree 3.

The functions

$$y(x) = x^4 - 2x^3 - 2x^2 + 4x + 1$$

and

$$y(x) = -x^4 + 2x^3 + 2x^2 - 4x - 1$$

are called ***quartic***, or ***fourth-degree***, functions.

Most quartic functions have graphs similar to those in Figure 5. The main difference between the graphs of quartic functions is where the turns occur.

The fourth finite differences in successive outputs are all 24. If the fourth finite differences in successive outputs are constant (and the first, second, and third finite differences in successive outputs vary), then the function has degree 4.

The graphs appear in Figure 5.

FIGURE 5

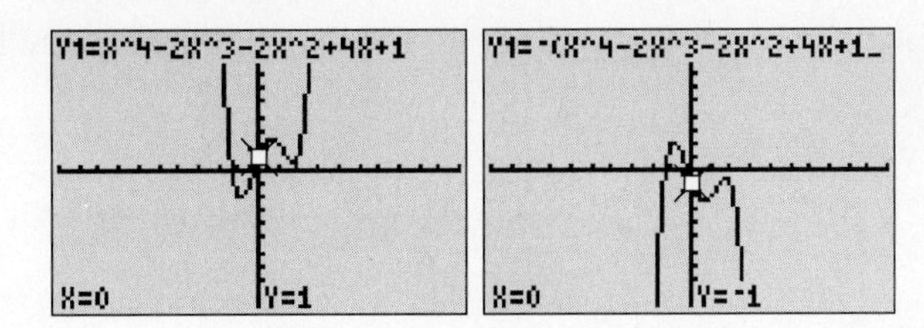

Now that we have investigated some of the key features and representations of both polynomial functions and exponential functions, we are ready to investigate how to model real-world data with such functions. This will be the focus of the remainder of the chapter.

STUDENT DISCOURSE

Pete: This section really helped me pull all the ideas together.

Sandy: I know. I constantly referred back to this section when I forgot something about one of the functions.

Pete: I do have a question.

Sandy: What's that?

Pete: We need exactly two points to create a linear function and three points to create a quadratic function, right?

Sandy: Yes.

Pete: How many points do I need to create a third- or a fourth-degree function?

Sandy: Look at the pattern—what do you think?

Pete: A quadratic that is degree 2 requires three points. So does a cubic that is degree 3 require four points?

Sandy: It makes sense. The general form of a cubic is $y = ax^3 + bx^2 + cx + d$. How many parameters do you see?

Pete: I guess there are four—a, b, c, and d.

Sandy: What connection do you see between the number of parameters and the number of points needed?

Pete: Well, both linear and exponential functions had two parameters and required two points. Quadratics have three parameters and require three points. So I guess the number of points required is determined by the number of parameters I have.

Sandy: Very well said—good job. And how does the number of parameters compare to the degree of the polynomial function?

Pete: Hmmm. I'm not sure.

Sandy: For example, a quadratic is degree 2 and has three parameters.

Pete: The degree seems to be one less than the number of parameters.

Sandy: Exactly. So if you had a quartic function, how many points would you need?

Pete: A quartic is degree 4, so I would have five parameters. So I need five points.

Sandy: Great job. you're ready for the rest of the chapter.

1. In the Glossary, list and define the words in this section that appear in ***italic boldface*** type.

2. Consider the points $(-3, 5)$ and $(6, 1)$.

a. Find the specific linear function determined by the two points.

b. What is the slope of the linear function through the two points?

c. Where does the graph of the function intersect the vertical axis?

3. Consider the points $(-3, 5)$ and $(6, 1)$.

a. Find the specific exponential function determined by the two points.

b. What is the base of the exponential function through the two points?

c. Where does the graph of the function intersect the vertical axis?

4. Consider the points $(-1, 8)$, $(0, 4)$, and $(1, 6)$.

a. Find the specific quadratic function determined by the three points.

b. What is the second finite difference in successive outputs?

c. Where does the graph of the function intersect the vertical axis?

5. Create a table of data that is

a. linear.

b. exponential.

c. quadratic.

d. cubic.

6. Solve each of the following equations.

a. $5 - 3x = 17$

b. $4t + 17 = -23$

c. $-3x + 2 = 0$

d. $5x - 3 = 17 + x$

e. $4x + 9 = 7x - 11$

f. $3 - 9p = 6p + 2$

7. Write the equation of an exponential function with base 1.3 that contains the point $(1, 65)$.

8. Consider the data in each of the following tables. Identify the data as linear, quadratic, or exponential. Create the specific algebraic function for the data and graph the function.

a.

TABLE 6 **Random Data I**

Input	–2	–1	0	1	2	3
Output	–10	0	4	2	–6	–20

b.

TABLE 7 **Random Data 2**

Input	–2	–1	0	1	2	3
Output	0.024	0.12	0.6	3	15	75

c.

TABLE 8 **Random Data 3**

Input	–2	–1	0	1	2	3
Output	6.6	4.9	3.2	1.5	–0.2	–1.9

9. Create an example in which the "=" expresses the equality of two functions. Solve the equation you wrote using the appropriate notation.

10. Create an example of an equation and an example of an inequality. Solve both of your examples.

11. Consider the functions $f(x) = 7x - 2$ and $g(x) = -x^2 + x + 4$.

a. When will the output $g(x)$ be equal to 2? Explain what you did.

b. When will the output $f(x)$ be equal to –15? Explain what you did.

c. When will the outputs of the two functions be the same? Explain what you did.

d. When will the outputs of $f(x)$ be less than the outputs of $g(x)$? Explain what you did.

12. Telephone Bill: Table 9 displays the charges by a local telephone company based on the number of calls made per month.

TABLE 9 Local Telephone Charges

Number of Calls	7	12	15	23	31	33	40
Charges ($)	12.65	13.65	14.25	15.85	17.45	17.85	19.95

a. Identify the data as linear, exponential, or quadratic. Defend your choice.

b. Create a specific algebraic model for the data.

c. Use the model to determine the cost of making 28 calls per month.

d. Use the model to determine the most calls that can be made if the phone charges must be less than $17.

REFLECTION

Write a paragraph comparing and contrasting the polynomial functions: linear, quadratic, cubic, quartic. Discuss numeric, symbolic, and geometric representations.

SECTION 4.4

Applying Linear Models to Data

Purpose

- Use linear regression to create a linear function, given two points.
- Fit linear models to data.
- Use the correlation coefficient and residuals to measure the "goodness of fit."
- Use regression lines and the standard error of estimate to answer questions about data.

We have spent much time developing an understanding of linear, exponential, and quadratic relationships. In this section, you will see how to use the calculator to quickly find a linear function given the minimum of two points. You already know how to do this without the calculator, but the model may be easier to find using the calculator especially if the numbers are difficult to work with.

After fitting a linear function to two points, we will look at how to create a linear function to approximate real-world data. Such data most often are not exactly linear or exponential or quadratic. Instead we use one of these types of functions to approximate the data. In this section, we look at data that appears visually to be approximately linear.

1. Consider the two points (–3, 7) and (–8, –1).

a. What is the slope of the line through the two points?

b. Derive the specific linear function that fits these two points.

2. Enter the two points (–3, 7) and (–8, –1) by placing –3 and –8 in an input list and 7 and –1 in an output list. Use your calculator's linear regression command to find the equation of the regression line (least squares line) for the line that contains the points (–3, 7) and (–8, –1). (Ask a buddy or your instructor if you do not know how to do this, or consult the manual.) Record the equation of the linear regression line.

3. How do your answers to Investigations 1b and 2 compare?

Given the points (–3, 7) and (–8, –1), we calculate the slope as follows:

$$\text{slope} = \frac{\Delta\,(\text{output})}{\Delta\,(\text{input})} = \frac{-1 - 7}{-8 - (-3)} = \frac{-8}{-5} = \frac{8}{5}$$

The parameter a has a value of $\frac{8}{5}$ in the general linear function $y = ax + b$. Given $y = \frac{8}{5}x + b$, we can find b by substituting one of the points for (x, y). We'll use the point (–3, 7).

Substitute $x = -3$ and $y = 7$: $\quad 7 = \frac{8}{5}(-3) + b$

Multiply on right side of "=": $\quad 7 = \frac{-24}{5} + b$

Add $\frac{24}{5}$ to both sides: $\quad 7 + \frac{24}{5} = b$

Write left side of "=" using common denominator:

$$\frac{35}{5} + \frac{24}{5} = b$$

Add fractions: $\quad \frac{59}{5} = b$

So the specific model is $\quad y = \frac{8}{5}x + \frac{59}{5}$.

After entering the points on the calculator and executing a linear regression command on the paired lists, we get the results displayed in Figure 1.

FIGURE 1

```
L1     | L2     | L3    3
-3     | 7      |
-8     | -1     |
------ | ------ |
L3(1)=
```

```
LinReg
 y=ax+b
 a=1.6
 b=11.8
```

```
1.6▸Frac
            8/5
11.8▸Frac
           59/5
```

Notice that the calculator returns decimal values for the parameters a and b. Converting each to fractions shows that we get the same function as we did without the calculator.

The resulting model can be graphed along with the points as a check (Figure 2).

FIGURE 2

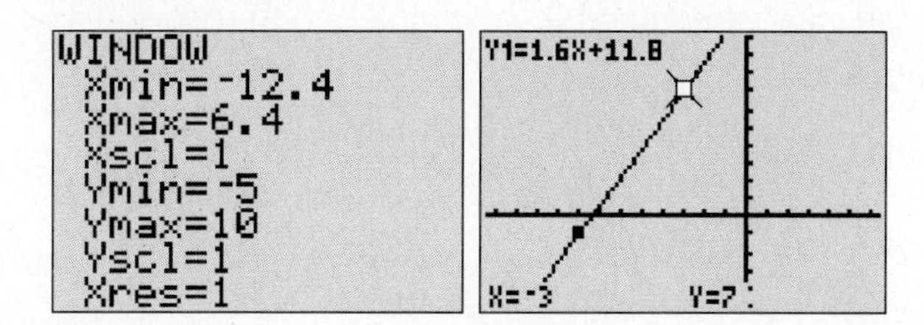

In summary, given two points, we can use the linear regression feature of the calculator to create the specific linear function that satisfies the two points.

Next we look at some actual data. In this case, we will first check to see if the data fits any of our models exactly. If not, we will try to find the model that provides the "best" approximation of the data.

Natural Gas Production: Natural gas is the world's third-largest energy source. By 2010, due to economic and environmental advantages, natural gas may be the world's largest energy source. We would like to determine what kind of relationship exists between the variables "year" and "world natural gas production."

Table 1 displays the world natural gas production (in millions of tons of oil equivalent) since 1988.

TABLE 1 World Natural Gas Production

Year	World Natural Gas Production, *NGP* (million tons of oil equiv.)
1988	1768
1989	1828
1990	1865
1991	1920
1992	1936
1993	1974

Source: U.S. Department of Energy, *Annual Energy Review 1992* (electronic database).

4. Let the variable x be zero when the year is 1988.

a. What is the value of x in 1989? Why?

b. What is the value of x in 1993? Why?

5. Using the data from Table 1 and the definition of x in Investigation 4, complete Table 2. For the last three columns, compute the finite differences in *NGP*, the finite ratios in *NGP*, and the second finite differences in *NGP*.

TABLE 2 World Natural Gas Production

x	World Natural Gas Production, *NGP* (million tons of oil equiv.)	Finite Differences, $\Delta(NGP)$	Finite Ratios, ®*NGP*	Second Finite Differences, $\Delta(\Delta(NGP))$

Source: U.S. Department of Energy, *Annual Energy Review 1992* (electronic database).

6. Are the data exactly linear, exponential, or quadratic based on the results of Table 2? Justify your answers.

7. Enter the data from the first two columns of Table 2 into your calculator. (Ask a friend or your instructor if you are not sure how to do this, or consult the manual.) Graph the data in an appropriate viewing window.

a. What does the input axis represent? What does the output axis represent?

b. Does it appear that the data could be approximated by a linear function? Why or why not?

It is common when the input to a function is a time unit, such as a year, to select a starting year as an ***arbitrary zero***. In this case, we chose 1988 as the arbitrary zero. Then, instead of using the actual year as input, we define an input variable, such as x, to represent the number of years since the starting year. We did this when we looked at the various salary offers beginning in Chapter 2.

We see both the table of data points and their graph in Figure 3.

FIGURE 3

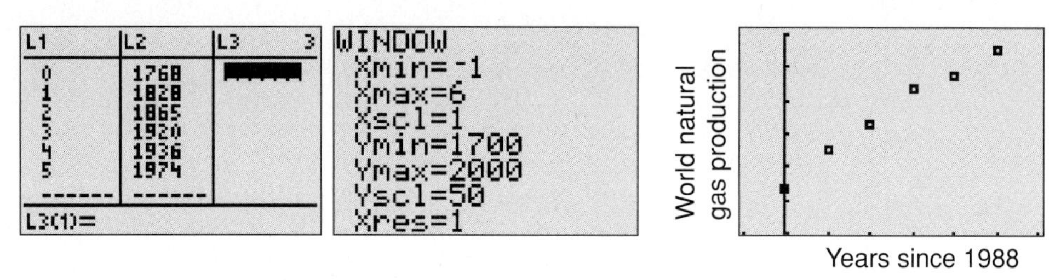

Note that L1 represents the years since 1988 and L2 represents the world natural gas production.

The graph in Figure 2 is called a ***scatter plot***. This is a graph of all the ordered pairs. Scatter plots are used to analyze the relationship between data. If there is a relationship between the variables, a pattern is often apparent in the scatter plot. The points in the graph in Figure 1 appear to lie approximately in a straight line. This suggests that we try a linear model to describe the data.

Note that the data are not exactly linear, exponential, or quadratic. This conclusion comes from Table 2, in which the finite differences, the finite ratios, and the second finite differences all vary.

We will use the calculator to find values of the parameters a and b for the line of best fit. Such a line is called the ***regression line*** for the data.

8. Use your calculator's linear regression command to find a best-fit regression line for the data. (Ask a buddy or your instructor if you do not know how to do this, or consult the manual.) Record the values of a, b, and the equation for the regression line.

9. Graph the regression line over the data points. Comment on how well the line approximates the data visually.

10. Associated with each regression line is a number between –1 and 1 inclusive called the correlation coefficient. Whenever you use the calculator to find the regression line, the calculator computes the correlation coefficient and stores it as variable r. Find this value and record below. (Ask a buddy or your instructor if you do not know how to do this, or consult the manual.)

Figure 4 displays the calculator output for the parameters of the regression line and the value of the correlation coefficient.

FIGURE 4

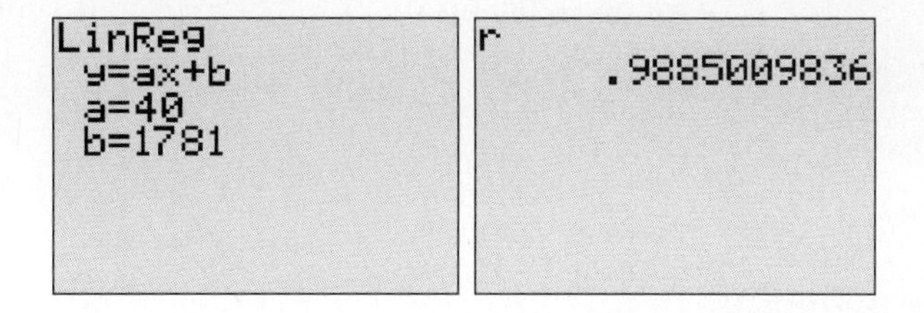

If we round to the nearest integer, we get $NGP(x) = 40x + 1781$ as the equation of the regression line.

The calculator returns a value of 0.9885009836 as a value for the variable r (the correlation coefficient).

Values of r that are close to 1 indicate a strong *positive* correlation between the input and output. This means that the quantities most likely exhibit a linear relationship and that the regression line has a positive slope.

Values of r that are close to -1 indicate a strong *negative* correlation between the input and output. This means that the quantities most likely exhibit a linear relationship and that the regression line has a negative slope.

Values of r near zero indicate little correlation and suggest that there is no linear relationship between the input and the output.

Typically, deciding whether there is a linear correlation or not between two lists of data depends on the value of r and the number of data points.

In the case of the World Natural Gas Production data, we used only six data points and we see a strong positive correlation since the value of r is near +1.

Statisticians express their confidence in an estimate as a percent. Most commonly, results are expressed with 95% certainty (other common levels of certainty are 90% and 99%). The values of r required for a given sample size and confidence level are found in statistical tables.

For a sample of size 6 the correlation coefficient must be at least 0.811 for a positive linear correlation to exist. Since $r = 0.9885009836$ for the World Natural Gas Production data, we can state with 95% certainty that there is a positive linear correlation between year and natural gas production.

Essentially, for a sample of size 6, there is no correlation if r is between -0.811 and 0.811 (Figure 5).

FIGURE 5

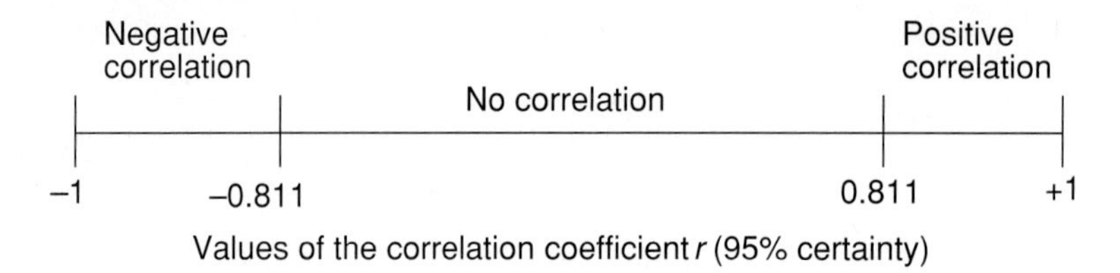

Figure 6 displays a graph of both the original scatter plot and the regression line.

FIGURE 6

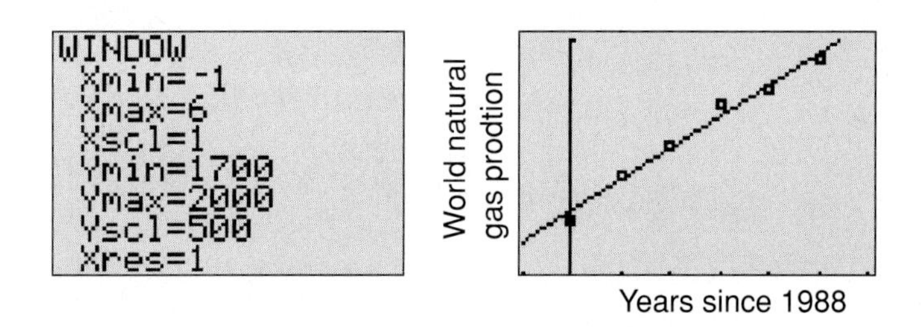

The model $NGP(x) = 40x + 1781$ visually fits the data quite well. The correlation coefficient suggests that there is a positive linear correlation.

We usually use such models to predict output values. To help you understand how the ***"goodness of fit"*** is measured and to help you understand the accuracy of predictions, we will look at something called residuals.

The vertical distance between the observed output and the predicted output is commonly used to measure the fit (Figure 7). These vertical distances are called ***residuals***. These are the differences between an actual (observed) output and an output predicted by a model.

FIGURE 7

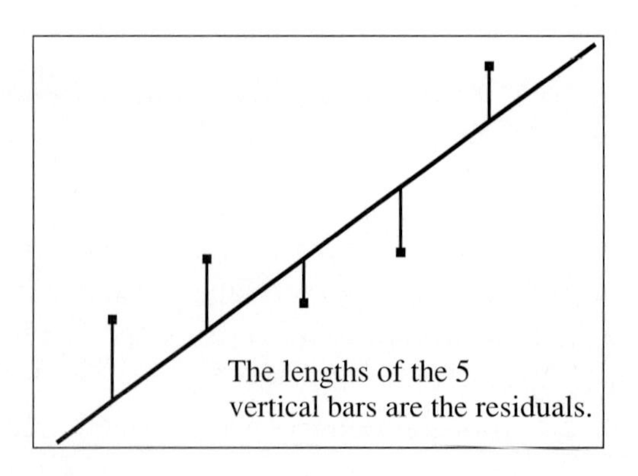

Let's use some notation as we investigate residuals.

Use NGP_O to represent the observed output, which is the output given in the data.

Use NGP_P to represent the predicted output, which is the output obtained from the regression line function. Then the residual is

$$NGP_O - NGP_P$$

for a given pair of outputs.

The regression line (also called the ***least squares line***) produces residuals in which the sum of the squares of the residuals is as small as possible. Squares are used to eliminate the effect of some residuals being negative.

Another important measure is called the standard error of the estimate. It is used to establish the reliability of predictions made from the regression line. The ***standard error of the estimate*** is the number that results from calculating the square root of the mean of the squares of the residuals.

Let's compute the residuals and the standard error of the estimate for the natural gas production data.

11. Complete Table 3. Use output values, rounded to the nearest integer, from the regression line for the third column. The first one is done for you.

TABLE 3 World Natural Gas Production

x	Observed, NGP_O	Predicted, NGP_P	Residual, $NGP_O - NGP_P$	Square of Residual
0	1768	1781	1768 − 1781 = −13	$(-13)^2 = 169$
1	1828			
2	1865			
3	1920			
4	1936			
5	1974			

Source: U.S. Department of Energy, *Annual Energy Review 1992* (electronics edition).

12. Find the standard error of the estimate as follows:

a. Find the mean of the squares of the residuals. That is, find the mean of the numbers in the rightmost column of Table 3.

b. Calculate the square root of your answer to part **a**. This number is called the standard error of the estimate.

Table 4 displays the results of computation of residuals using the regression line and rounding the predicted outputs to the nearest integer (parameters not rounded).

TABLE 4 World Natural Gas Production

x	Observed, NGP_O	Predicted, NGP_P	Residual, $NGP_O - NGP_P$	Square of Residual
0	1768	1781	−13	169
1	1828	1821	7	49
2	1865	1862	3	9
3	1920	1902	18	324
4	1936	1942	−6	36
5	1974	1982	−8	64

Source: U.S. Department of Energy, *Annual Energy Review 1992* (electronics database).

Summing the squares of the residuals, we get 651. The mean is $\frac{651}{6} = 108.5$. The square root of the mean is $\sqrt{108.5} \approx 10.4$.

So the standard error of the estimate is approximately 10.4. This value is important when using the regression line to predict values of output given a value for input.

Given an input that is between the smallest and largest input in the data, the predicted output is usually a good estimate of the actual output (this is called ***interpolating***).

However, given an input that is outside the interval between the smallest and largest input in the data, the predicted output may not be a good estimate of the actual output (this is called ***extrapolating***).

However, on the basis of statistical principles, when extrapolating, we can generally say with 95% confidence that the predicted output will be within 2 standard errors in the estimate of the actual output (Figure 8).

FIGURE 8

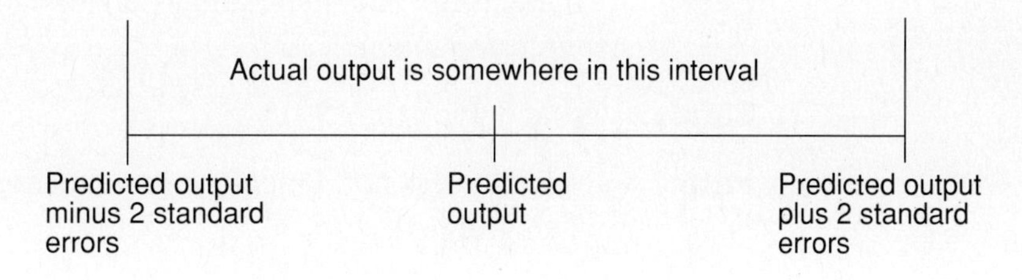

Let's investigate how to use these ideas.

13. a. Natural gas could be the most important fossil fuel by 2010. Use the regression line to predict the world's natural gas production in 2010.

b. Calculate an interval centered by your answer to part **a** by subtracting and adding 2 standard deviations of error in the estimate to the predicted output.

c. Use the interval in part **b** to describe what the actual natural gas production might be in 2010.

In 2010 the value of the input variable x is 22. So we evaluate the function at 22.

$$\begin{aligned} NGP(22) &= 40(22) + 1781 \\ &= 880 + 1781 \\ &= 2661. \end{aligned}$$

Thus the world's natural gas production in 2010 is predicted to be about 2661 million tons of oil equivalent.

Since the standard error in the estimate is about 10.4, we want an interval from $2661 - 2(10.4)$ up to $2661 + 2(10.4)$.

$$2661 - 2(10.4) = 2661 - 20.8 = 2640.2 \approx 2640$$

$$2661 + 2(10.4) = 2661 + 20.8 = 2681.8 \approx 2682$$

Therefore, we are 95% confident that the world's natural gas production (million tons of oil equivalent) will be between 2640 and 2682 in the year 2010.

STUDENT DISCOURSE

Pete: That was fun and interesting. Can all data be modeled using regression lines?

Sandy: What do you think?

Pete: I would guess not. I mean lots of data doesn't lie in a straight line when graphed.

Sandy: That's right.

Pete: What do you do then?

Sandy: There are other regression models that can be used. What do you think some might be?

Pete: I'll bet exponential is one.

Sandy: That's right. In fact, you will look at some other models in the next two sections.

Pete: By the way, why do we use two measures of "fit"—r and the residuals?

Sandy: Well, r can help you measure the strength of the linear relationship.

Pete: I understand that.

Sandy: Residuals lead to the standard error in the estimate. They help you measure how accurate your predictions are. Also, the residuals can be computed regardless of the type of model being used. Since r only measures the degree of linear correlation, it cannot be used when working with other polynomial models.

Pete: You mean like quadratics or cubics.

Sandy: Exactly. You'll see that when you do the last section of the chapter.

1. In the Glossary, list and define the words in this section that appear in ***italic boldface*** type.

2. Paper Production: The data in Table 5 represent world paper production (million tons) by year.

TABLE 5 World Paper Production

Year	World Paper Production (million tons)
1988	227
1989	232
1990	240
1991	243
1992	245
1993	247

Sources: FAO, *Forest Products Yearbook*, 1950–1992; *Pulp and Paper*, January 1994.

a. Select an arbitrary zero and re-create Table 5 using, as input, the number of years since the arbitrary zero.

b. Construct a scatter plot. Would a linear model approximate the data? Why or why not?

c. Find the regression line.

d. Compute the correlation coefficient. What does it tell you about the linear relationship?

e. Compute the standard error of the estimate.

f. Estimate the world paper production this year. Write a sentence describing your estimate.

g. Would you expect world paper production to continue to increase indefinitely? Why or why not?

3. Cigarette Smoking: Table 6 lists the percentages of total smokers in the United States over the age of 18 by year.

TABLE 6 Cigarette Smoking

Year	Total Smokers (%)
1965	42.4
1974	37.1
1979	33.5
1983	32.1
1985	30.1
1987	28.8
1990	25.5
1992	26.5
1993	25.0

Source: U.S. Bureau of the Census, *Statistical Abstract of the United States: 1996* 116th ed. (Washington, D.C., 1996), p. 145.

a. Select an arbitrary zero and re-create Table 6 using, as input, the number of years since the arbitrary zero.

b. Construct a scatter plot. Would a linear model approximate the data? Why or why not?

c. Find the regression line.

d. Compute the correlation coefficient. What does it tell you about the linear relationship?

e. Compute the standard error of the estimate.

f. Estimate the percentage of U.S. adults over the age of 18 who smoke cigarettes this year. Write a sentence stating your estimate.

College Student Objectives: On the basis of a survey of first-time fall-semester full-time U.S. college freshman, the Higher Education Research Group reported the percentage of students in the sample who identified their essential or important objectives. Two of these objectives were "be very well-off financially" and "develop a meaningful philosophy of life." The data appear in Table 7.

TABLE 7 College Student Objectives

Year	Well-off Financially (%)	Meaningful Philosophy of Life (%)
1970	39	76
1975	50	64
1980	63	50
1982	69	47
1983	69	44
1984	71	45
1985	71	43
1986	73	41
1987	76	39

Source: U.S. Bureau of the Census, *Statistical Abstract of the United States: 1989,* 109th ed. (Washington, D.C., 1989).

4. Let the number of years since 1970 be the input and let the percentage of students who wish to be very well-off financially be the output.

a. Construct a scatter plot. Does the data appear approximately linear?

b. Use your calculator to find a best-fit regression function. Comment on the "goodness of fit" visually.

c. Find the regression line.

d. Compute the correlation coefficient. What does it tell you about the linear relationship? (*Note*: With nine data points the absolute value of r must be at least 0.666 for a linear correlation to exist.)

e. Compute the standard error of the estimate.

f. Use your model to predict the percentage of students who wish to be very well-off financially this year.

5. Let number of years since 1970 be the input and let the percentage of students who wish to develop a meaningful philosophy of life be the output.

a. Construct a scatter plot. Does the data appear approximately linear?

b. Find the regression line.

c. Compute the correlation coefficient. What does it tell you about the linear relationship?

d. Compute the standard error of the estimate.

e. Use your model to predict the percentage of students who wish to develop a meaningful philosophy of life this year.

6. Consider the question: Is there a relationship between the percentage of students who wish to be very well-off financially and the percentage of students who wish to develop a meaningful philosophy of life? Let's explore.

a. Construct a scatter plot with input the percentage of students who wish to be very well-off financially and output the percentage of students who wish to develop a meaningful philosophy of life. Does the data appear to be approximately linear?

b. Find the regression line.

c. Compute the correlation coefficient. What does it tell you about the linear relationship?

d. Compute the standard error of the estimate.

e. Suppose 80% of students wish to be very well-off financially. Use your model to predict the percentage of students who wish to develop a meaningful philosophy of life.

f. Comment on the relationships between objectives explored in this exploration.

Cars and Gasoline: Table 8 on page 284 lists the number of passenger-car registrations (millions) and the gasoline consumption (thousand barrels per day) in the United States during selected years.

7. Is there a linear relationship between

a. the year and passenger-car registrations? If your answer is yes, find the regression line for the data and justify why it is a reasonable model. If no, explain why.

b. the year and gasoline consumption? If yes, find the regression line for the data and justify why it is a reasonable model. If no, explain why.

c. passenger-car registrations and gasoline consumption? If yes, find the regression line for the data and justify why it is a reasonable model. If no, explain why.

TABLE 8 Car Registrations and Gasoline Consumption

Year	Passenger-Car Registrations (millions)	Gasoline Consumption (thousand barrels per day)
1978	116.6	7555
1980	121.7	6820
1984	127.9	6850
1986	135.4	7229
1987	137.3	7359
1990	145.0	7474

Source: World Resources Institute, *1992 Environmental Almanac* (Boston: Houghton Mifflin, 1992), p. 67.

8. **Popcorn Volume and Mass:** A chemistry class is given 12 samples of popped popcorn. They measure both the volume and the mass of each sample. The data appear in Table 9 on page 285.

   a. Construct a scatter plot. Does the data appear to be approximately linear? Why or why not?

   b. Find the regression line.

   c. Compute the correlation coefficient. What does it tell you about the linear relationship? (*Note:* With 12 data points the absolute value of r must be at least 0.576 for a linear correlation to exist.)

   d. Compute the standard error of the estimate.

   e. Predict the mass of a sample if the volume is 1.5 ml. Write a sentence describing your prediction and your confidence in the prediction.

TABLE 9 Popcorn Volume and Mass

Volume (ml)	Mass (g)
4.5	37.14
5.0	47.61
1.0	12.92
2.0	24.64
6.0	50.43
1.7	16.52
5.0	44.04
0.8	8.28
8.0	63.34
2.7	20.87
4.0	39.19
3.0	27.658

9. We can say with 68% confidence that a predicted output will be within 1 standard deviation of error in the estimate of the actual output. Estimate the world natural gas production in 2010 with 68% confidence instead of 95% confidence.

10. Consider the world natural gas production data and the corresponding regression line. If you use the year as input and the residuals as output, the fit is considered good if the residuals cluster around the input access with no apparent pattern. Graph the residuals for the world natural gas production and describe the graph.

11. Use linear regression to find the linear function that contains the two given points.

a. (5, –2) and (–4, 18) **b.** (6, 9) and (11, –2)

c. (–8, –2.3) and (5.9, 6.7) **d.** (8.1, 2.9) and (11.7, –9.5)

12. A linear function has a rate of change of $-\frac{2}{7}$. What happens to:

a. the output if the input increases by 7?

b. the input if the output increases by 2?

c. the output if the input decreases by 14?

d. the input if the output increases by 14?

13. Natural Gas Production: Find out why the largest increase in the production of natural gas occurred in 1991. How might world events affect your predictions?

REFLECTION

Search through a newspaper and find a set of data that is approximately linear. Write two interesting questions about the data. Construct a linear regression model and use the model to answer the questions, if possible. If not possible, explain why.

SECTION 4.5

Applying Exponential Models to Data

Purpose

- Use exponential regression to create a function, given two points.
- Fit exponential models to data.
- Introduce the reversibility of exponentiation.
- Use exponential models to answer questions about data.

In Section 4.4, we looked at regression lines as good estimates of real-world data. In this section, we look at data that appears to be exponential, rather than linear. We will use the calculator to return best-fit exponential functions for data and use both the correlation coefficient and the standard error to analyze the regression model and answer questions about the data.

1. Consider the two points $(0, 12)$ and $(1, 11.4)$.

a. What is the finite ratio of the two outputs?

b. Derive the specific exponential function that fits these two points.

2. Enter the two points $(0, 12)$ and $(1, 11.4)$ by placing 0 and 1 in an input list and 12 and 11.4 in an output list. Use your calculator's exponential regression command to find the equation of the exponential functions for the curve that contains the points $(0, 12)$ and $(1, 11.4)$. (Ask a buddy or your instructor if you do not know how to do this, or consult the manual.) Record the equation of the exponential function.

3. How do your answers to Investigations 1b and 2 compare?

To derive the function we must determine the values of the parameters a and b in the standard exponential equation. Given the points (0, 12) and (1, 11.4), we calculate the finite ratio in outputs as follows:

$$\text{Finite ratio in outputs} = \frac{11.4}{12} = 0.95.$$

The parameter b has a value of 0.95 in the general exponential function $y = a(b)^x$. We know that a is 12 since 12 is the output when the input is 0.

So the specific model is $y = 12(0.95)^x$.

After entering the points on the calculator and executing a exponential regression command on the paired lists, we get the results displayed in Figure 1.

FIGURE 1

The resulting model can be graphed along with the points as a check (Figure 2).

FIGURE 2

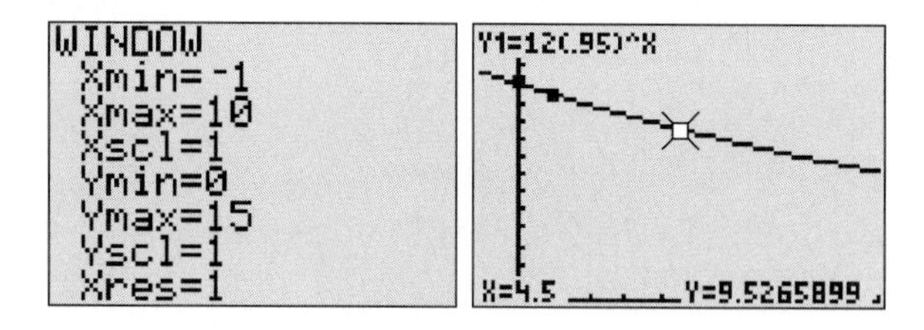

In summary, given two points, we can use the exponential regression feature of the calculator to create the specific exponential function that satisfies the two points.

Next we look at some actual data. In this case, we will first check to see if the data fits any of our models exactly. If not, we will try and find the model to provides the "best" approximation of the data.

Outstanding Credit: The amount of installment credit in the United States continues to rise. Table 1 displays the amount of money (in billions of dollars) in installment credit by year

TABLE 1 Installment Credit Outstanding

Year	Installment Credit Outstanding ($ billions)
1986	572
1987	609
1988	663
1989	724
1990	734
1991	728
1992	731
1993	790
1994	903
1995	1025

Source: U.S. Bureau of the Census, *Statistical Abstract of the United States: 1996,* 116th ed. (Washington, D.C., 1996), p. 516.

4. Let the variable x be zero when the year is 1986.

a. What is the value of x in 1986? Why?

b. What is the value of x in 1994? Why?

5. Using the data from Table 1 and the definition of x in Investigation 4, complete Table 2. For the last three columns, compute the finite differences in C, the finite ratios in C, and the second finite differences in C.

TABLE 2 Installment Credit Outstanding

x	Installment Credit Outstanding C ($ billions)	Finite Differences, $\Delta(C)$	Finite Ratios, ®C	Second Finite Differences, $\Delta(\Delta(C))$

6. Are the data exactly linear, exponential, or quadratic based on the results of Table 2? Justify your answer.

7. Enter the data from the first two columns of Table 2 into your calculator. (Remember how to do this from the previous section? If not, ask someone or consult your manual.) Graph the data in an appropriate viewing window.

a. What does the input axis represent? What does the output axis represent?

b. Does it appear that the data could be approximated by a linear function? Why or why not?

c. Does it appear that the data could be approximated by an exponential function? Why or why not?

Again we have selected an arbitrary zero from which our input will be measured. In this case, we chose 1986 as the arbitrary zero.

We see both a partial table of data points and their graph in Figure 3.

FIGURE 3

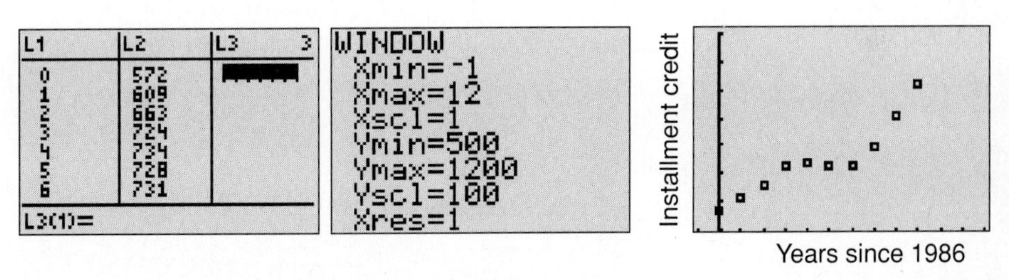

Note that L1 represents the years since 1986 and L2 represents installment credit outstanding.

The points in the graph in Figure 3 certainly do not appear to lie in a straight line. The rate of change in the output in the last three years is steadily increasing. Instead of a linear model, this suggests an exponential model.

Note that the data are not exactly linear, exponential, or quadratic. This conclusion comes from Table 2, in which the finite differences, the finite ratios, and the second finite differences all vary.

Let's try to approximate the data with an exponential model. Recall that the exponential model has the form $y(x) = a(b)^x$, where a and b are parameters that must be found. The parameter b is the base and represents a common ratio between successive outputs.

Just as with linear data, many calculators have a feature that automatically determines a best-fit exponential model for data. Such a model is called an ***exponential regression model.***

8. Use your calculator's exponential regression command to find a best-fit exponential regression equation for the data. (Ask a buddy or your instructor if you do not know how to do this, or consult the manual.) Record the exponential regression model.

9. Graph the exponential regression function over the data points. Comment on how well the curve approximates the data visually.

10. Find the value of the correlation coefficient r and record below. (Ask a buddy or your instructor if you do not know how to do this, or consult the manual.)

Figure 4 displays the calculator output for the parameters of the exponential regression equation, the correlation coefficient, and the graph.

FIGURE 4

If we round a to the nearest integer and b to 3 decimal places, we get $C(x) = 579(1.055)^x$ as the equation of the exponential regression equation.

The calculator returns a value of 0.9481203649 as a value for the variable r (the correlation coefficient).

For a sample of size 10 the correlation coefficient must be at least 0.632 for a positive correlation to exist according to the statistical tables. Since $r =$ 0.9481203649 for the installment credit outstanding data, we can state with 95% certainty that there is a positive exponential correlation between year and installment credit outstanding.

A comment is in order on how the calculator is fitting an exponential function to the data. All the regression techniques discussed thus far apply only to linear situations. The calculator is using logarithms to transform the data into a linear form (this is called linearizing the data). The calculator then applies linear regression technique in order to find the correlation coefficient. We will discuss logarithms in Chapter 8 and will discuss how the linearizing works.

Let's compute the residuals and the standard error of the estimate for the installment credit outstanding data.

11. Complete Table 3 using a rounded integer value for a and a value rounded to three decimal places for b. Use output values, rounded to the nearest integer, from the exponential regression function for the third column. The first one is done for you.

TABLE 3 Installment Credit Outstanding

Year Since 1986	Observed C ($ billions)	Predicted C ($ billions)	Residual	Square of Residual
0	572	579	−7	49
1	609			
2	663			
3	724			
4	734			
5	728			
6	731			
7	790			
8	903			
9	1025			

12. Find the standard error of the estimate as follows:

a. Find the mean of the squares of the residuals. That is, find the mean of the numbers in the rightmost column of Table 3.

b. Calculate the square root of your answer to part **a**. This number is called the standard error of the estimate.

Table 4 displays the results of computation of the residuals using the regression line and rounding the predicted outputs to the nearest integer.

TABLE 4 Installment Credit Outstanding

Year Since 1986	Observed C ($ billions)	Predicted C ($ billions)	Residual	Square of Residual
0	572	579	−7	49
1	609	609	−2	4
2	663	645	18	324
3	724	680	44	1936
4	734	718	16	256
5	728	758	−30	900
6	731	800	−69	4761
7	790	844	−54	2916
8	903	891	12	144
9	1025	941	84	7056

Summing the squares of the residuals yields 18,346. The mean is $\frac{18{,}346}{10} = 1834.6$. The square root of the mean is $\sqrt{1834.6} \approx 42.8$.

So the standard error of the estimate is approximately 42.8.

Let's wrap up the section by posing and answering some questions about the data.

13. a. Use the exponential regression function to predict the number of dollars of installment credit outstanding in the year 2000.

b. Calculate an interval centered by your answer to part **a** by subtracting and adding 2 standard errors in the estimate to the predicted output.

c. Use the interval in part **b** to state what the actual number of dollars of installment credit outstanding in the year 2000 will be.

In the year 2000 the value of the input variable x is 14. So we evaluate the function at 14.

$$\begin{aligned} C(14) &= 579(1.055)^{14} \\ &= 579(2.1160915) \\ &\approx 1225. \end{aligned}$$

Thus the installment credit outstanding in the year 2000 is predicted by the model to be about \$1225 billion. This estimate may seem to be low if we look at the graph.

Since the standard error in the estimate is about 42.8, we want an interval from $1225 - 2(42.8)$ up to $1225 + 2(42.8)$.

$$1225 - 2(42.8) = 1225 - 85.8 \approx 1139$$

$$1225 + 2(42.8) = 1225 + 85.8 \approx 1311$$

Therefore, we are 95% confident that the installment credit outstanding in the year 2000 will be between \$1139 billion and \$1311 billion in the year 2000. That's a pretty wide range, but it's the best we can do with the data available.

STUDENT DISCOURSE

Pete: It seems to me that we must use exponential regression when the output grows faster and faster. Is that correct?

Sandy: If the output increases as the input increases, that's true. But remember that there are exponential functions in which the output is decreasing.

Pete: Hmmm. I don't remember.

Sandy: Think back to the graphs of exponential functions. Weren't there two types?

Pete: Oh yeah. There was one type that curved downward rather than upward.

Sandy: That's right. Do you remember when that happened?

Pete: It had something to do with the size of the base.

Sandy: Go on.

Pete: I think the base was smaller than 1. If that was the case, the output decreased. That's right. I remember the problem about the drug in the bloodstream.

Sandy: Very good.

Pete: So in that case the output would initially decrease quickly and then the rate of decrease would slow down. Is that right?

Sandy: Why do you think the decrease in outputs slows down?

Pete: Because the curve seems to level out almost to horizontal.

Sandy: That seems right to me.

1. In the Glossary, list and define the words in this section that appear in ***italic boldface*** type.

2. **World Population:** Table 5 displays the world's population, 1982–1993.

   a. Calculate a best-fit exponential regression model for this data.

   b. Write a question that requires you to evaluate the model. Use the best-fit model to answer your question.

   c. Write a question that requires you to solve an equation. Use the best-fit model to answer your question.

TABLE 5 Worldwide Population

Year	Population (billion)
1982	4.614
1983	4.695
1984	4.775
1985	4.856
1986	4.942
1987	5.029
1988	5.117
1989	5.206
1990	5.295
1991	5.381
1992	5.469
1993	5.557

Source: Lester Brown, Hal Kane, and David Malin Roodman, *Vital Signs 1994,* Worldwatch Institute (New York: W. W. Norton, 1994), p. 99.

3. **AIDS Cases:** Table 6 displays global estimates of AIDS cases, 1982–1993.

   a. Select an arbitrary zero and re-create Table 6 using, as input, the number of years since the arbitrary zero.

   b. Construct a scatter plot. Would an exponential model approximate the data? Why or why not?

TABLE 6 Worldwide AIDS Cases

Year	AIDS Cases (1000)
1982	7
1983	21
1984	65
1985	143
1986	279
1987	493
1988	813
1989	1275
1990	1892
1991	2701
1992	3657
1993	4820

Source: Lester Brown, Hal Kane, and David Malin Roodman, *Vital Signs 1994*, Worldwatch Institute (New York: W. W. Norton, 1994), p. 103.

c. Find the exponential regression function.

d. Compute the correlation coefficient. What does it tell you about the exponential relationship?

e. Compute the standard error of the estimate.

f. Estimate the number of AIDS cases worldwide this year. Write a sentence describing your estimate.

4. Credit Card Charges: Table 7 shows, in billions of dollars, how credit-card charges have grown in recent years.

TABLE 7 Credit Card Charges ($ billions)

Year	Bank Card	All Credit Cards
1980	54	206
1983	80	240
1986	142	335
1989	216	529
1990	250	563
1991	260	659

Source: Credit card companies, Federal Reserve Board, Bankcard Holders of America.

a. Determine a best-fit model for the relationship between years since 1980 and bank-card charges. How good is the model?

b. Determine a best-fit model for the relationship between years since 1980 and charges on all credit cards. How good is the model?

c. Determine a best-fit model for the relationship between bank-card charges and charges on all credit cards. How good is the model?

d. What year was your arbitrary zero? What were the inputs for the year?

5. Use your models from the previous exploration to answer these questions. Justify all answers.

a. Predict the charges on all credit cards this year.

b. What will the charges on all credit cards be when bank-card charges reach $500 billion?

c. What will the bank-card charges be when the charges on all credit cards reach $1 trillion?

6. Use exponential regression to find the exponential function that contains the two given points.

a. $(5, 2)$ and $(-4, 18)$

b. $(6, 9)$ and $(11, 3)$

c. $(-8, 2.4)$ and $(5.9, 6.7)$

d. $(8.1, 2.9)$ and $(11.7, 11.2)$

7. Use linear regression to find the linear function that contains the two given points.

a. $(8, 32)$ and $(-5, 59)$

b. $(7.9, 4.3)$ and $(-2.8, -1.9)$

8. Solve the following equations for x.

a. $5 + 3x = 7.2$

b. $4.9 = 2.3 - 9.7x$

c. $3x + 11 = 5 - 4x$

d. $4.8 - x = 3.7x + 2$

9. Solve each of the following for y. Identify the slope.

a. $5x + 7y = 19$

b. $4y - 32x = -9$

c. $2x - y = 23$

d. $5 - 3y = 11x$

e. $x = 9 + 3y$

10. Write the equations of two linear functions that are

a. parallel

b. perpendicular

REFLECTION

Search through a newspaper and find a set of data that is approximately exponential. Write two interesting questions about the data. Construct an exponential regression model and use the model to answer the questions, if possible. If not possible, explain why.

SECTION 4.6

Applying Polynomial Models to Data

Purpose

- Fit nonlinear polynomial models to data.
- Use residuals to measure the "goodness of fit."
- Use nonlinear polynomial models to answer questions about data.

MAKING CONNECTIONS

So far in this chapter we have looked at data that can be modeled by either linear or exponential functions. In this section, we look at general polynomial functions as models for data. Unfortunately, we are unable to use a correlation coefficient to measure the fit to the data. Instead we must rely on the visual feedback from the graphs and on residuals and the standard error of the estimate. This section concludes our study of regression models until Chapter 8, when we look at exponential regression one more time.

1. Consider the three points $(-1, 7)$, $(0, -2)$, and $(1, 5)$, which are on a parabola with equation $y = ax^2 + bx + c$. Derive the specific quadratic function that fits these three points.

2. Enter the three points $(-1, 7)$, $(0, -2)$, and $(1, 5)$, by placing –1, 0, and 1 in an input list and 7, –2, and 5 in an output list. Use your calculator's quadratic regression command to find the equation of the quadratic function for the parabola that contains the points $(-1, 7)$, $(0, -2)$, and $(1, 5)$. (Ask a buddy or your instructor if you do not know how to do this, or consult the manual.) Record the specific quadratic function.

3. How do your answers to Investigations 1 and 2 compare?

Given the points $(-1, 7)$, $(0, -2)$, and $(1, 5)$, we calculate the second finite difference in outputs in Table 1.

TABLE 1

x	y	Δ(y)	Δ(Δ(y))
–1	7	–9	16
0	–2	7	
1	5		

The parameter a has a value of $\frac{1}{2}(16) = 8$ in the general quadratic function $y = ax^2 + bx + c$.

The parameter c has a value of –2 (the output when the input is 0).

So the model is $y = 8x^2 + bx - 2$. Now we need to find b.

We can find b by substituting one of the points for (x, y). We'll use the point $(1, 5)$.

Substitute $x = 1$ and $y = 5$:	$5 = 8(1)^2 + b(1) - 2$
Simplify on right side of "=":	$5 = 6 + b$

Subtract 6 to both sides: $-1 = b$

So the specific model is $y = 8x^2 - x - 2$.

After entering the points on the calculator, executing a quadratic regression command on the paired lists, and graphing the quadratic regression function, we get the results displayed in Figure 1.

FIGURE 1

In summary, given three points, we can use the quadratic regression feature of the calculator to create the specific quadratic function that satisfies the three points.

Now we look at some real-world data and try and use a polynomial function as a model.

Promoting Racial Understanding: The percentage of college freshmen who say that helping to promote racial understanding is essential or very important is displayed, by year, in Table 2 on page 304.

4. Let the arbitrary zero occur in 1977. If x is zero in 1977, then what is x in 1988? Why?

5. Enter the data into your calculator. Construct a scatter plot in an appropriate viewing window. Do the data seem to fit either a linear or exponential model? Why or why not?

6. What is different about the behavior of the data in Table 2 compared to others we have investigated?

TABLE 2 Percentage of College Freshmen Who Feel Promoting Racial Understanding Is Very Important.

Year	Promoting Racial Understanding Is Important (%)
1977	38
1979	34
1981	31
1983	30
1987	29
1988	32
1989	35
1992	42

Source: UCLA.

7. Create both linear and exponential regression equations for the data. Do either appear to be good predictors of the data? Why or why not?

Figure 2 displays the data along with the scatter plot.

FIGURE 2

L1	L2	L3
0	38	
2	34	
4	31	
6	30	
10	29	
11	32	
12	35	

L3(1)=

WINDOW
Xmin=-2
Xmax=16
Xscl=2
Ymin=20
Ymax=50
Yscl=10
Xres=1

The data points do not seem to lie on either a linear or an exponential curve. One thing that is different about the behavior of the data is that the inputs neither increase steadily nor do they decrease steadily. The outputs first decrease and then increase. This pattern is not typical of either linear or exponential models.

The fact that this set of data is not modeled by linear or exponential models is finalized when we look at the correlation coefficients. In both cases, the correlation coefficient is less than 0.2. This suggests little, if any, linear or exponential relationship between input and output.

Can we model the data at all? If yes, we need to look at other types of functions. In particular, we look at ***polynomial functions***.

The data in Table 2 decrease and then increase. It is possible that the data could be modeled by a quadratic function since such functions behave similarly. Thus we want to create a function

$$Race(x) = ax^2 + bx + c,$$

where the parameters a, b, and c must be found in order to determine the specific model.

Some calculators will automatically determine a best-fit quadratic regression model for data.

Let's investigate.

8. Use your calculator's quadratic regression command to find a best-fit quadratic model for the data in Table 2. (Ask a buddy or your instructor if you are unsure how to do this, or consult the manual.) Round the values of a, b, and c to the nearest one-hundredth and record the model.

9. Use Table 3 to compute the residuals and their squares for your model.

10. Calculate the standard error of the estimate.

Use the quadratic model to do the following investigation.

TABLE 3 Percentage of College Freshmen Who Feel Promoting Racial Understanding Is Very Important.

Years Since 1977	Observed, $Race_O$ (%)	Predicted, $Race_P$ (%)	Residual, $Race_O - Race_P$	Square of Residual
0	38			
2	34			
4	31			
6	30			
10	29			
11	32			
12	35			
15	42			

Source: UCLA.

11. Predict the percentage of college freshmen who say that helping to promote racial understanding is essential or very important for this year. Provide an interval in which you are 95% confident that the actual data value lies in. Write a sentence stating your conclusion.

The parameters for the best-fit quadratic model of the racial understanding data along with the graph of the model appear in Figure 3.

FIGURE 3

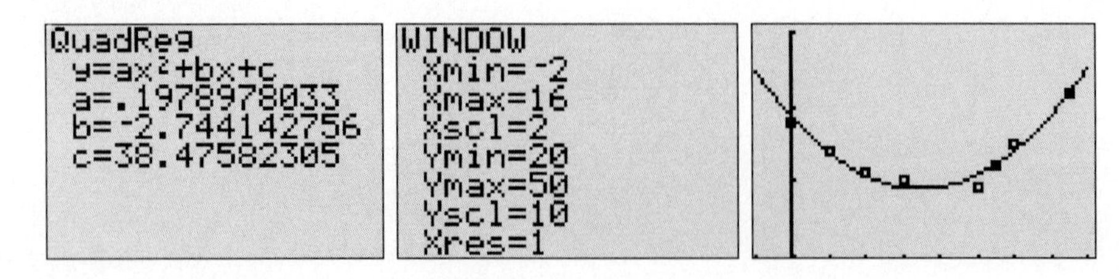

If we round the value of each parameter to the nearest one-hundredth, we get the following quadratic model.

$$Race(x) = 0.20x^2 - 2.74x + 38.48$$

While the fit seems to be pretty good, we can use residuals to find out exactly how good. Table 4 displays the residuals for this quadratic model using rounded values.

TABLE 4 Percentage of College Freshmen Who Feel Promoting Racial Understanding Is Very Important.

Years Since 1977	Observed, $Race_O$ (%)	Predicted, $Race_P$ (%)	Residual, $Race_O - Race_P$	Square of Residual
0	38	38.48	−0.48	0.2304
2	34	33.80	0.20	0.04
4	31	30.72	0.28	0.0784
6	30	29.24	0.76	0.5776
10	29	31.08	−2.08	4.3264
11	32	32.54	−0.54	0.2916
12	35	34.40	0.60	0.36
15	42	42.38	−0.38	0.1444

Source: UCLA.

The standard error in the estimate is $\sqrt{\frac{6.0488}{8}} \approx 0.87$.

Figure 4 contains a graph of the residuals as a function of the input.

FIGURE 4

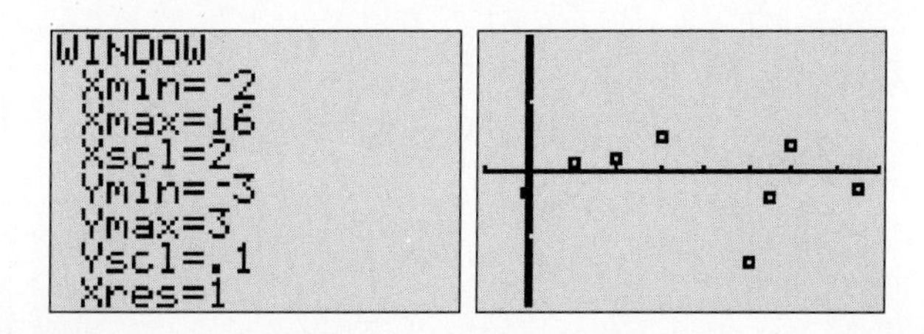

The residuals seem randomly distributed. There is no apparent pattern in the graphs of the residuals. This evidence suggests a good fit, at least within the range of the data.

If we wish to estimate the percentage of college freshmen who say that helping to promote racial understanding is essential or very important during 1997, we calculate

$$\begin{aligned} Race(20) &= 0.20(20)^2 - 2.74(20) + 38.48 \\ &\approx 64. \end{aligned}$$

Since the standard error of the estimate is about 0.87, twice the standard error is 1.74 (less than 2), roughly we can conclude with 95% confidence that the actual percentage is between 62 and 66.

Let's conclude this section with a set of data that might be modeled by a higher-degree polynomial.

U.S. Unemployment Rates: Table 5 displays the unemployment rate in the United States for selected years.

12. Choose a reasonable year as an arbitrary zero. Enter the data in your calculator and construct a scatter plot. What kind of polynomial would seem to best model the data? Why?

13. Use your calculator to find the best-fit regression model for the data. Record both the parameters and the model. Round each parameter to three decimal places.

A scatter plot for the unemployment data appears in Figure 5.

FIGURE 5

L1	L2	L3
0	7	
1	6.1	
2	5.5	
3	5.2	
4	5.6	
5	6.9	
6	7.3	

L3(1)=

WINDOW
Xmin=-1
Xmax=10
Xscl=2
Ymin=4
Ymax=8
Yscl=1
Xres=1

TABLE 5 U.S. Unemployment Rates

Year	Unemployment (%)
1986	7.0
1987	6.1
1988	5.5
1989	5.2
1990	5.6
1991	6.9
1992	7.3
1993	6.9
1994	6.4

Sources: 1993 World Fact Book, U.S. Commerce Department.

The shape of the data most resembles a cubic function. Let's model the data with a third-degree function. The parameters of the regression model and the graph of the regression model appear in Figure 6.

FIGURE 6

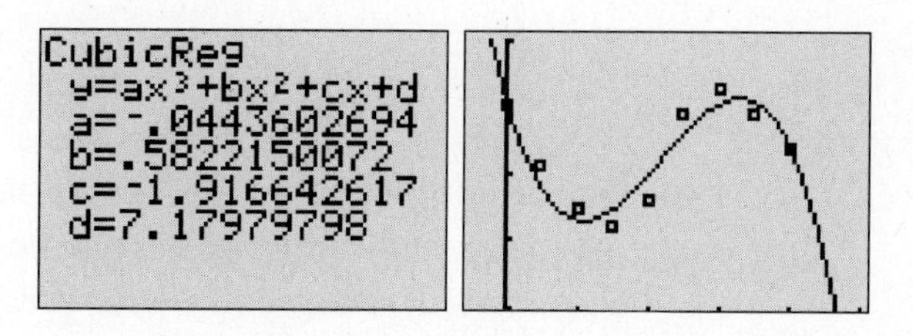

Visually this seems like a pretty good fit though we would be hesitant about predicting future trends using this model.

STUDENT DISCOURSE

Pete: Why wouldn't you want to predict future trends with the cubic model? It looks like a pretty good fit.

Sandy: What does the output represent?

Pete: Percentage of people unemployed.

Sandy: Right. And what happens to the graph of the regression function as the input increases past 7?

Pete: The curve drops very fast.

Sandy: And what would that mean in the problem situation?

Pete: That the unemployment rate would drop quickly to zero.

Sandy: That's right. Is that realistic?

Pete: No, I guess not. At some point, the unemployment rate would level off or, more than likely, increase.

Sandy: So the model could be used to make predictions during the years that span the input data. But . . .

Pete: But it's not a good predictor of the future. Right?

Sandy: That's the way I see it.

1. In the Glossary, list and define the words in this section that appear in ***italic boldface*** type.

2. Do you believe the regression model (page 307) for promoting racial understanding data is a good predictor of future trends? Why or why not? What would this model predict for the year 2010? Is this reasonable? Why or why not? What do you think caused the decrease in promotion of racial understanding in the 1980s?

3. Create several examples of a fifth-degree polynomial function. Predict the shape of the graphs and justify your prediction. Graph each example on your calculator and modify your prediction if necessary.

4. Create several examples of a sixth-degree polynomial function. Predict the shape of the graphs and justify your prediction. Graph each example on your calculator and modify your prediction if necessary.

5. Nuclear Warheads: Table 6 displays global nuclear arsenals in thousands by year.

TABLE 6 World Nuclear Arsenals

Year	Nuclear Warheads (1000s)
1970	39.695
1975	52.325
1980	61.480
1985	68.590
1987	68.835
1989	63.650
1990	60.240
1991	55.775
1992	52.875

Source: Lester Brown, Hal Kane, and David Malin Roodman, *Vital Signs 1994*, Worldwatch Institute (New York: W. W. Norton, 1994), p. 115.

a. Construct a scatter plot to determine the "shape" of the data. Which type of function would best model the data? Why?

b. Use your calculator to find a best-fit regression model for the data. Record the parameters and the regression model.

c. Use your model to predict the world nuclear arsenals in 1995. How accurate is this prediction?

d. If the model remained accurate, in what year would all nuclear warheads be eliminated? Is this reasonable? Justify your answers.

6. Do you believe the regression model (page 309) for the U.S. unemployment rate data is a good long-term predictor of unemployment? Why or why not? If the model remained accurate, what would the unemployment rate be in 2000? Is this reasonable? Justify your answer.

7. Professional Sports Salaries: Table 7 displays the average salary of professional sports figures.

TABLE 7 Professional Sport Salaries

Year	National Basketball Assn. ($100,000)	National Football League ($100,000)	Major League Baseball ($100,000)	National Hockey League ($100,000)
1980	1.9	0.9	1.8	1.0
1982	2.2	1.0	2.2	1.2
1984	3.2	1.9	3.2	1.5
1986	4.8	2.0	4.0	1.7
1988	6.4	2.2	4.2	1.9
1990	8.1	3.8	6.0	2.4

Sources: NFL Players Association, NBA Players Association, Major League Baseball Players Association, NHL Players Association.

Use a scatter plot to predict a good model, if any, for each of the following pairs of data. Find an appropriate best-fit regression model for each pair. Discuss the accuracy of your model.

a. Input: years since 1980; output: NBA salaries.

b. Input: years since 1980; output: NFL salaries.

c. Input: years since 1980; output: MLB salaries.

d. Input: years since 1980; output: NHL salaries.

e. Input: NBA salaries; output: NFL salaries.

f. Input: NBA salaries; output: MLB salaries.

g. Input: NBA salaries; output: NHL salaries.

h. Input: NFL salaries; output: MLB salaries.

i. Input: NFL salaries; output: NHL salaries.

j. Input: MLB salaries; output: NHL salaries.

8. Given three **non-*colinear*** points with different imput values, there is a parabola that contains all three points. Write down three different ordered pairs.

a. Use your calculator to create a function containing the three points. Check your answers.

b. Write one more ordered pair. Create a quadratic regression model for the four points. Is the model an exact fit for the data? How is this problem different from part **a**?

9. Given any four non-colinear points, there is a cubic function that contains all four points. Write down four different ordered pairs. Use your calculator to create a function containing the four points. Check your answers.

10. Given any five non-colinear points, there is a quartic function that contains all five points. Write down five different ordered pairs. Use your calculator to create a function containing the five points. Check your answers.

11. Use linear regression to find the linear function that contains the two given points.

a. $(11, 8)$ and $(2, 1)$ **b.** $(-8, -5)$ and $(3, -6)$

12. Use exponential regression to find the exponential function that contains the two given points.

a. $(1, 7)$ and $(4, 13)$ **b.** $(7, 23)$ and $(4, 9)$

13. Use quadratic regression to find the quadratic function that contains the three given points.

a. $(-2, 15)$, $(3, -4)$, and $(1, -5)$ **b.** $(5, 11)$, $(-8, 1)$, and $(-4, -2)$

14. Use cubic regression to find the cubic function that contains the four given points.

a. $(5, -2)$, $(2, 9)$, $(-1, -3)$, and $(-4, 18)$

b. $(6, 9)$, $(-2, 4)$, $(-8, -1)$, and $(11, -2)$

15. Use quartic regression to find the quartic function that contains the five given points.

a. $(5, -2)$, $(0, 8)$, $(-3, -1)$, $(7, 19)$, and $(-4, 18)$

b. $(6, 9)$, $(1, -1)$, $(-3, -5)$, $(-7, 12)$, and $(11, -2)$

REFLECTION

Search through a newspaper and find a set of data that is approximately quadratic. Write two interesting questions about the data. Construct a quadratic regression model and use the model to answer the questions, if possible. If not possible, explain why.

SECTION 4.7

Data: Creating Models and Answering Questions: Review

Purpose

- Reflect on the key concepts of Chapter 4.

In this section you will reflect upon the mathematics in Chapter 4—what you've done and how you've done it.

1. State the most important ideas in this chapter. Why did you select each?

2. Identify all the mathematical concepts, processes, and skills you used to investigate the problems in Chapter 4.

There are a number of important ideas that you might have listed, including evaluating, solving, simplifying, expression, equation, function, linear regression model, exponential regression model, polynomial regression model, correlation coefficient, residuals, standard error of the estimate and confidence interval.

REVIEW

1. Complete the following statements:

a. A mathematical expression is . . .

b. A mathematical statement is . . .

2. Create examples of the following.

a. Simplifying an expression

b. Evaluating a function

c. Solving an equation

3. Complete the following sentences.

a. An example of an equation is . . .

b. An example of an inequality is . . .

4. Consider the quadratic function $y(x) = 3x^2 - x + 5$.

a. Write a question that is answered by evaluating the function. Do the required evaluation.

b. Write a question about the function that is answered by solving an equation. Write the equation. Solve the equation.

5. Consider the function $y(x) = \frac{2}{5}x - 11$.

a. What happens to the output if the input increases by 5? What did you do to answer the question?

b. What change in the input causes the output to increase by 2? What did you do to answer the question?

c. What happens to the output if the input increases by 20? What did you do to answer the question?

d. What change in the input causes the output to decrease by 6? What did you do to answer the question?

6. Write an example of:

a. a linear function with a slope of $\frac{7}{6}$.

b. an exponential function with a vertical intercept of $(0, 351)$.

c. an exponential function in which the base is 0.7.

d. a quadratic function with a vertical intercept of $(0, -8)$ and with a second finite difference in consecutive outputs of 12.

e. two linear functions that are parallel. One function should have a vertical intercept of $(0, 7)$ and the other should have a vertical intercept of $(0, -9)$.

f. two linear functions that are perpendicular. One function should have a slope of $-\frac{3}{7}$.

7. Consider the functions $f(x) = x^2 - 5x + 3$ and $g(x) = 2x - 3$:

a. When will the output of $g(x)$ be equal to 15? Explain what you did.

b. When will the output of $f(x)$ be equal to 2? Explain what you did.

c. When will the output of $f(x)$ be equal to -7? Explain what you did.

d. When will the outputs of the two functions be the same? Explain what you did.

e. When will the outputs of $f(x)$ be less than the outputs of $g(x)$? Explain what you did.

8. Use linear regression to find the linear function that contains the two given points.

a. $(-8, 1)$ and $(4, -3)$ **b.** $(-1, 0)$ and $(4.5, -2.3)$

9. Use exponential regression to find the exponential function that contains the two given points.

a. $(4, 29)$ and $(-1, 15)$ **b.** $(3.9, 1.4)$ and $(1.2, 5.3)$

10. Use quadratic regression to find the quadratic function that contains the three given points.

a. $(5, -2)$, $(-5, 14)$, and $(9, 0)$

b. $(1.2, -5.1)$, $(-8.4, -1.7)$, and $(3.4, 8.9)$

11. Use cubic regression to find the cubic function that contains the four given points.

a. $(-1, 16)$, $(5, -2)$, $(-5, 14)$, and $(9, 0)$

b. $(-7.5, 4.3)$, $(1.2, -5.1)$, $(-8.4, -1.7)$, and $(3.4, 8.9)$

12. Use quartic regression to find the quartic function that contains the five given points.

a. $(-8, -9)$, $(-1, 16)$, $(5, -2)$, $(-5, 14)$, and $(9, 0)$

b. $(9.6, 11.2)$, $(-7.5, 4.3)$, $(1.2, -5.1)$, $(-8.4, -1.7)$, and $(3.4, 8.9)$

13. Solve the following equations for the variable.

a. $2.3p - 7.9 = -8.2$ **b.** $5.1 = 0.2t + 5.7$

c. $9 - x = 2 + x$

d. $3.5 - 7.1x = 5.2x + 9.7$

14. Solve each of the following for y. Identify the slope.

a. $11x + 7y = 23$

b. $2x = 5y - 9$

15. A linear function has a rate of change of $\frac{5}{4}$. What happens to:

a. the output if the input increases by 4?

b. the input if the output increases by 5?

c. the output if the input decreases by 12?

d. the input if the output increases by 30?

16. Consider the data in each of the following tables. Identify the data as linear, quadratic, or exponential based on the scatter plot. Create the specific algebraic function for the data and graph the function.

a.

TABLE 1 **Random Data 1**

Input	−2	−1	0	1	2	3
Output	−41	−18	−7	−8	−21	−46

b.

TABLE 2 **Random Data 2**

Input	−2	−1	0	1	2	3
Output	−2.8	−2.2	−1.6	−1	−0.4	0.2

c.

TABLE 3 **Random Data 3**

Input	−2	−1	0	1	2	3
Output	112.5	45	18	7.2	2.88	1.152

Tuition and Fees: Table 4 displays data on the average annual tuition and fees charged by community colleges and public four-year colleges in the United States by year.

TABLE 4 **Tuition and Fees**

Year	Two-Year College Tuition and Fees ($)	Four-Year College Tuition and Fees ($)
1985	584	1386
1987	660	1651
1988	706	1726
1989	730	1846
1990	756	2035
1991	824	2159
1992	937	2410
1993	1025	2604
1994	1125	2820
1995	1194	2982

Source: U.S. Bureau of the Census, *Statistical Abstract of the United States: 1996,* 116th ed. (Washington, D.C., 1996), p. 607.

17. Let the year be the input and the two-year college tuition and fees be the output.

a. Construct a scatter plot. Identify the data as linear, quadratic, or exponential based on the scatter plot.

b. Create the specific algebraic function for the data and graph the function.

c. Use your regression function to predict the average tuition and fees at two-year colleges this year. Write a sentence describing and defending your answer.

18. Let the year be the input and the four-year college tuition and fees be the output.

a. Construct a scatter plot. Identify the data as linear, quadratic, or exponential based on the scatter plot.

b. Create the specific algebraic function for the data and graph the function.

c. Use your regression function to predict the average tuition and fees at four-year colleges this year. Write a sentence describing and defending your answer.

19. Let the two-year college tuition and fees be the input and the four-year college tuition and fees be the output.

a. Construct a scatter plot. Identify the data as linear, quadratic, or exponential based on the scatter plot.

b. Create the specific algebraic function for the data and graph the function.

c. Use your regression function to predict the average tuition and fees at four-year colleges assuming the average tuition and fees at two-year colleges is $1300. Write a sentence describing and defending your answer.

CONCEPT MAP

Construct a Concept Map centered on one of the following:

a. mathematical statement

b. solve

c. standard error of the estimate

d. linear model

e. exponential model

f. polynomial regression model

REFLECTION

Select one of the important ideas you listed in Investigation 1. Write a paragraph to Izzy explaining your understanding of this idea. How has your thinking changed as a result of studying this idea?

5 Considering Several Models Simultaneously

SECTION 5.1

Comparing Models: An Introduction to Systems of Equations

Purpose

- Compare two or more models.
- Introduce the concept of a system of equations.
- Solve systems numerically, graphically, and algebraically.

We have spent much time in this course studying linear relationships between two variables. In this chapter, we look at problems that involve not one, but several relationships between variables. Each relationship between the variables will be expressible as an equation. Therefore, we will study problems in which more than one equation is needed—in fact, there will be as many equations as there are relationships. When investigating problems involving several relationships between the same variables, we are working with a ***system of equations***. The word "system" indicates more than one equation. One goal, when looking at systems of equations, is to determine what solutions the equations in the system have in common.

This chapter is devoted to studying both how to create systems of equations from problem situations and techniques for finding the common solutions to the equations in the system. In the process, we will investigate two algebraic solving techniques, a technique using tables, and a technique using graphs. Finally, you will learn about an important algebraic quantity—a matrix—as a way to represent and eventually answer questions about linear systems.

Comparing Annual Salaries: Tom accepted a job in public relations at CDs-R-Us. His salary function is $y(x) = 2000x + 33{,}000$, where Y is his annual salary after x years.

Tom's friend, Sarah, accepted a job at the same time in public relations at Super Dog Burger Bits.

Her salary function is $y(x) = 23364.48598(1.07)^x$, where Y is her annual salary after x years.

1. Who makes more in year 5? Justify your answer.

2. Who makes more in year 20? Justify your answer.

3. Is there a year when the annual salaries will be approximately the same? Why?

To calculate the salary in year 5, we evaluate the salary functions at 5.

$$y(5) = 2000(5) + 33{,}000 = 43{,}000$$

Tom's salary in year 5 is \$43,000.

$$y(5) = 23364.48598(1.07)^5 \approx 32769.90$$

Sarah's salary in year 5 is approximately \$32,769.90. So, in year 5, Tom is earning more than Sarah.

To calculate the salary in year 20, we evaluate the salary functions at 20.

$$y(20) = 2000(20) + 33{,}000 = 73{,}000$$

Tom's salary in year 20 is \$73,000.

$$y(20) = 23364.48598(1.07)^{20} \approx 90413.19$$

Sarah's salary in year 20 is approximately \$90,413.19. Sarah will earn more than Tom in year 20.

The previous answers suggest that there must be a year (between 5 and 20) when the two salaries will be approximately equal.

In this case, we are considering the two salary functions simultaneously. Whenever a problem involves two or more mathematical relationships involving equations, the mathematical model is called a ***system of equations***.

Solving such a model involves finding all values of the variables that make all equations in the system true.

So a system of equations is a list of two or more equations. The solution to a system is the set of common solutions to each equation in the system.

For example, if a system consists of two linear equations with two variables, the solution to the system is the set of all ordered pairs that make both equations true. Looking at a table, we want to find the input–output pair in which the input produces the same output for both equations. On a graph, we are looking for the intersection point of the two lines.

In the previous Investigation, we looked at the system

$$\begin{cases} y(x) = 2000x + 33{,}000 \\ y(x) = 23364.48598(1.07)^x. \end{cases}$$

Where are the two outputs approximately the same? Let's investigate.

4. Use a table to numerically determine when the two salaries will be approximately equal. Briefly describe what you did.

5. Use a graph to graphically determine when the two salaries will be approximately equal. Briefly describe what you did.

6. Write *one* mathematical statement that expresses algebraically that the two salaries are equal. Can you solve this statement algebraically? Why or why not?

7. Tom is very proud that his starting salary was so much higher than Sarah's. Write a mathematical statement that expresses algebraically the years when Tom's salary will exceed Sarah's salary. Solve this problem any way you wish. Describe what you did and state your answer.

DISCUSSION

Figure 1 contains both a numerical and graphical solution to the question of when the two salaries will be approximately equal.

FIGURE 1

X	Y1	Y2
14	61000	60246
14.1	61200	60655
14.2	61400	61067
14.3	61600	61481
14.4	61800	61899
14.5	62000	62319
14.6	62200	62742

X=14.4

WINDOW
Xmin=0
Xmax=20
Xscl=2
Ymin=0
Ymax=100000
Yscl=25000
Xres=1

Intersection
X=14.354686 Y=61709.371

In year 14, the two salaries will be approximately equal according to both the table and the graph.

To express the problem algebraically, we are saying the outputs are equal. Thus we write an equation that sets the two processes equal.

$$2000x + 33{,}000 = 23364.48598(1.07)^x$$

This is an example of the substitution approach to solving a system. In this case, we have substituted the salary from CDs-R-Us for the salary from Super Dog Burger Bits. The result is an equation in only *one* variable. If possible, we'd like to solve for this variable.

This equation mixes linear and exponential expressions. Solving for the single variable x is not possible by any technique familiar to you. Thus our best bets were the table or the graph.

We represent the years when Tom's annual salary exceeded Sarah's with the inequality

$$2000x + 33{,}000 > 23364.48598(1.07)^x.$$

However, the answer to the inequality is dependent on the answer to the equation

$$2000x + 33{,}000 = 23364.48598(1.07)^x.$$

Since the two salaries are approximately equal in year 14 and Tom started out earning more, his salary must exceed Sarah's for all years up to year 14. The solution to the inequality is

$$x \le 14.$$

The equality to 14 is included since Tom's annual salary in year 14 was still slightly more than Sarah's.

Let's look at a problem in which both models are linear. We might be able to solve this algebraically.

8. Concert Sales: CDs-R-Us is branching out. They just purchased an outdoor concert facility. Seating is available in the pavilion or on the lawn. The price of pavilion tickets is $25 while the price of lawn tickets is $10. The average crowd size is 6000. To remain solvent, a concert must bring in $75,000 in total ticket sales. To meet this guideline, let's determine how many of each kind of seat must be sold.

a. How many different quantities must be found to answer this question? List each and assign each a variable name.

b. How many relationships are described in the problem statement? What quantities (money, how many of something, etc.) do they relate?

Two quantities must be found: the number of pavilion tickets that must be sold and the number of lawn tickets that must be sold. Use variables p and l, respectively, to represent these quantities.

Two relationships are described in the problem. One is the average crowd size of 6000. This is a relationship between the number of pavilion tickets sold and the number of lawn tickets sold. The other is the total ticket revenue of $75,000. This is a relationship between the income from the sale of pavilion tickets and the income from the sale of lawn tickets.

With this in place, let's find the algebraic models.

9. a. Using the variables p and l, write an equation stating the relationship between number of lawn tickets, number of pavilion tickets, and average number of total tickets sold.

b. What domain restrictions must be placed on the variables p and l? Why?

c. What is the unit on each term in this equation?

10. a. Using the variables p and l, write an equation stating the relationship between the income from lawn tickets, the income from pavilion tickets, and the total income.

b. What is the unit on each term in this equation?

The two relationships in this problem lead us to a system of equations. The two equations along with reference names are as follows.

Ticket equation: $p + l = 6000$

Income equation: $25p + 10l = 75{,}000$

The unit in the **Ticket equation** is the number of tickets sold.

The unit in the **Income equation** is dollars.

It is very important that every term in an individual equation represent the same unit.

Solving this system means finding a value for p and l that makes both equations true. Let's try a ***substitution method*** for solving system.

We need to rewrite one of the equations so that a variable is isolated (alone) on one side of the "=". This is the variable we will substitute for in the other equation.

11. Which equation would you choose to rewrite? Why?

12. Which variable would you choose to isolate? Why?

13. Solve the equation you chose for the variable you chose to isolate.

It probably is easiest to solve the **Ticket equation** for a variable since the coefficients on the variables are both 1. There is no real advantage to isolating one variable over the other in this case. We generally try to isolate a variable with a numerical coefficient of 1.

For consistency, let's isolate p in the **Ticket equation.**

Ticket equation:	$p + l = 6000$
Subtract l from both sides:	$p + l - l = 6000 - l$
Simplify:	$p = 6000 - l$

Now we need to substitute into the **Income equation**.

14. Use your answer to Investigation 13 to substitute in the **Income equation**.

15. Solve the resulting equation.

16. How would you find the value of the other variable? Find this value.

17. State your answer with justification to the Board of Directors of CDs-R-Us.

Here's the solution process:

Income equation:	$25p + 10l = 75{,}000$
Substitute $6000 - l$ for p:	$25(6000 - l) + 10l = 75{,}000$
Distribute 25:	$150{,}000 - 25l + 10l = 75{,}000$
Simplify:	$150{,}000 - 15l = 75{,}000$

Subtract 150,000 from both sides:

$$150{,}000 - 15l - 150{,}000 = 75{,}000 - 150{,}000$$

Simplify:	$-15l = -75{,}000$
Divide both sides by -15:	$\dfrac{-15l}{-15} = \dfrac{-75{,}000}{-15}$
Simplify:	$l = 5000.$

Thus 5000 lawn tickets must be sold to satisfy the conditions of the problem. Since we have two relationships between l and p, we should be able to substitute the value for l to find the value of p. It is usually easiest to use the "simplest" relationship. In this case, it would be the **Ticket equation.**

Ticket equation:	$p + l = 6000$
Substitute 5000 for l:	$p + 5000 = 6000$
Subtract 5000 from both sides:	$p + 5000 - 5000 = 6000 - 5000$
Simplify:	$p = 1000$

So 1000 pavilion tickets must be sold to satisfy the conditions of the problem. As a double check, you could substitute your answers into the **Income equation**.

We could solve this problem numerically or graphically too. To do so, we must express each relationship as a function. This means defining one variable in terms of the other. We have already expressed p as a function of l in the **Ticket equation** ($p = 6000 - l$).

18. Rewrite the **Income equation** so that it expresses p as a function of l.

19. Use your calculator to solve the system numerically. State what you did. Does your answer match the algebraic solution?

20. Use your calculator to solve the system graphically. State what you did. Does your answer match the algebraic and tabular solutions?

To express p as a function of l in the **Income equation**, we must isolate p.

Income equation:	$25p + 10l = 75{,}000$
Subtract $10l$ from both sides:	$25p + 10l - 10l = 75{,}000 - 10l$
Simplify:	$25p = 75{,}000 - 10l$
Divide both sides by 25:	$\frac{25p}{25} = \frac{75{,}000}{25} - \frac{10l}{25}$

Simplify: $p = 3000 - 0.4l$

Now we can display a table for both $p = 6000 - l$ and $p = 3000 - 0.4l$ or graph both. Recall that calculators use only x for the input and y for the output, so y replaces p and x replaces l on the calculator. Figure 2 displays the numerical and graphical solutions.

FIGURE 2

X	Y1	Y2
4700	1300	1120
4800	1200	1080
4900	1100	1040
5000	1000	1000
5100	900	960
5200	800	920
5300	700	880

X=5000

WINDOW
Xmin=0
Xmax=6000
Xscl=1000
Ymin=-1000
Ymax=5000
Yscl=1000
Xres=1

TICKETS
INCOME
Intersection
X=5000 Y=1000

Both approaches support the solution obtained algebraically. Note that equations solved algebraically can use variables that help us recall what the variables stand for.

There are several other algebraic techniques available to attack systems. One method, called ***elimination***, will be explored in the next section. Other techniques involve using an ***array*** of numbers called a ***matrix***. We will investigate these in detail later in the chapter.

Before concluding the section, let's look at how systems might be used in conjunction with a data analysis problem.

21. Median Annual Income: Table 1 on page 332 displays the median annual income for selected years for year-round, full-time U.S. workers who are at least 25 years old. Let 1980 be the arbitrary zero.

a. Construct a scatter plot that displays men's annual income by year since 1980. Predict a type of function that is a good model for the data.

b. Construct a scatter plot that displays women's annual income by year since 1980. Predict a type of function that is a good model for the data.

c. Find a regression equation for men's annual income as a function of years since 1980. Round each parameter two decimal places.

TABLE 1 Median Annual Income

Year	Men ($)	Women ($)
1980	20,297	12,156
1981	21,689	13,259
1982	22,857	14,477
1983	23,891	15,292
1984	25,497	16,169
1985	26,365	17,124
1986	27,335	17,675
1987	28,313	18,531

Source: Statistical Abstract of the United States: 1990.

d. Find a regression equation for women's annual income as a function of years since 1980. Round each parameter two decimal places.

The scatter plots for the **Median Annual Income** data appear in Figure 3.

FIGURE 3

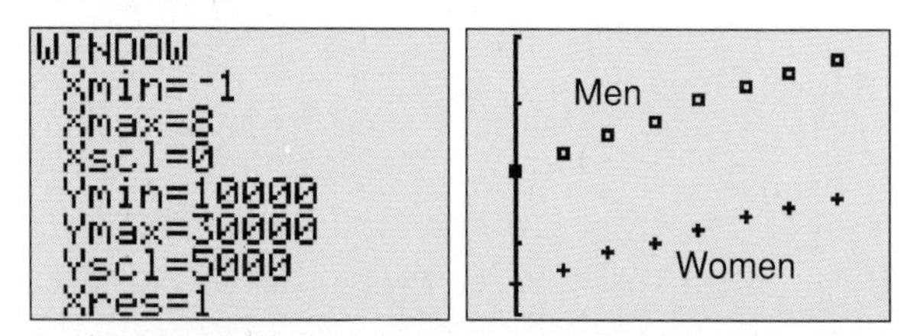

Both scatter plots are approximately linear. We use a linear regression model to describe the relationship in the data. The best-fit models, with parameters rounded to the nearest one-hundredth appear below.

Men's income: $M(x) = 1148.48x + 20510.83$

Women's income: $W(x) = 899.08x + 12438.58$

An interesting question is: When will women attain pay equity with men? This implies treating the above two regression models as a system.

22. Use substitution to determine when women will attain pay equity with men. Discuss your answer in terms of reasonableness.

23. Solve the system graphically. Does your graphic solution support the algebraic solution?

To solve the system by substitution, we set the outputs of the two functions equal to each other.

Outputs equal: $M(x) = W(x)$

Substitute processes: $1148.48x + 20510.83 = 899.08x + 12438.58$

Subtract $899.08x$ from both sides to collect x terms on one side:

$$1148.48x + 20510.83 - 899.08x = 899.08x + 12438.58 - 899.08x$$

Simplify: $249.40x + 20510.83 = 12438.58$

Subtract 20510.83 from both sides:

$$249.40x + 20510.83 - 20510.83 = 12438.58 - 20510.83$$

Simplify: $249.40x = -8072.25$

Divide both sides by 249.32: $\dfrac{249.40x}{249.40} = \dfrac{-8072.25}{249.40}$

Simplify: $x \approx -32.37$

Since the domain of both regression models is the whole numbers, this value for x does not make sense in the problem context. Thus there is no realistic solution to the question of when women will attain pay equity. These models suggest that it will never happen.

Figure 4 displays a graphical interpretation of this result.

FIGURE 4

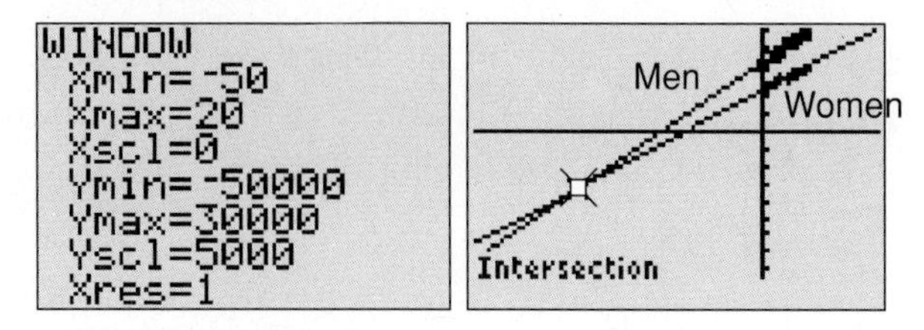

The two graphs intersect in Quadrant III where both the input and output are negative. Although mathematically significant, this has no significance to the problem under discussion since we are not concerned when the income of men and women was negative.

STUDENT DISCOURSE

Pete: So, if I have two linear functions and I find where they intersect, I have solved a system of equations. Is that right?

Sandy: That is the way I understand it.

Pete: What if the graphs of the two lines don't intersect?

Sandy: Good question. What do you think?

Pete: I would guess that there aren't any solutions to the system.

Sandy: Exactly. When might this happen?

Pete: What do you mean?

Sandy: When won't the graphs intersect?

Pete: Hmmm . . . . let's see. I guess the two lines must be parallel.

Sandy: And how would you know this from looking at their equations?

Pete: Parallel lines have the same slope. So I guess the slopes must be different if the system has one solution.

Sandy: If two lines have the same slope, are they always parallel?

Pete: I guess so. What are you getting at?

Sandy: What if they also have the same vertical intercept?

Pete: So the lines have the same slope and share a point. They must be the same line.

Sandy: That's right. And if they are the same line, how many intersection points are there?

Pete: A lot. I'd say an infinite number because they intersect at all points on the line.

Sandy: That's right.

Pete: Let me summarize. If two lines have the same slope, they are either parallel or are the same line. If parallel, there are no solutions to the system and if they are the same line, there are an infinite number of solutions to the system.

Sandy: Very good. I think you have it.

1. In the Glossary, define the words in this section that appear in ***italic boldface*** type.

2. Both the **Men's income** regression model and the **Women's income** regression model are linear. Thus they have a constant rate of change.

a. What is the rate of change of the **Men's income** model? Interpret its meaning in context of the data.

b. What is the rate of change of the **Women's income** model? Interpret its meaning in context of the data.

c. How could you use your answers to parts **a** and **b** to determine that women will never reach pay equity if these models are correct?

3. Hardware Sale: A hardware store marks down every item in the store 15%. Consider the relationship between the original price and the sale price. The difference between the sale price and the original price of a hammer is $2.00. We wish to determine both the original price and the sale price of the hammer.

a. Write the system of equations for this problem.

b. Solve the system algebraically. Check your answer numerically and graphically.

c. Write a complete sentence describing your results.

4. **An 80-foot dog pen:** We have 80 feet of fence and wish to create a rectangular dog pen. Express the length as a function of the width. The length must be twice the width. The task is to determine the length and the width.

   a. Write the system associated with the problem.

   b. Solve the system algebraically. Check your answer numerically and graphically.

   c. Write a complete sentence describing your results.

5. **Men and Women in Education:** Nationally, the ratio of men to women in elementary education is 2:7. Express the number of men in elementary education as a function of the number of women in elementary education. The total number of elementary education students at our state university is 853.

   a. Write the system associated with the problem.

   b. Solve the system algebraically. Check your answer numerically and graphically.

   c. Write a complete sentence describing your results.

6. Two different systems were solved graphically in Section 4.2. Identify them by writing down the systems and describing the answers to the systems. Why weren't these systems solved algebraically?

7. **Grand Canyon Hike:** Sue Sierra, an avid hiker, left at sunrise from the South Rim of the Grand Canyon and hiked down the Kaibab Trail to the Phantom Ranch on the bank of the Colorado River. She reached the ranch at sunset. After spending a peaceful night at the ranch, she hiked back up to the rim along the same trail the next day. Again she left at sunrise and arrived at sunset. Prove there is a spot on the trail that Sue occupied on both trips at precisely the same time of day.

8. Flu Epidemic: The graph in Figure 5 tracks the progress of a flu epidemic in a local community. The graph labeled S represents the number of people susceptible to infection on a given day. The graph labeled I represents the number of people infected on a given day. The graph labeled R represents the number of people who recovered on a given day.

FIGURE 5

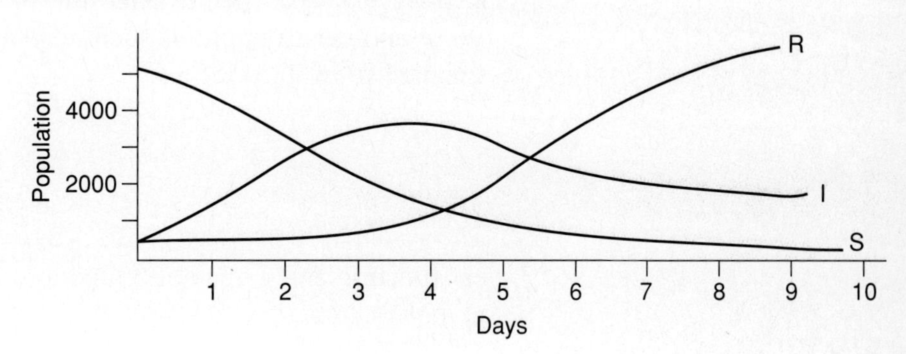

a. Inherent in this graph are three different systems that involve exactly two models. Identify the systems.

b. Identify the approximate solution to each of the systems identified in part **a**. Interpret the solutions in terms of the problem situation.

c. One system involves all three of the models. Does this system have a solution? Why or why not?

9. Solve the following systems algebraically. Check results using both a table and a graph. If there is no solution, explain why.

a. $x - 5y = 7$
$4x + 3y = 17$

b. $7x - 9y = -2$
$4x + y = 11$

c. $3x = 4 - 11y$
$y = 3x + 5$

d. $5x - 3y = 0$
$8x + 2y = 14$

e. $5x - 3y = 11$
$10x = 6y + 5$

f. $4x - 2y = 12$
$y = 2x - 6$

10. One of the systems in Exploration 9 has no solution. Explain how you could determine this by looking at the equations. Write another example of such a system. Write both equations in the standard form $Ax + By = C$.

11. One of the systems in Exploration 9 has an infinite number of solutions. Explain how you could determine this by looking at the equations. Write another example of such a system. Write both equations in the standard form $Ax + By = C$.

12. The system of equations $y = 5x - 3$ and $y = x^2 - 3x - 9$ involves a linear function and a quadratic function. Use a graph to find the solution to the system.

REFLECTION

Describe the difference between a system of equations and an equation. Create examples that illustrate your description. What does it mean to solve a system of equations? What does it mean to solve an equation? Refer to your examples in your explanation.

SECTION 5.2

More Manipulations of Systems of Equations

Purpose

- Investigate the effect of multiplying an equation by a constant.
- Investigate the effect of adding two equations.
- Introduce the elimination technique for solving systems of equations.

In the last section, we introduced the idea of a system of equations—in particular, a linear system of equations. We used the substitution technique to find the solution to each system algebraically. The solutions can also be approximated using a table or a graph. For the table, we locate the case where the outputs of the two functions are the same. On the graph, we locate the intersection point of the two graphs.

Occasionally, a system will not have a solution. We begin this section by looking at how the parameters of linear functions help us to decide the type of solution a system will have. Then we will look at a second algebraic technique, called elimination, for solving systems.

1. Each of the following systems is written using slope-intercept form of a linear equation. Find the solution to each system, if the solution exists.

a. $y = 3x + 2$
$y = 5x - 3$

b. $y = 3x + 2$
$y = 3x - 5$

c. $y = 3x + 2$
$y = 3x + 2$

2. Each equation in Investigation 1 is in the form $y = ax + b$, where a and b are parameters. Explain how you can use the parameters to determine if a system has no solutions, one solution, or infinitely many solutions.

3. Each linear equation is written in standard form $Ax + By = C$, where A, B, and C are parameters. Solve each for y. Identify the slope and the y value of the y-intercept for each function.

a. $3x + 4y = 8$

b. $5x + 2y = -1$

c. $6x - 5y = 7$

d. $3x - 4y = -5$

4. Complete Table 1 using the answers from Investigation 3.

TABLE 1 Parameters and Slopes

$Ax + By = C$	A	B	C	Slope	y Value of y-Intercept
$3x + 4y = 8$	3	4	8		
$5x + 2y = -1$					
$6x - 5y = 7$					
$3x + 4y = -5$					

5. Given the standard linear function $Ax + By = C$, explain to your friend Izzy how you can use the parameters to determine the slope.

6. How many solutions will each of the following systems have? Why?

a. $\begin{cases} 3x - 2y = 15 \\ 6x - 4y = -2 \end{cases}$

b. $\begin{cases} 4p + 7t = 3 \\ 4p - 7t = 3 \end{cases}$

c. $\begin{cases} 5x + 2y = -6 \\ -15x - 6y = 18 \end{cases}$

Given a system of two linear equations with two variables, the system may have:

- exactly one ordered pair as a solution. This occurs when the graphs intersect at exactly one point. Any two lines will intersect at exactly one point if they have different slopes. For example, the system $\begin{cases} y = 3x + 2 \\ y = 5x - 3 \end{cases}$ has exactly one solution since the slopes of the two lines, 3 and 5, are different.

- no solutions when the graphs do not intersect. The lines are parallel so they have the same slope. However, they must have different vertical intercepts. For example, the system $\begin{cases} y = 3x + 2 \\ y = 3x - 5 \end{cases}$ has no solutions since both lines have a slope of 3 and since the two lines have different y-intercepts, $(0, 2)$ and $(0, -5)$.

- an infinite number of solutions if the two equations represent the same function. This means that the two lines have the same slope and the same vertical intercept. For example, the system $\begin{cases} y = 3x + 2 \\ y = 3x + 5 \end{cases}$ has an infinite number of solutions since both lines have the same slope and the same y-intercept. They are the same function.

Given a linear function in standard form, we solve for y to identify the slope and y-intercept. In general, if we have $Ax + By = C$:

Subtract Ax from both sides: $By = -Ax + C$

Divide both sides by B: $y = -\frac{A}{B}x + \frac{C}{B}$

So, $-\frac{A}{B}$ is the slope of $Ax + By = C$ and $\frac{C}{B}$ is the y value of the y-intercept.

Let's look at a couple of systems.

Given $\begin{matrix} 3x - 2y = 15 \\ 6x - 4y = -2 \end{matrix}$, the slope of the first equation is $-\frac{3}{-2} = \frac{3}{2}$ and the slope of the second equation is $-\frac{6}{-4} = \frac{3}{2}$. Both equations have the same slope. The y-intercepts are different since $\frac{C}{B} = \frac{15}{-2}$ for the first equation and $\frac{C}{B} = \frac{-2}{-4} = \frac{1}{2}$ for the second equation. Since the slopes are the same and the y-intercepts are different, we conclude that the lines are parallel and that there is no solution to the system.

Given $\begin{matrix} 4p + 7t = 3 \\ 4p - 7t = 3 \end{matrix}$, the slope of the first equation is $-\frac{4}{7}$ and the slope of the second equation is $-\frac{4}{-7} = \frac{4}{7}$. Since the slopes are different, the system has exactly one solution.

Given $\begin{matrix} 5x + 2y = -6 \\ -15x - 6y = 18 \end{matrix}$, for the first equation we have $-\frac{A}{B} = -\frac{5}{2}$ and $\frac{C}{B} = \frac{-6}{2} = -3$. or the second equation, we have $-\frac{A}{B} = -\frac{-15}{-6} = -\frac{5}{2}$ and $\frac{C}{B} = \frac{18}{-6} = -3$. Since both the slopes and y-intercepts are the same, these equations represent the same function and the system has an infinite number of solutions.

Let's return to an earlier problem and look at another solution technique. The following problem was initially investigated in Section 5.1.

Concert Sales: CDs-R-Us is branching out. They just purchased an outdoor concert facility. Seating is available in the pavilion or on the lawn. The average price of pavilion tickets is \$25, while the average price of lawn tickets is \$10. The average crowd size is 6000. To remain solvent, a concert must bring in \$75,000 in total ticket sales. To meet this guideline, let's determine how many of each kind of seat must be sold.

The system that models this problem is

Ticket equation: $p + l = 6000$

Income equation: $25p + 10l = 75{,}000$

We solved this system numerically, graphically, and algebraically using substitution in the last section. We will look at another algebraic approach in this section.

First let's investigate the effect of certain operations on equations.

7. Solve $p + l = 6000$ for the variable p.

8. Multiply both sides of the $p + l = 6000$ by a positive number. Then solve the result for the variable p. Is this equation equivalent to the original?

9. Multiply both sides of the $p + l = 6000$ by a negative number. Then solve the result for the variable p. Is this equation equivalent to the original?

10. Does a relationship between two variables, expressed as an equation, change when the equation is multiplied by a number? Justify your answer.

11. Does multiplying both sides of an equation in two variables, such as $p + l = 6000$, change the graph of the equation? Why or why not?

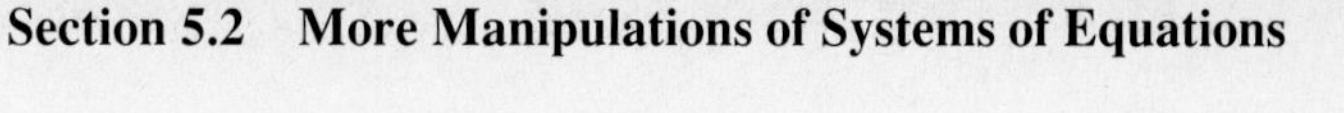

DISCUSSION

Investigations 8–11 were designed to convince you that multiplying an equation by a nonzero number results in an ***equivalent equation***. This means the equation has the same set of solutions,

First note that $p = 6000 - l$ is equivalent to $p + l = 6000$. The same quantity was subtracted from both sides of the original model.

Let's multiply $p + l = 6000$ by 17. Due to the ***distributive property of multiplication over addition***, this is the same as multiplying each term by 17.

Original model:	$p + l = 6000$
Multiply both sides by 17:	$17p + 17l = 102{,}000$
Solve the result for p as follows.	
Subtract $17l$ from both sides:	$17p + 17l - 17l = 102{,}000 - 17l$
Simplify:	$17p = 102{,}000 - 17l$
Divide both sides by 17:	$\frac{17p}{17} = \frac{102{,}000}{17} - \frac{17l}{17}$
Simplify:	$p = 6000 - l$

Since the result is equivalent to $p + l = 6000$, we know multiplying an equation by a constant seems to result in an equation equivalent to the original. A similar result occurs if you multiply both sides by a negative number.

This result is not surprising since we have been using a similar step to solve linear equations. Let's look at a different operation on equations.

INVESTIGATION

12. Begin with the original system $p + l = 6000$ and $25p + 10l = 75{,}000$. Add the two equations together term by term. Write the resulting equation.

13. Solve each of the original equations for p. Also solve the equation from the last investigation for p. This expresses p as a function of l in each equation.

14. Graph all three functions from the last investigation. What do you notice about the graphs?

Adding the two equations results in $26p + 11l = 81{,}000$. Solve for p.

Subtract $11l$ from both sides: $26p + 11l - 11l = 81{,}000 - 11l$

Simplify: $26p = 81{,}000 - 11l$

Divide both sides by 26: $\frac{26p}{26} = \frac{81{,}000}{26} - \frac{11l}{26}$

Simplify: $p = \frac{81{,}000}{26} - \frac{11l}{26}$

The other two functions have been solved for p previously.

The graph of all three appears in Figure 1.

FIGURE 1

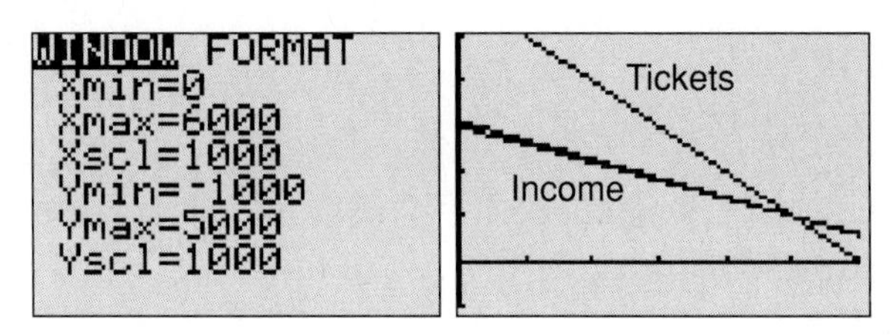

There are three functions graphed in Figure 1, but two are very close together. (That's the reason for the darker line near Income.)

The important point is that all three graphs intersect at the same point. Thus solving either of the systems

$$\begin{cases} p + l = 6000 \\ 26p + 11l = 81{,}000 \end{cases} \qquad \begin{cases} 25p + 10l = 75{,}000 \\ 26p + 11l = 81{,}000 \end{cases}$$

results in the same solution as solving the original system.

Let's look at one more example.

INVESTIGATION

15. Multiply the **Ticket equation** by –15.

16. Add the answer to Investigation 15 to the **Income equation**.

17. Graph the original two equations and your answer to Investigation 16. What is true about the graphs?

Multiplying the **Ticket equation** by –15 results in $-15p - 15l = -90{,}000$.

Add $-15p - 15l = -90{,}000$ and $25p + 10l = 75{,}000$ to get:

$$10p - 5l = -15{,}000$$

Solve this for p as follows.

Add $5l$ to both sides: $10p - 5l + 5l = -15{,}000 + 5l$

Simplify: $10p = -15{,}000 + 5l$

Divide both sides by 10: $\frac{10p}{10} = -\frac{15{,}000}{10} + \frac{5l}{10}$

Simplify: $p = -1500 + 0.5l$

The graph of all three functions appears in Figure 2.

FIGURE 2

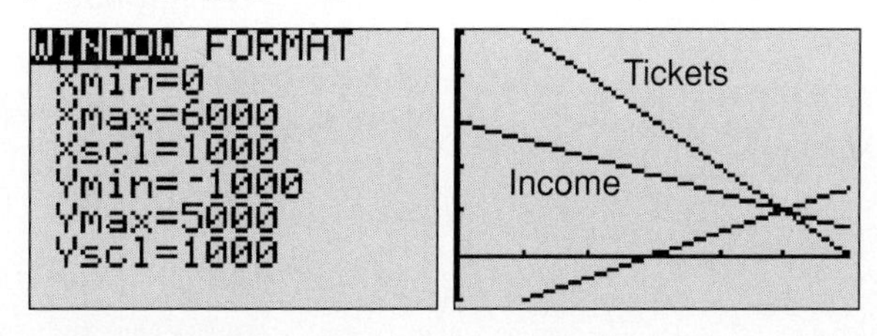

Again notice that all three equations intersect at one point. This point is the solution to the system determined by any two of the three equations.

To sum up, suppose we obtain an equation from the original system by multiplying equations by nonzero constants and adding equations. The resulting equation can be paired with either of the original equations to create an equivalent system.

This idea is the basis of the ***elimination method*** for solving systems. We will now investigate this technique.

18. First, focus on eliminating a variable. Select one of the two variables in the original system to eliminate. Write the variable and its numerical coefficient in the **Income equation**.

19. Multiply the **Ticket equation** by the opposite of the numerical coefficient identified in Investigation 18. Write the resulting equation.

20. Add the equation in Investigation 19 to the **Income equation**. Something desirable happens. What is it? Solve this equation for the variable.

21. Does the graph of the equation obtained by addition intersect the graphs of the original equations at the same point?

22. Return to the original system and use a similar approach to eliminate the other variable. Solve the resulting equation.

Suppose we choose to eliminate the variable l. The numerical coefficient of l in $25p + 10l = 75{,}000$ is 10. Thus, we multiply the **Ticket equation** $p + l = 6000$ by -10. The new system appears below.

$$-10p - 10l = -60{,}000$$
$$25p + 10l = 75{,}000$$

Add the equations: $15p = 15{,}000$

Solve for p: $p = 1000$

This solution matches the solution previously found by other methods.

Suppose we choose to eliminate p. The numerical coefficient of p in $25p + 10l = 75{,}000$ is 25. We multiply the **Ticket equation** $p + l = 6000$ by -25. The new system appears below.

$$-25p - 25l = -150{,}000$$
$$25p + 10l = 75{,}000$$

Add the equations: $-15l = -75{,}000$

Solve for l: $l = 5000$

We again get a match with a previous answer.

Summarizing, the elimination technique uses multiplication to write an equivalent system in which the numerical coefficients of one of the variables are opposites. Adding the two equations results in an equation in one variable which is easy to solve.

STUDENT DISCOURSE

Pete: So how do I know when to use the old substitution procedure and this new elimination procedure?

Sandy: How do you begin the substitution process?

Pete: I have to solve one of the equations for one of the variables, if I don't already have an equation solved for a variable.

Sandy: What problems might occur when you do this?

Pete: I know I don't like it when I have to divide and create fractions. In fact, I usually write the fractions as decimals.

Sandy: But doesn't that cause a problem?

Pete: What do you mean?

Sandy: If you write the fractions as decimals, you may end up rounding and lose some accuracy.

Pete: I can see that. Anyway, I don't like using substitution if I have to use fractions.

Sandy: How can you tell if a fraction will arise in the process?

Pete: If I have to divide. So I need to find a variable I can solve for without dividing.

Sandy: That's right. You need a coefficient of 1 on one of the variables.

Pete: And if that doesn't occur?

Sandy: Then elimination may be a better choice.

Pete: Can't I use elimination regardless?

Sandy: Sure. If both equations are written in standard form, elimination may be the least complicated algebraic method.

Pete: Thanks. I'll keep these points in mind as I work on some problems.

1. In the Glossary, define the words in this section that appear in ***italic boldface*** type.

2. If we multiply an equation on both sides by zero, is the result an equation that is equivalent to the original equation? Justify your answer.

3. **Hot Dog Stand Profits:** A hot dog stand makes a 25¢ profit on each hot dog sold and 15¢ profit on each bag of french fries sold. The stand breaks even on all other items sold. The stand has a weekly overhead cost of \$600. The week's profit must be \$400. For every three hot dogs sold, two orders of french fries are sold. We'd like to find the number of hot dogs and the number of bags of french fries sold.

   a. Write the system associated with the problem.

   b. Solve the system algebraically. Check your answer numerically and graphically.

   c. Write a complete sentence describing your results.

4. **Super Dog Burger Bits** produces two kinds of dry dog food: chunks and minichunks. The cost to manufacture a 25-pound bag of chunks is \$4 for materials and \$3 for labor. The cost to manufacture a 25-pound bag of minichunks is \$5 for materials and \$2 for labor. The total materials budget in this division for one week is \$10,000. The total labor budget in this division for one week is \$5000.

   a. Create a table that displays the information visually.

   b. Write an equation representing the total weekly materials cost based on the number of bags of chunks and the number of bags of minichunks produced.

   c. Write an equation representing the total weekly labor cost based on the number of bags of chunks and the number of bags of minichunks produced.

d. Solve the system defined by parts **b** and **c** to determine how many bags of chunks and how many bags of minichunks can be produced in one week. Write a sentence stating your answer.

5. Consider the system of equations

$$3x - 5y = 7$$
$$-6x + 10y = 3.$$

What happens when you try to solve this system? What does the graph of this system look like?

6. Consider the system of equations

$$10x + 5y = 30$$
$$y = -2x + 6.$$

What happens when you try to solve this system? What does the graph of the system look like?

7. Is it possible to create a dog pen with an area of 20 square feet and a perimeter of 20 feet? Justify your answer.

8. Identify the slope and y-intercept of the following equations by inspection (without writing anything).

a. $5x - 11y = 23$

b. $x + 17y = 17$

c. $x + y = -2$

d. $4x + 3y = 0$

9. Use inspection to determine if the following systems have no solution, exactly one solution, or an infinite number of solutions.

a. $y = 7x - 9$
$y = 7x + 1$

b. $4x - 7y = 2$
$4x + 7y = 2$

c. $5x + 2y = 7$
$10x + 4y = 7$

d. $x - 2y = -1$
$-4x + 8y = 4$

10. If you feel you need the practice, solve the following systems algebraically. Along with the solution, indicate whether you feel the substitution approach or the elimination approach is easier to use.

a. $2x - y = 7$
$5x + 7y = 2$

b. $4x + 9y = -11$
$x + 5 = 3y$

c. $9x + 4y = 7$
$5x - 3y = -15$

d. $x = 2y - 7$
$y = 8 - 5x$

e. $5t + 3c = -1$
$3c = 2 - 5t$

f. $a + b = -23$
$4a - 4b = 11$

g. $-3t - 4q = 15$
$6t = q - 1$

h. $y = \frac{3}{4}x - \frac{5}{2}$
$x = \frac{4}{3}y + 1$

REFLECTION

Write a paragraph comparing and contrasting the substitution method versus the elimination method for solving systems of equations. Include examples of systems best solved by elimination and examples of systems best solved by substitution.

SECTION 5.3

Organizing Information to Create Systems

Purpose

- Identify the number of variables and the number of equations based on a problem statement.
- Represent problem situations in a table format.
- Determine when systems will have many, one, or no solution(s).

Now that you know several techniques for solving systems of equations, we will look at situations in which a system of equations is an appropriate model. Given a problem situation, it is crucial to identify the number of unknown quantities. This indicates how many variables we need. Additionally, it is important to determine the number of relationships given. Each relationship results in an equation. Writing equations for each relationship using the variables that have been identified results in the system that we must solve.

Multiple Part-Time Jobs: You decided not to take the job offers at CDs-R-Us, Super Dog Burger Bits, or Virtual Reality Experiences. They just offer too much security for you. Instead you have lined up four part-time jobs that pay a flat amount each week. You will work only two jobs per week. You plan to try every job combination (pair of jobs) possible over the subsequent weeks.

1. Represent each job with a letter.

2. Using these letters, record all possible job combinations (pairs of jobs).

3. How many weeks must you work to experience a different pair of jobs each week?

DISCUSSION

Since there are four jobs, you could use the letters A, B, C, and D to represent them. There are six possible job combinations, so it will take 6 weeks to experience every combination.

Suppose you have kept track of your total earnings each week, but not how much each job paid. Every week you work on a given job, you are paid the same flat amount, which happens to be a whole number. Your total earnings in dollars (ordered from least to most) for each week are as follows: \$289, \$293, \$298, \$315, \$320, and \$324.

The question we eventually wish to answer is: "How much did each job pay?" Rather than try and answer this immediately, let's explore the structure of the problem in more depth.

INVESTIGATION

4. Create enough variables to represent the weekly amount paid by each job.

5. How many different equations can you write to represent how much you were paid weekly? How many variables are in each equation?

DISCUSSION

Since there were four different jobs, you need four variables to represent the weekly amount paid by each job. Use a, b, c, and d to represent these amounts.

Since there were 6 weeks you could write six different equations to represent the total earned each week. The problem is that you don't know which two jobs' salaries will add up to which amount.

We need to investigate the structure of the problem a little further.

INVESTIGATION

6. Each job pays a different amount. You have listed four variables for this amount. What might be a good thing to do to the variables to better organize the problem?

7. What restrictions can you place on the domain of each variable?

8. The smallest amount earned in a week was \$289. Write an equation involving some of the variables to represent this fact. How did you choose the variables for this equation?

9. The largest amount earned in a week was \$324. Write an equation involving some variables to represent this fact. How did you choose the variables for this equation?

10. Write any other equations you know to be correct. Explain why these are correct models for a part of the problem.

Since the weekly salaries must be ordered, it makes some sense to order the variables from smallest to largest. We will assume that the list a, b, c, and d is ordered from least amount to greatest amount.

We know that the variables must represent whole numbers and that no variable can exceed the largest total salary. This creates an initial limitation on the domain of the variables.

The ordering of the variables allows us to write several models.

The smallest weekly total must come from adding the smallest two salaries.

$$\text{So } a + b = \$289.$$

The largest weekly total must come from adding the largest two salaries.

$$\text{So } c + d = \$324.$$

We can assume that the second-smallest total comes from adding the smallest salary to the third-smallest salary.

$$\text{Thus } a + c = \$293.$$

Likewise we can assume that the second-largest total results from adding the largest salary to the third-largest salary.

$$\text{So } b + d = \$320.$$

Beyond this, it is rather tough to write specific relationships. For example, we don't know if $a + d$ is larger or smaller than $b + c$. Thus we don't know which to set equal to \$298, the next-largest total salary. While we need a total of six relationships, we are limited to only four right now. We will return to this problem as we delve further into systems. In the meantime give it some thought and consider investigating additional relationships using a table.

It is important to note that the number of objects we need to find represents the number of variables in the system. The number of relationships represents the number of equations in the system. Let's delve further into these structural ideas about systems by looking at some new and old problems.

11. Elementary School Consumer Math: Consider this standard elementary school problem: If I had \$2 more, I would have \$10. How much money do I have?

a. How many variables are there in this problem?

b. How many relationships are specified in this problem?

c. How many solutions are there to this problem?

12. Recall the **Concert Sales** problem: CDs-R-Us is branching out. They just purchased an outdoor concert facility. Seating is available in the pavilion or on the lawn. The average price of pavilion tickets is \$25, while the average price of lawn tickets is \$10. To remain solvent, concerts must bring in an average of \$75,000 in total ticket sales. How many of each type of ticket must be sold?

a. How many variables are there in this problem? Why?

b. How many relationships are specified in this problem? Why?

c. How many solutions are there to this problem. Why?

13. Use the problem described in Investigation 12, but add the restriction that the average number of tickets sold is 6000.

a. How many variables are there in this problem? Why?

b. How many relationships are specified in this problem? Why?

c. How many solutions are there to this problem. Why?

14. Indoor Concert Sales: CDs-R-Us is doing so well with their outdoor concert facility that they have purchased a small concert venue to highlight up-and-coming performers. To encourage people to attend concerts featuring relative unknowns, they have priced the tickets very inexpensively. In this case, tickets are priced at \$6, \$8, and \$10. Unfortunately, the group that performed last night had a poor reputation. Only 50 people showed up, which meant ticket revenues of \$450. How many of each type ticket were purchased?

a. How many variables are there in this problem?

b. What restrictions must be placed on the domain of the variables?

c. How many relationships are specified in this problem? Write equations representing these relationships.

Systems of equations are classified by the number of variables, the number of equations, and the type of equations. In the preceding investigation all equations were first-degree. We will focus more on the number of variables and the number of equations.

Investigation 11, involving **Elementary School Consumer Math,** results in one variable and one equation, $x + 2 = 10$. There is only one solution, 8, to this problem. Technically, this would not be referred to as a system since it involves only one relationship.

Similarly, Investigation 12 involves only one relationship—the one involving income from ticket sales $25p + 10l = 75{,}000$. But while involving one relationship, the problem requires two variables. If we ignore the problem restriction, there are an infinite number of solutions to this equation. Any ordered pair that satisfies the equation is a solution. This means any point on the graph, which is a straight line, is a solution.

It is very common to have an infinite number of solutions when the number of variables exceeds the number of equations. The actual number of solutions to this problem is dependent on the domain restrictions on the variables. We know they must be whole numbers. A table would help us explore the number of solutions to this problem. There are quite a few.

Investigation 13 adds a restriction and thus the equation $p + l = 6000$. Now we have a system with two variables and two equations. We know from a previous section that this system has exactly one solution because the slopes of the graphs are different.

Finally, Investigation 14 presents a problem with three variables—say x, y, and z—which represent the number of tickets sold at each of the three prices. However, there are only two relationships.

Tickets sold:	$x + y + z = 50$
Receipts from tickets:	$6x + 8y + 10z = 450$

Minimal restrictions on the domain of the variables include the fact that they must be whole numbers less than 50.

We need to investigate this problem further in order to determine how many solutions it has.

We will use an elimination technique on the system

$$x + y + z = 50$$
$$6x + 8y + 10z = 450.$$

15. a. Use multiplication and addition to eliminate y from the system.

b. Solve the resulting equation for z in terms of x. Enter this expression into your calculator.

16. a. Use multiplication and addition to eliminate z from the system.

b. Solve the resulting equation for y in terms of x. Enter this expression into your calculator.

17. Display a table of values for the variables x, y, and z. Is there more than one solution to this problem?

There are more variables than equations in the system representing the **Indoor Concert Sales** problem. This requires us to define two variables, say y and z, as functions of the remaining variable x.

To eliminate y, we multiply the tickets equation by –8 and add the resulting equation to the income equation.

Multiply:	$-8x - 8y - 8z = -400$
	$6x + 8y + 10z = 450$
Add equations:	$-2x + 2z = 50$
Add $2x$ to both sides:	$-2x + 2x + 2z = 50 + 2x$
Simplify:	$2z = 50 + 2x$
Divide both sides by 2:	$\frac{2z}{2} = \frac{50}{2} + \frac{2x}{2}$
Simplify:	$z = 25 + x$

We have an equation expressing a relationship between variables x and z.

To eliminate z, we multiply the tickets equation by –10 and add the resulting equation to the income equation.

Multiply:	$-10x - 10y - 10z = -500$
	$6x + 8y + 10z = 450$
Add equations:	$-4x - 2y = -50$
Add $4x$ to both sides:	$-4x - 2y + 4x = -50 + 4x$
Simplify:	$-2y = -50 + 4x$
Divide both sides by –2:	$\frac{-2y}{-2} = \frac{-50}{-2} + \frac{4x}{-2}$
Simplify:	$y = 25 - 2x$

This provides a relationship between x and y.

A table display of solutions satisfying the relationships appears in Figure 1.

FIGURE 1

Y1=25-2X
Y2=25+X
Y3=
Y4=
Y5=
Y6=
Y7=
Y8=

X	Y1	Y2
0	25	25
1	23	26
2	21	27
3	19	28
4	17	29
5	15	30
6	13	31

X=6

X	Y1	Y2
7	11	32
8	9	33
9	7	34
10	5	35
11	3	36
12	1	37
13	-1	38

X=13

$y = 25 - 2x \quad z = 25 + x$

Note that this system has 13 different answers: all values for x from zero to 12 inclusive. If we were not limited to whole numbers there would be an infinite number of answers. The multiple answers are primarily due to the fact that there are more variables than relationships between variables.

We conclude this section by introducing you to the idea of a *matrix*. A ***matrix*** is a rectangular array (arrangement) of objects. In our case, the objects will be numbers. We investigate a method of displaying information from a system in a table.

Recall the system for the **Concert Sales** problem.

$$p + l = 6000$$
$$25p + 10l = 75{,}000.$$

a. Complete Table 1 by writing only the coefficients of the variables in the appropriate space.

TABLE 1 Coefficient Matrix for Concert Sales

	Pavilion	Lawn
Tickets		
Income		

b. Complete Table 2 by writing the values of the constants from the system.

TABLE 2 Constant Matrix for Concert Sales

Total Tickets	
Total Income	

18. Refer to the system for the **Indoor Concert Sales** problem.

Tickets sold: $x + y + z = 50$

Receipts from tickets: $6x + 8y + 10z = 450$

a. Create a table similar to Table 1 for this system.

b. Create a table similar to Table 2 for this system.

We often use a table of numbers called a ***matrix*** to represent a problem's information. A ***matrix*** is essentially a rectangular array of numbers. The plural of matrix is ***matrices***. Unlike tables, matrices do not contain row or column headings. Rather, the position of a number in a matrix determines what it means. Let's look at matrices for the three systems we have discussed in this section.

When a matrix contains only the coefficients of the variables from a system of equations, it is called a ***coefficient matrix.***

The system for the **Concert Sales** problem is

$$p + l = 6000$$
$$25p + 10l = 75{,}000.$$

The two equations are written in a form that has like variable terms aligned on one side of the equation. The constant terms are on the other side of the equation. This form is necessary to represent the system using matrices.

The coefficient matrix for the **Concert Sales** problem is

$$\begin{bmatrix} 1 & 1 \\ 25 & 10 \end{bmatrix}$$

Notice that the columns represent variables and the rows represent relationships.

Matrices are often classified by their size. This is determined by stating the number of rows followed by the number of columns. The coefficient matrix we just wrote is 2 by 2, written 2×2, because it has two rows and two columns.

When a matrix contains the constants of a system, it is called a ***constant matrix***. (What a surprise?) The constant matrix for the **Concert Sales** problem is the 2×1 following matrix.

$$\begin{bmatrix} 6000 \\ 75{,}000 \end{bmatrix}$$

The coefficient matrix for the **Indoor Concert Sales** problem is the 2×3 matrix

$$\begin{bmatrix} 1 & 1 & 1 \\ 6 & 8 & 10 \end{bmatrix}$$

The constant matrix for this problem is the 2×1 matrix

$$\begin{bmatrix} 50 \\ 450 \end{bmatrix}$$

To conclude, notice that the number of variables in a system is determined by how many quantities we are asked to find. The number of equations is determined by the number of relationships described in the problem.

STUDENT DISCOURSE

Pete: It docsn't seem too hard to write the systems in the matrix form, but why would you want to do that?

Sandy: Once you know a little about matrices, you'll learn a way to use matrices to solve linear systems.

Pete: Wait! You said matrices. I thought we were talking about a matrix.

Sandy: Sorry. Matrices is just the plural of matrix.

Pete: Oh, okay. Anyway, I already know lots of ways to solve systems. Why do I need another?

Sandy: Graphing and looking at a table to estimate solutions are fine if the system involves only two equations and two variables. Elimination and substitution also are good techniques as long as there aren't too many equations (three or fewer) or too many variables (three or fewer).

Pete: I've mainly worked with two equations and two variables so far.

Sandy: That's right, but what if there were five variables and five equations, for example?

Pete: That sounds pretty complicated.

Sandy: True and it would be difficult to solve such a system by any of the techniques you have learned so far. Even three equations and three unknowns requires care and patience to solve.

Pete: So what about matrices?

Sandy: Using a calculator that can manipulate matrices, you can solve any size system that consists of linear equations as long as the calculator can store a matrix of that size. Most calculators handle at least 10×10 matrices and the newer ones can handle over 100 rows and columns. The technique is the same for each and the only difference in time and effort is how long it takes to enter the numbers in the matrix into the calculator.

Pete: That's interesting. Will I learn about that in this course?

Sandy: You'll learn about matrices in the next section and how to use them to solve linear systems in the following section.

Pete: Sounds good. I'll look forward to that.

1. In the Glossary, define the words in this section that appear in ***italic boldface*** type.

2. Hardware Sale: A hardware store marks down every item in the store 15%. Consider the relationship between the original price and the sale price.

a. How many variables are in this problem? What are they?

b. How many relationships are described?

c. How many solutions are there to this problem?

Add the condition that the difference between the sale price and original price of a hammer is $2.00. We wish to determine both original price and the sale price of the hammer.

d. How many variables are in this problem? What are they?

e. How many relationships are described?

f. How many solutions are there to this problem?

g. Write a coefficient matrix and a constant matrix for this system.

3. An 80-foot Dog Pen: We have 80 feet of fence and wish to create a rectangular dog pen. Express the length as a function of the width.

a. How many variables are in this problem? What are they?

b. How many relationships are described?

c. How many solutions are there to this problem?

Add the restriction that the length must be twice the width. The task is to determine the length and the width.

d. How many variables are in this problem? What are they?

e. How many relationships are described?

f. How many solutions are there to this problem?

g. Write a coefficient matrix and a constant matrix for this system.

4. Men and Women in Education: The ratio of men to women in elementary education is 2:7. Express the number of men in elementary education as a function of the number of women in elementary education.

a. How many variables are in this problem? What are they?

b. How many relationships are described?

c. How many solutions are there to this problem?

Add the restriction that the total number of elementary education students is 853.

d. How many variables are in this problem? What are they?

e. How many relationships are described?

f. How many solutions are there to this problem?

g. Write a coefficient matrix and a constant matrix for this system.

5. Hot Dog Stand Profits: A hot dog stand makes a 25¢ profit on each hot dog sold and 15¢ profit on each bag of french fries sold. The stand breaks even on all other items sold. The stand has a weekly overhead cost of $600. The week's profit must be $400. Express the number of bags of french fries sold as a function of the number of hot dogs sold.

a. How many variables are in this problem? What are they?

b. How many relationships are described?

c. How many solutions are there to this problem?

Add the condition that for every three hot dogs sold, two orders of french fries are sold.

d. How many variables are in this problem? What are they?

e. How many relationships are described?

f. How many solutions are there to this problem?

g. Write a coefficient matrix and a constant matrix for this system.

6. Super Dog Burger Bits produces two kinds of dry dog food: chunks and minichunks. The cost to manufacture a 25-pound bag of chunks is $4 for materials and $3 for labor. The cost to manufacture a 25-pound bag of minichunks is $5 for materials and $2 for labor. The total materials budget in this division for one week is $10,000. The total labor budget in this division for one week is $5000.

a. How many variables are in this problem? What are they?

b. How many relationships are described? Write an equation for each relationship.

c. How many solutions are there to this problem?

d. Write a coefficient matrix and a constant matrix for this system.

7. Athletic Shoes: The athletic shoe manufacturer NikBok manufactures three types of athletic shoes: walking, tennis, and aerobic. The manufacture of each pair of walking shoes costs $8 in materials, $10 in labor, and $3 in overhead. The manufacture of each pair of tennis shoes costs $6 in materials, $10 in labor, and $4 in overhead. Finally, each pair of aerobic shoes costs $7 in materials, $9 in labor, and $5 in overhead to produce. The company budgets $5200 for materials, $7000 for labor, and $3000 for overhead each day. You, as production supervisor, must determine the approximate number of each type of shoe that can be manufactured in a day.

a. How many variables are in this problem? What are they?

b. How many relationships are described? Write an equation for each relationship.

c. Write a coefficient matrix and a constant matrix for this system.

8. Multiple Part-Time Jobs: Suppose there are three total jobs. You will work only two jobs per week. You plan to try every job combination (pair of jobs) possible over the subsequent weeks. Your total earnings in dollars (ordered from least to most) for each week are as follows: $353, $370, and $379.

a. How many variables are in this problem? What are they?

b. How many relationships are described? Write an equation for each relationship.

c. Write a coefficient matrix and a constant matrix for this system.

9. The relationship between the number of jobs and the number of possible combinations of two jobs is interesting to explore.

a. Complete Table 3 by determining the number of possible combinations of two jobs given the number of jobs.

TABLE 3 Counting Combinations

Number of Jobs	Number of Pairs
1	
2	
3	
4	6
5	
6	
7	
8	

b. Table 3 can be thought of as data. Analyze the data to determine the best model. Find a function that will calculate the number of pairs of objects given the number of objects.

c. Use your answer to part **b** to determine the number of possible pairs given 10 part-time jobs.

10. Find the slope and the y-intercept of the following linear functions.

a. $5x - 9y = 23$

b. $x + 2y = 21$

11. Write the coefficient matrix and the constant matrix for each of the following linear systems.

a. $9x + 7y = 23$
$2x + y = -19$

b. $4x + 10y = 2$
$2x + 5y = -1$

c. $4x - 2y + 7z = 4$
$x + y - 3z = -8$

d. $5x - y = 21$
$3x + 2y = 7$
$x - y = 1$

e. $y = 2x + 4$
$3x - 7y + 9 = 0$

f. $3x - 2y + 5z = 21$
$x + 2y + 7z = -1$
$2x + 3z = 14$

g. $x - z = 2$
$y + 3z = 0$
$4x + 7y = -8$

h. $4x - y + 2z + 8w = 17$
$x + 2y + 7z - w = -7$
$3x + 8z + 7z = 2$

12. Solve the systems given in parts **a**, **b**, and **e** of Exploration 11.

13. Solve the system given in part **c** of Exploration 11 by expressing both x and y in terms of z.

14. Use a graph or a table to solve the following systems.

a. $3x + 2y = 7$
$y = 3x^2 + x - 4$

b. $y = x^2 + x + 1$
$y = -2x^2 + 3x + 7$

REFLECTION

Create a problem that contains three variables and four relationships. Write the system that represents your problem.

SECTION 5.4

Coded Messages and the Algebra of Matrices

Purpose

- Investigate matrix operations in a problem context.

MAKING CONNECTIONS

Matrices were introduced briefly in the last section. Before you learn how matrices can be used to solve systems of equations, you need to learn a little about the algebra of matrices. Just as we can perform operations on numbers, there are operations that can be performed on matrices. There are also some special types of matrices, such as identity and inverse matrices, that you will learn about. To introduce operations with matrices, we will look at how matrices might be used to encode and decode privileged information.

Cryptography is the science of written information unintelligible to all but those intended to read it. One technique is to send an encrypted (scrambled) message that has been generated by an ***encoding key.*** The receiver decodes the message using an inverse transformation of the encoding key. The security of the message is directly related to the secrecy of the encoding key.

Let's start the encryption process. Initially, we convert English into numerical data using the letter–number correspondence.

The first message we will encode is:

One person can make a difference.

1. Using Table 1 on page 368, write the sequence of numbers that represent the message.

TABLE 1 Letter–Number Correspondence: One person can make a difference.

A	B	C	D	E	F	G	H	I	J	K	L	M
1	2	3	4	5	6	7	8	9	10	11	12	13

N	O	P	Q	R	S	T	U	V	W	X	Y	Z
14	15	16	17	18	19	20	21	22	23	24	25	26

2. Thus far, do you think this is a good scheme for writing a secret message? Why or why not? How could you improve the scheme?

The sequence of numbers for the message is

15, 14, 5, 16, 5, 18, 19, 15, 14, 3, 1, 14, 13, 1,
11, 5, 1, 4, 9, 6, 6, 5, 18, 5, 14, 3, 5.

This code would be easy to break, especially for someone who knows how to analyze the frequency of occurrence of letters.

It is common to apply a linear transformation to the data generated from Table 1. This is accomplished by creating a linear function and using the position value as input. The output mod 26 would represent the encoded letter. Mod 26 means that the output of the linear function would be divided by 26 and the remainder of this division would be the encoded number.

Instead of using a linear transformation, we will use matrix multiplication to do the encoding.

Oh no! We don't know how to multiply matrices yet. We better collect some data on the calculator and figure out how this works.

3. Use your calculator to find the product, if it exists. (Ask a classmate or your instructor if you are unsure how to do this, or consult the manual.) If the calculator gives an error when you try to find the product, the product does not exist.

a. $\begin{bmatrix} 3 & 2 \end{bmatrix} \cdot \begin{bmatrix} 5 \\ 4 \end{bmatrix}$

b. $\begin{bmatrix} 5 \\ 4 \end{bmatrix} \cdot \begin{bmatrix} 3 & 2 \end{bmatrix}$

c. $\begin{bmatrix} 3 & 2 \end{bmatrix} \cdot \begin{bmatrix} 5 & 6 \\ 4 & 7 \end{bmatrix}$

d. $\begin{bmatrix} 5 & 6 \\ 4 & 7 \end{bmatrix} \cdot \begin{bmatrix} 3 & 2 \end{bmatrix}$

e. $\begin{bmatrix} 2 & 3 \end{bmatrix} \cdot \begin{bmatrix} 4 & 6 & 8 \\ 5 & 7 & 9 \end{bmatrix}$

f. $\begin{bmatrix} 4 & 6 & 8 \\ 5 & 7 & 9 \end{bmatrix} \cdot \begin{bmatrix} 2 & 3 \end{bmatrix}$

g. $\begin{bmatrix} 3 & 2 \\ 5 & 1 \end{bmatrix} \cdot \begin{bmatrix} 5 & 4 & 2 \\ 3 & 7 & 4 \end{bmatrix}$

h. $\begin{bmatrix} 5 & 4 & 2 \\ 3 & 7 & 4 \end{bmatrix} \cdot \begin{bmatrix} 3 & 2 \\ 5 & 1 \end{bmatrix}$

i. $\begin{bmatrix} 5 & 4 & 2 \\ 3 & 7 & 4 \end{bmatrix} \cdot \begin{bmatrix} 1 & 3 & 4 & 1 \\ 5 & 0 & 6 & 2 \\ 2 & 1 & 0 & 3 \end{bmatrix}$

4. For each matrix product in the preceding exercise, record in Table 2 the dimensions of both given matrices along with the dimensions of the product matrix. We started the first one for you.

TABLE 2 Matrix Dimensions

Problem	Number of Rows (1st Matrix)	Number of Columns (1st Matrix)	Number of Rows (2nd Matrix)	Number of Columns (2nd Matrix)	Number of Rows (Product Matrix)	Number of Columns (Product Matrix)
1a	1	2	2	1	1	1
1b						
1c						
1d						
1e						
1f						
1g						
1h						
1i						

5. Study Table 2 carefully. Look for relationships between the dimensions of the two original matrices that would allow you to predict the size of the product matrix.

a. What condition(s) on the dimensions of the two matrices must hold if their product is to be defined?

b. Explain how to determine the dimensions of the product matrix from the dimensions of the two original matrices.

6. a. The element $c_{1,2}$ (the element in position row 1, column 2) is in the product matrix. Describe how it is computed from the original two matrices.

b. The element $c_{i,j}$ is an element of the product matrix. Describe how this element is computed from the original two matrices. In other words how is any element in a product matrix computed?

Matrices can be named using capital letters. Your calculator may use this convention. We may wish to refer to a specific element of a matrix. For example, if we refer to matrix $\begin{bmatrix} 2 & -8 & 9 \\ 0 & 6 & -1 \end{bmatrix}$ as A, then an element of this matrix would be referred to by using a lowercase a followed by two subscripts. The first subscript would represent the row the element is in. The second subscript represents the column the element is in.

Thus $a_{2,1} = 0$, the entry in row 2, column 1. Also the entry 9 would be referenced by $a_{1,3}$.

The product of two matrices exists only if the number of columns in the first matrix equals the number of rows in the second matrix.

The product $\begin{bmatrix} 2 & 3 \end{bmatrix} \cdot \begin{bmatrix} 4 & 6 & 8 \\ 5 & 7 & 9 \end{bmatrix}$ exists since the first matrix has two columns and the second matrix has two rows.

Meanwhile, the product $\begin{bmatrix} 4 & 6 & 8 \\ 5 & 7 & 9 \end{bmatrix} \cdot \begin{bmatrix} 2 & 3 \end{bmatrix}$ is not defined since the first matrix has three columns while the second matrix has only one row.

This last example suggests that $\begin{bmatrix} 2 & 3 \end{bmatrix} \cdot \begin{bmatrix} 4 & 6 & 8 \\ 5 & 7 & 9 \end{bmatrix} \neq \begin{bmatrix} 4 & 6 & 8 \\ 5 & 7 & 9 \end{bmatrix} \cdot \begin{bmatrix} 2 & 3 \end{bmatrix}$.

So the order we multiply two matrices is important unlike when we multiply real numbers. Formally, we say that matrix multiplication is not commutative.

The size of the product matrix is determined by the number of rows in the first matrix and the number of columns in the second matrix.

The product of $\begin{bmatrix} 2 & 3 \end{bmatrix} \cdot \begin{bmatrix} 4 & 6 & 8 \\ 5 & 7 & 9 \end{bmatrix}$ will have one row (the number of rows in $\begin{bmatrix} 2 & 3 \end{bmatrix}$) and three columns (the number of columns in $\begin{bmatrix} 4 & 6 & 8 \\ 5 & 7 & 9 \end{bmatrix}$). It is a 1×3 matrix.

Matrix multiplication consists of multiplying the rows of the first matrix by the corresponding elements in the columns of the second matrix and then adding the products.

For example, the $c_{1,2}$ element in a product matrix would come from multiplying the elements in row 1 of the first matrix by the corresponding elements of column 2 in the second matrix and adding the products.

For example, the first row of $\begin{bmatrix} 2 & 3 \end{bmatrix}$ is $\begin{bmatrix} 2 & 3 \end{bmatrix}$. The second column of $\begin{bmatrix} 4 & 6 & 8 \\ 5 & 7 & 9 \end{bmatrix}$ is $\begin{bmatrix} 6 \\ 7 \end{bmatrix}$. The element $c_{1,2}$ is $2(6) + 3(7) = 12 + 21 = 33$.

Matrix multiplication is not too hard, but it can be confusing and tedious. Fortunately, we have some technology to assist us in these computations.

Let's explore a few more ideas about matrices before we return to our encoding problem.

A few definitions are in order.

A ***square matrix*** has the same number of rows as columns The ***order*** of a square matrix is the number of rows or columns in the matrix.

For example, $\begin{bmatrix} 2 & 7 & 0 \\ -8 & 7 & 1 \\ -8 & 0 & -5 \end{bmatrix}$ is a square matrix of order 3, but $\begin{bmatrix} 2 & 5 & -9 \\ 0 & 3 & 5 \end{bmatrix}$ is not a square matrix since the number of rows and columns are not equal.

Associated with each square matrix is an ***identity matrix***. An identity matrix is square, has ones down the main diagonal (upper left to lower right) and zeros everywhere else.

For example, the identity matrix of order 3 is $\begin{bmatrix} 1 & 0 & 0 \\ 0 & 1 & 0 \\ 0 & 0 & 1 \end{bmatrix}$.

7. a. Create any 2×2 matrix. Multiply your 2×2 matrix by its identity matrix. What is the answer?

b. Create a 3×3 matrix. Predict the result when you multiply it by its identity matrix. Perform the multiplication and modify your prediction if necessary.

c. Create a 4×4 matrix. Predict the result when you multiply it by its identity matrix. Perform the multiplication and modify your prediction if necessary.

8. If any square matrix is multiplied by its identity matrix, what is the product matrix?

9. Some square matrices have inverses. If A represents a square matrix, then A^{-1} represents the inverse of matrix A.

a. Create a 2×2 square matrix. Find its inverse, if it exists, on your calculator and write the matrix and its inverse below. If the inverse does not exist, alter your matrix until you have a matrix that has an inverse.

b. Find the product of the matrix you created in part **a** and its inverse. Record the product.

c. Create a 3×3 square matrix and find its inverse, if it exists, using your calculator. If the inverse does not exist, alter your matrix until you have a matrix that has an inverse. Record your matrix and the inverse.

d. Predict the product of the matrix you created in part **c** and its inverse. Find the product and modify your prediction, if necessary.

e. Write a general statement describing the product of a square matrix and its inverse.

10. Associated with every square matrix is a number called its determinant. Your calculator will probably compute the *determinant* of a square matrix automatically. (Ask someone how to do this if you do not know how.)

a. Find the determinant of the 2×2 matrix you wrote in Investigation 9**a**.

b. Find the determinant of the 3×3 matrix you wrote in Investigation 9**c**.

The product of any square matrix and its corresponding identity matrix is always the original square matrix. This result is analogous to multiplying a real number by 1.

For example, if A is a square matrix and I is its identity matrix, then

$$AI = A.$$

The product of a square matrix and its inverse is always the corresponding identity matrix. This is equivalent to multiplying a real number by its multiplicative inverse.

For example, if A is a square matrix and I is its identity matrix, then

$$AA^{-1} = I.$$

For every square matrix, there is a special number called the ***determinant***. A determinant *det* is a function with input a square matrix and output that number. The domain of *det* is the set of all square matrices (Figure 1).

FIGURE 1

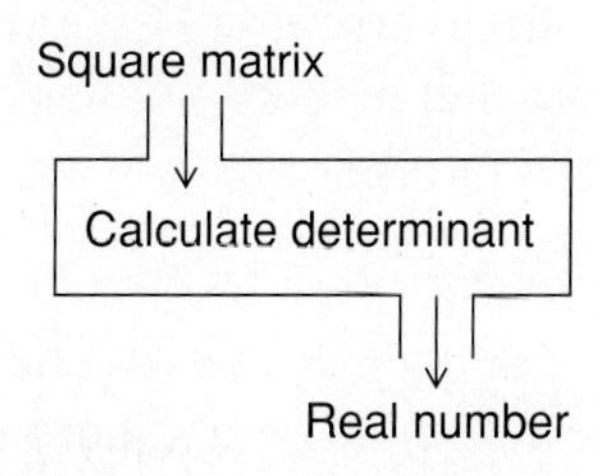

The use of a determinant makes it convenient to express formulas and has proven useful in many branches of mathematics. The calculation of the determinant involves all possible products in which exactly one element in each row and column are added or subtracted. It is relatively easy to calculate the determinant of a 2 × 2 matrix.

If $A = \begin{bmatrix} a_{11} & a_{12} \\ a_{21} & a_{22} \end{bmatrix}$, then the determinant of A, written $det(A)$, is $a_{11}a_{22} - a_{21}a_{12}$. In words, we multiply the elements in the upper left to lower right diagonal. Then we subtract the product of the elements in the other diagonal.

For example, $det\left(\begin{bmatrix} 3 & 7 \\ 5 & -4 \end{bmatrix}\right) = 3(-4) - 5(7) = -12 - 35 = -47$.

For larger matrices, the procedure is more complex. For a 3 × 3 matrix there are six products and for a 4 × 4 matrix there are 24 products. The confusing part is figuring out whether each product is added or subtracted. If you are interested, you might look it up in a book on matrices. Fortunately, we can use the calculator to find determinants.

One important point: a square matrix has an inverse only if the determinant of the square matrix is not zero. Thus we have an easy way to determine if a matrix has an inverse by checking to see whether the detminant is zero or not.

For example, matrix D = $\begin{bmatrix} 8 & 6 \\ 4 & 3 \end{bmatrix}$ does not have an inverse since its determinant is $8(3) - 4(6) = 0$ (Figure 2). When we ask the calculator to find the

FIGURE 2

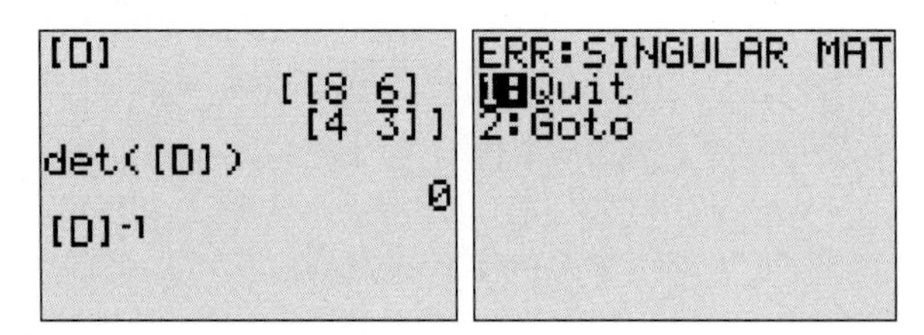

inverse of matrix [D], the calculator responds with an error message. A matrix that does not have an inverse is called a ***singular matrix*** (see the error message in Figure 2). Note that technology has put the name of the matrix in brackets to distinguish it from other types of variables.

Now we have sufficient background to encode our message. To do so, we need a matrix to serve as an encryption key. The encryption key matrix must have an inverse and therefore must be a square matrix. For our coding, let's use a 3 × 3 matrix. This is complex enough without being too complex. To encode the message, we write the message as a matrix and compute the product encryption key matrix times the message matrix. Let's see how this works.

Recall that the sequence of numbers for the message is

15, 14, 5, 16, 5, 18, 19, 15, 14, 3, 1, 14, 13, 1, 11, 5, 1, 4,
9, 6, 6, 5, 18, 5, 14, 3, 5.

11. Suppose the encryption key matrix has name K and the message matrix has name M. We will need the product KM. If K is 3×3, what do you know about the dimensions of the M if we must be able to find the product KM?

12. Let's record the message in a matrix M. Using the restriction on size you stated in Investigation 11, fill matrix M *column by column* with the sequence of numbers for the message. If you run out of numbers, fill the remainder of the column with zeros.

Since the encryption key K has three columns, the message matrix M must have three rows. So the number of rows in the message matrix must match the number of columns in the encryption key.

The sequence of numbers for the message is

15, 14, 5, 16, 5, 18, 19, 15, 14, 3, 1, 14, 13, 1, 11, 5,
1, 4, 9, 6, 6, 5, 18, 5, 14, 3, 5.

The message matrix M is

$$\begin{bmatrix} 15 & 16 & 19 & 3 & 13 & 5 & 9 & 5 & 14 \\ 14 & 5 & 15 & 1 & 1 & 1 & 6 & 18 & 3 \\ 5 & 18 & 14 & 14 & 11 & 4 & 6 & 5 & 5 \end{bmatrix}.$$

Now let's create an encryption key and encode the message.

13. Write a 3×3 matrix K as an encryption key. Be sure it has an inverse.

14. Compute the product KM. Write the product matrix.

15. The product matrix from Investigation 14 is the coded matrix. Write the sequence of numbers that represents the coded message.

16. To check our work, let's all use the same encryption key. Multiply encryption key $K = \begin{bmatrix} 1 & 2 & 4 \\ 3 & -2 & 8 \\ 7 & 6 & -5 \end{bmatrix}$ times the message matrix M. Write the product matrix.

17. Write the sequence of numbers that represents the coded message from Investigation 16.

Now let's decode a message. The encoded message is

19, 37, 26, 62, 42, 116, 135, 177, 166, 122, 162,
41, 100, 188, 8, 68, 4, 276.

18. Record this information by column in a matrix with three rows. Remember to enter the first three numbers in the first column and so on.

19. To decode, we multiply the inverse of $K = \begin{bmatrix} 1 & 2 & 4 \\ 3 & -2 & 8 \\ 7 & 6 & -5 \end{bmatrix}$ times the matrix in Investigation 18. Write the resulting matrix.

20. Write the sequence of numbers that represents the decoded message.

21. Convert the sequence of numbers to a sequence of letters to read the message. Write the message.

The message "One person can make a difference" was encoded initially as the matrix $M = \begin{bmatrix} 15 & 16 & 19 & 3 & 13 & 5 & 9 & 5 & 14 \\ 14 & 5 & 15 & 1 & 1 & 1 & 6 & 18 & 3 \\ 5 & 18 & 14 & 14 & 11 & 4 & 6 & 5 & 5 \end{bmatrix}$.

Multiplying encryption key $K = \begin{bmatrix} 1 & 2 & 4 \\ 3 & -2 & 8 \\ 7 & 6 & -5 \end{bmatrix}$ matrix times M, we get

$$\begin{bmatrix} 1 & 2 & 4 \\ 3 & -2 & 8 \\ 7 & 6 & -5 \end{bmatrix} \begin{bmatrix} 15 & 16 & 19 & 3 & 13 & 5 & 9 & 5 & 14 \\ 14 & 5 & 15 & 1 & 1 & 1 & 6 & 18 & 3 \\ 5 & 18 & 14 & 14 & 11 & 4 & 6 & 5 & 5 \end{bmatrix} = \begin{bmatrix} 63 & 98 & 105 & 61 & 59 & 23 & 45 & 61 & 40 \\ 57 & 182 & 139 & 119 & 125 & 45 & 63 & 19 & 76 \\ 164 & 52 & 153 & -43 & 42 & 21 & 69 & 118 & 91 \end{bmatrix}.$$

Thus the encoded message is sent as 63, 57, 164, 98, 182, 52, 105, 139, 153, 61, 119, –43, 59, 125, 42, 23, 45, 21, 45, 63, 69, 61, 19, 118, 40, 76, 91.

To decode 19, 37, 26, 62, 42, 116, 135, 177, 166, 122, 162, 41, 100, 188, 8, 68, 4, 276, first write as a 3×6 matrix

$$\begin{bmatrix} 19 & 62 & 135 & 122 & 100 & 68 \\ 37 & 42 & 177 & 162 & 188 & 4 \\ 26 & 116 & 166 & 41 & 8 & 276 \end{bmatrix}.$$

Multiply the inverse of K times the matrix containing the message to obtain

$$\begin{bmatrix} 1 & 2 & 4 \\ 3 & -2 & 8 \\ 7 & 6 & -5 \end{bmatrix}^{-1} \begin{bmatrix} 19 & 62 & 135 & 122 & 100 & 68 \\ 37 & 42 & 177 & 162 & 188 & 4 \\ 26 & 116 & 166 & 41 & 8 & 276 \end{bmatrix} = \begin{bmatrix} 5 & 8 & 21 & 8 & 12 & 18 \\ 1 & 15 & 19 & 15 & 4 & 25 \\ 3 & 6 & 19 & 21 & 20 & 0 \end{bmatrix}.$$

The decoded message is 5, 1, 3, 8, 15, 6, 21, 19, 19, 8, 15, 21, 12, 4, 20, 18, 25, 0.

Translating to alphabetic characters, we get

Each of us should try.

The complete message: "One person can make a difference. Each of us should try." is a quote from a speech by Robert F. Kennedy.

One final note—you can make up any encryption key you like as long as the encryption key matrix has an inverse. The two people communicating this way must share the encryption key along with the message. If the encryption key falls into the wrong hands, watch out!

STUDENT DISCOURSE

Pete: That was fun, though I have to admit that matrix multiplication is a bit tricky.

Sandy: I agree. I know how it works but its easy to make a mistake.

Pete: I guess that's why it helps to have the calculator to do the complicated ones. By the way, we learned how to multiply matrices. Is it possible to add and subtract them too?

Sandy: Sure it is. In fact, you'll use your calculator to learn how to add matrices in the explorations.

Pete: I have another question. Are matrices used a lot in mathematics?

Sandy: Well, I don't have a lot of background, but I did ask my math instructor about that.

Pete: And?

Sandy: It seems that matrices are a good way to organize and display information in a concise format. Matrix operations are then used to transform the data in some way. As a result, matrices appear in many areas of mathematics.

Pete: So I'll probably run into them again?

Sandy: Most likely.

1. In the Glossary, define the words in this section that appear in ***italic boldface*** type.

2. Use your calculator to create some examples to help you determine how to add two matrices. Once you think you understand how to add matrices, write a statement to a friend explaining how to add two matrices. In your statement include any restrictions on the matrices in order for the sum to be defined.

3. A matrix can be multiplied by a real number. This is called ***scalar multiplication***. Use your calculator to create some matrices and multiply them by a real number. Once you understand what is happening, write a statement explaining to a friend how to multiply a matrix by a real number. Include in your statement any restrictions on the matrices in order for the scalar product to be defined.

4. Create a square matrix that has an inverse and a square matrix that does not have an inverse. Compute the determinant of each. How can the determinant help you determine if the matrix inverse exists?

5. Consider matrix $E = \begin{bmatrix} 1 & 3 \\ -5 & 7 \end{bmatrix}$ and matrix $F = \begin{bmatrix} 0 & -5 \\ 3 & 7 \end{bmatrix}$. Is $EF = FE$? Why or why not? What important property does this example illustrate?

6. Write down a short message. Select an encryption key and encode it. Give the coded message and the encryption key to a friend and ask your friend to decode it.

7. The message 107, 133, –14, 89, 133, 37, 88, 146, –75, 50, 52, 9, 59, 157, –90 was encoded using the following encryption key.

$$\begin{bmatrix} 5 & 2 & -1 \\ 3 & 8 & 1 \\ 0 & 5 & -6 \end{bmatrix}$$

Decode the message.

8. We have used encryption keys that have been size 3×3. Could we use a 4×4 encryption key? Explain how that would alter how we would encode and decode.

9. $\begin{bmatrix} 2 & 5 & 7 \\ 4 & -3 & 7 \\ 8 & 1 & -4 \end{bmatrix} \cdot \begin{bmatrix} x \\ y \\ z \end{bmatrix} = \begin{bmatrix} -15 \\ 67 \\ 43 \end{bmatrix}$ is a matrix representation of a system of equations. Write the system of equations. (*Hint*: Perform the matrix multiplication.)

10. Matrix multiplication can be used to scale and rotate drawings.

a. Plot the points P(–4, 1), E(–1, 6), N(4, 7), T(6, 2), and A(3, –2). Connect successive points with straight line segments. Finally, connect A to P. This creates a five-sided polygon with the given points as vertices.

b. Write the matrix for this polygon. This is done by creating a matrix with two rows. The first row contains the x values. The second row contains the corresponding y values.

c. Multiply $\begin{bmatrix} 3 & 0 \\ 0 & 3 \end{bmatrix}$ by the polygon matrix. Sketch the new polygon. How is it different from the original polygon?

d. Multiply $\begin{bmatrix} 2 & 0 \\ 0 & 5 \end{bmatrix}$ by the polygon matrix. Sketch the new polygon. How is it different from the original polygon?

e. Multiply $\begin{bmatrix} -1 & 0 \\ 0 & 1 \end{bmatrix}$ by the polygon matrix. Sketch the new polygon. How is it different from the original polygon?

f. Multiply $\begin{bmatrix} 1 & 0 \\ 0 & -1 \end{bmatrix}$ by the polygon matrix. Sketch the new polygon. How is it different from the original polygon?

g. Multiply $\begin{bmatrix} 0 & 1 \\ 1 & 0 \end{bmatrix}$ by the polygon matrix. Sketch the new polygon. How is it different from the original polygon?

11. Solve the following systems.

a. $3x - y = 17$
$x = 4 - 3y$

b. $3x + 5y = 17$
$4x - 2y = -19$

c. $3x = 2y - 1$
$9x - 6y = 2$

d. $x + 2y - z = 7$
$3x - 5y + z = 2$

12. Write the coefficient matrix and the constant matrix for each system in Exploration 11.

13. Given the matrix $Fib = \begin{bmatrix} 1 & 1 \\ 1 & 0 \end{bmatrix}$:

a. Complete the table for the first eight powers of *Fib*.

TABLE 3 Powers of *Fib*

n	Fib^n
1	
2	
3	
4	
5	
6	
7	
8	

b. Discuss the pattern in the numbers in each power of *Fib*.

c. The numbers in row 1, column 1, of the powers of *Fib* form a very important sequence we learned earlier in the text. Identify the sequence.

14. Given the following graphs of two functions, estimate the solutions to the system.

a.

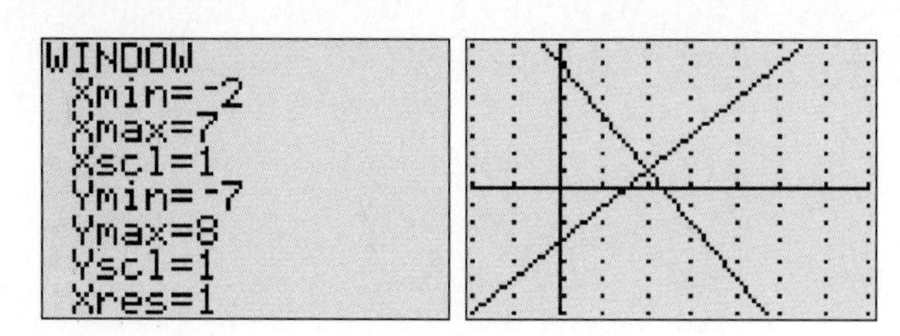

b.

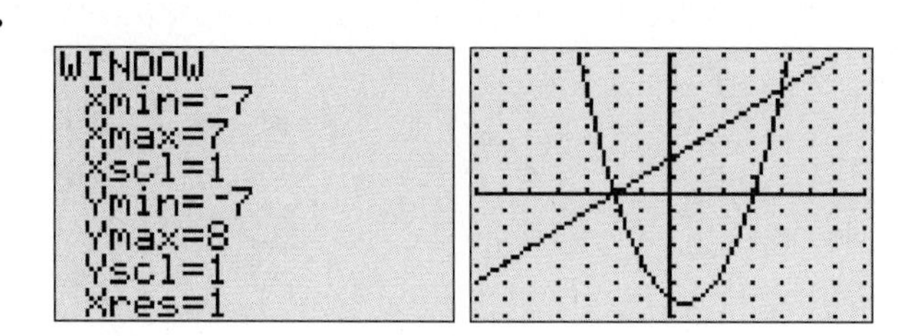

15. Perform the following matrix operations, if possible. If not possible, explain why.

a. $\begin{bmatrix} 3 \\ 7 \\ -2 \end{bmatrix} + \begin{bmatrix} 4 \\ -8 \\ 0 \end{bmatrix}$

b. $\begin{bmatrix} 3 & -4 \\ 0 & 7 \end{bmatrix} - \begin{bmatrix} 4 & -1 \\ 0 & 6 \end{bmatrix}$

c. $\begin{bmatrix} 3 & 5 \end{bmatrix} + \begin{bmatrix} 7 \\ -1 \end{bmatrix}$

d. $\begin{bmatrix} 3 & -5 \\ 6 & 2 \end{bmatrix} \begin{bmatrix} 4 & 7 & 0 \\ -8 & -6 & 4 \end{bmatrix}$

e. $\begin{bmatrix} 4 & 7 & 0 \\ -8 & -6 & 4 \end{bmatrix} \begin{bmatrix} 3 & -5 \\ 6 & 2 \end{bmatrix}$

f. $\begin{bmatrix} 3 & 0 & -4 \end{bmatrix} \begin{bmatrix} 7 & 9 \\ -2 & 0 \\ 5 & -6 \end{bmatrix}$

g. $det\left(\begin{bmatrix} 3 & 8 \\ -2 & 4 \end{bmatrix}\right)$

h. $\begin{bmatrix} 3 & 8 \\ -2 & 4 \end{bmatrix}^{-1}$

i. $det\left(\begin{bmatrix} 4 & -5 & 6 \\ 7 & -2 & 0 \\ 7 & 9 & 2 \end{bmatrix}\right)$

j. $\begin{bmatrix} 4 & -5 & 6 \\ 7 & -2 & 0 \\ 7 & 9 & 2 \end{bmatrix}^{-1}$

REFLECTION

Write a letter discussing all you know about the algebra of matrices. Address the letter to a friend who is trying to learn this material.

SECTION 5.5

Matrices and Linear Systems: An Auspicious Partnership

Purpose

- Use matrix equations to solve systems.
- Use matrices to create polynomial models from data.

In the preceding sections, you have learned how to solve systems using algebraic manipulation, tables, and graphs. In addition, you have learned a bit about the algebra of matrices. This section shows how matrices can be used to solve linear systems for which there is one unique solution. To begin let's return to a problem that was represented using matrices in Section 5.3.

Athletic Shoes: The athletic-shoe manufacturer NikBok manufactures three types of athletic shoes: walking, tennis, and aerobic. The manufacture of each pair of walking shoes costs \$8 in materials, \$10 in labor, and \$3 in overhead. The manufacture of each pair of tennis shoes costs \$6 in materials, \$10 in labor, and \$4 in overhead. Finally, each pair of aerobic shoes costs \$7 in materials, \$9 in labor, and \$5 in overhead to produce. The company budgets \$5200 for materials, \$7000 for labor, and \$3000 for overhead each day. You, as production supervisor, must determine the approximate number of each type of shoe that can be manufactured in a day.

Let's see how matrices can help solve the problem.

1. Create a table displaying the information stated in the problem. Use the types of shoes as column headings and the different relationships as row headings.

2. Use the variable w to represent the number of pairs of walking shoes manufactured in a day, the variable t to represent the number of pairs of tennis shoes manufactured in a day, and the variable a to represent the number of pairs of aerobic shoes manufactured in a day.

a. Write an equation that represents the total cost of materials on a given day.

b. Write an equation that represents the total cost of labor on a given day.

c. Write an equation that represents the total cost of overhead on a given day.

3. Using the equations from Investigation 2, write a system of equations that represents the **Athletic Shoes** problem.

4. a. Write the coefficient matrix for this system.

b. Write the constant matrix for the system.

c. A variable matrix has one column and as many rows as there are variables in a given problem. The variables are listed down the column in the matrix. Write the variable matrix for this problem.

d. Write the matrix representation of the system. (*Hint*: See Exploration 8 in Section 5.4.)

Table 1 displays the information given in the **Athletic Shoe** problem.

Each day the company budgets \$5200 for materials, \$7000 for labor, and \$3000 for overhead. This information reading across the rows expresses the three critical relationships in this problem.

TABLE 1 Cost of Manufacturing Athletic Shoes

	Walking Shoes	Tennis Shoes	Aerobic Shoes
Material cost per pair of shoes ($)	8	6	7
Labor cost per pair of shoes ($)	10	10	9
Overhead cost per pair of shoes ($)	3	4	5

The system of equations that represents the problem is

$$\begin{aligned} 8w + 6t + 7a &= 5200 \\ 10w + 10t + 9a &= 7000 \\ 3w + 4t + 5a &= 3000 \end{aligned}$$

where the variable w represents the number of pairs of walking shoes manufactured in a day, the variable t represents the number of pairs of tennis shoes manufactured in a day, and the variable a represents the number of pairs of aerobic shoes manufactured in a day.

The 3×3 coefficient matrix $[A]$, the 3×1 constant matrix $[C]$, and the 3×1 variable matrix $[X]$ all appear below. (Note that we have adopted the use of brackets around the variable which names the matrix.)

$$[A] = \begin{bmatrix} 8 & 6 & 7 \\ 10 & 10 & 9 \\ 3 & 4 & 5 \end{bmatrix} \qquad [C] = \begin{bmatrix} 5200 \\ 7000 \\ 3000 \end{bmatrix} \qquad [X] = \begin{bmatrix} w \\ t \\ a \end{bmatrix}$$

The matrix equation is $[A] \cdot [X] = [C]$.

Let's investigate techniques for solving such a matrix equation. To do so, we first look at how we solve linear equations.

5. Solve the equation $2x = 6$ for x. Be sure to write each step in the process.

6. If you used a step involving division, what would be an equivalent step involving multiplication?

7. Use similar steps to solve the matrix equation $[A] \cdot [X] = [C]$ for matrix $[X]$.

8. What condition must be placed on matrix $[A]$ in order to solve for matrix $[X]$?

The solution to the matrix equation $[A] \cdot [X] = [C]$ is the solution to the system of equations. Here's a little background on how to solve a matrix equation.

To solve the equation $2x = 6$, we must isolate x. One way of accomplishing this is by multiplying both sides of the equation by the multiplicative inverse of the coefficient of x.

The coefficient of x is 2. The multiplicative inverse of 2 is 2^{-1} which can be written as the fraction $\frac{1}{2}$.

Multiplying both sides by 2^{-1}, we get

$$2^{-1}(2x) = 2^{-1}(6).$$

Using the associative property of multiplication we get

$$(2^{-1}2)x = \frac{1}{2}(6).$$

The product of a number and its inverse is 1 and $1x = x$. So the result is

$$x = 3.$$

The algebra of matrices is similar. Think of the coefficient matrix $[A]$ as the 2 in the example, the variable matrix $[X]$ as the x in the example, and the constant matrix $[C]$ as the 6 in the example. We need to solve for matrix $[X]$ just as we had to solve for x in the example.

To isolate $[X]$, multiply both sides by the inverse of $[A]$, denoted by $[A]^{-1}$. This can be done only if the inverse of the coefficient matrix exists. The determinant of the coefficient matrix must be nonzero.

If the inverse of $[A]$ exists, we get

$$[A]^{-1}([A] \cdot [X]) = [A]^{-1}[C].$$

Using the associative property of multiplication of matrices we get

$$([A]^{-1}[A])[X] = [A]^{-1}[C].$$

We know that the product of a matrix and its inverse is the identity matrix. Let's call this matrix $[I]$. Simplifying the last equation yields

$$[I] \cdot [X] = [A]^{-1}[C].$$

But the identity matrix times any matrix yields the original matrix resulting in

$$[X] = [A]^{-1}[C].$$

Thus, to solve a linear system using matrices, we multiply the inverse of the coefficient matrix times the constant matrix. The resulting matrix is a one-column matrix listing the values for the variable successively down the column. Remember that matrix multiplication is not commutative. The order we use for multiplication affects the product. The solution matrix in the above derivation is $[A]^{-1}[C]$, not $[C][A]^{-1}$.

Now we find the solution to the original problem.

9. Enter matrix $[A] = \begin{bmatrix} 8 & 6 & 7 \\ 10 & 10 & 9 \\ 3 & 4 & 5 \end{bmatrix}$ and the matrix $[C] = \begin{bmatrix} 5200 \\ 7000 \\ 3000 \end{bmatrix}$ on your calculator.

10. Find the determinant of matrix $[A]$. Based on the answer, will you be able to find a unique solution to the system? Why?

11. Calculate $[A]^{-1} \cdot [C]$. A 3×1 matrix should appear. This matrix represents the ordered triple that is the solution to the system. Record the answers for w, t, and a.

12. Write a complete sentence describing the answer in terms of the original athletic shoe problem.

The determinant of $[A]$ is 44. The fact that the determinant is not zero suggests that the inverse of matrix $[A]$ exists and that the system has one unique solution.

Calculating the solution matrix, we get

$$[A]^{-1} \cdot [C] \approx \begin{bmatrix} 245.45 \\ 168.18 \\ 318.18 \end{bmatrix}$$

These are the approximate values for the variables w, t, and a. Of course, the domain set for each variable is the whole numbers since we must manufacture whole pairs of shoes.

We conclude that NikBok can manufacture about 245 pairs of walking shoes, 168 pairs of tennis shoes, and 318 pairs of aerobic shoes in a given day.

Let's return to the problem of working two part-time jobs per week.

Multiple Part-Time Jobs: You decided not to take the job offers at CDs-R-Us, Super Dog Burger Bits, or Virtual Reality Experiences. They just offer too much security for you. Instead you have lined up four part-time jobs that pay a flat amount each week. You will work only two jobs per week. You plan to try every job combination (pair of jobs) possible over the subsequent weeks. Your total earnings in dollars (ordered from least to most) for each week are as follows: $289, $293, $298, $315, $320, and $324.

We used a, b, c, and d to represent the amount earned weekly on each job, ordered from least to most. Remember that you are paid the same flat amount each week you work a given job and that amount is a whole number.

13. If there are four jobs and six combinations of two jobs per week, what size system of equations is needed to represent the problem?

14. There are two possible systems of equations for this problem. Write each system.

15. What is the minimum number of equations you must have to solve this system using matrix techniques? Why? (*Hint*: Consider the number of variables.)

There must be four different jobs if we work two jobs per week for six weeks, using all possible combinations of two jobs.

Let the letters a, b, c, and d represent the weekly pay of the four jobs arranged from least to most.

We know the following equations must be true.

$$a + b = 289$$
$$a + c = 293$$
$$b + d = 320$$
$$c + d = 324$$

These equations must be true due to the ordering of the variables. There are two other equations, however. The question is: What are they?

The two possibilities are as follows.

$$\begin{matrix} a + d = 298 \\ b + c = 315 \end{matrix} \quad \text{or} \quad \begin{matrix} b + c = 298 \\ a + d = 315 \end{matrix}$$

There are two possible systems involving six equations and four variables.

$$\begin{matrix} a + b = 289 \\ a + c = 293 \\ a + d = 298 \\ b + c = 315 \\ b + d = 320 \\ c + d = 324 \end{matrix} \quad \text{or} \quad \begin{matrix} a + b = 289 \\ a + c = 293 \\ a + d = 315 \\ b + c = 298 \\ b + d = 320 \\ c + d = 324 \end{matrix}$$

Regardless of which we choose, we have four variables. So we need a minimum of four equations. Let's try some matrix equations involving the four unknowns with four selected equations. Keep in mind we expect whole number answers.

16. Select any four of the six equations.

a. Write a matrix equation involving the four equations and four unknowns.

b. Use the determinant to decide if the system you wrote has a solution. State your conclusion.

c. If your system has a solution, find the solution. Is it a correct solution to the problem? (*Hint*: Remember that the solutions must be whole numbers.)

d. If you do not have a correct solution yet, choose four other equations, write the corresponding system, and attempt to solve. Continue until you have a solution to the problem.

17. Write a conclusion describing the solution to the problem. Explain why you know your solution is correct.

Probably the most obvious system to select is the one in which we are sure of all four equations.

$$a + b = 289$$
$$a + c = 293$$
$$b + d = 320$$
$$c + d = 324$$

This results in the matrix equation.

$$\begin{bmatrix} 1 & 1 & 0 & 0 \\ 1 & 0 & 1 & 0 \\ 0 & 1 & 0 & 1 \\ 0 & 0 & 1 & 1 \end{bmatrix} \cdot \begin{bmatrix} a \\ b \\ c \\ d \end{bmatrix} = \begin{bmatrix} 289 \\ 293 \\ 320 \\ 324 \end{bmatrix}$$

Unfortunately, the determinant of the coefficient matrix is zero.

Suppose we replace the last equation with $b + c = 315$.
We get the system

$$a + b = 289$$
$$a + c = 293$$
$$b + d = 320$$
$$b + c = 315$$

The matrix equation is

$$\begin{bmatrix} 1 & 1 & 0 & 0 \\ 1 & 0 & 1 & 0 \\ 0 & 1 & 0 & 1 \\ 0 & 1 & 1 & 0 \end{bmatrix} \cdot \begin{bmatrix} a \\ b \\ c \\ d \end{bmatrix} = \begin{bmatrix} 289 \\ 293 \\ 320 \\ 315 \end{bmatrix}$$

The determinant of the coefficient matrix is 2. Thus we can solve this system.

$$\begin{bmatrix} 1 & 1 & 0 & 0 \\ 1 & 0 & 1 & 0 \\ 0 & 1 & 0 & 1 \\ 0 & 1 & 1 & 0 \end{bmatrix}^{-1} \cdot \begin{bmatrix} 289 \\ 293 \\ 320 \\ 315 \end{bmatrix} = \begin{bmatrix} 133.5 \\ 155.5 \\ 159.5 \\ 164.5 \end{bmatrix}$$

These answers are not whole numbers, so we don't have the correct solution yet.

Let's replace $b + c = 315$ with $b + c = 298$.

The system is

$$\begin{aligned} a + b &= 289 \\ a + c &= 293 \\ b + d &= 320 \\ b + c &= 298 \end{aligned}$$

The matrix equation is

$$\begin{bmatrix} 1 & 1 & 0 & 0 \\ 1 & 0 & 1 & 0 \\ 0 & 1 & 0 & 1 \\ 0 & 1 & 1 & 0 \end{bmatrix} \cdot \begin{bmatrix} a \\ b \\ c \\ d \end{bmatrix} = \begin{bmatrix} 289 \\ 293 \\ 320 \\ 298 \end{bmatrix}$$

The solution is

$$\begin{bmatrix} 1 & 1 & 0 & 0 \\ 1 & 0 & 1 & 0 \\ 0 & 1 & 0 & 1 \\ 0 & 1 & 1 & 0 \end{bmatrix}^{-1} \cdot \begin{bmatrix} 289 \\ 293 \\ 320 \\ 298 \end{bmatrix} = \begin{bmatrix} 142 \\ 147 \\ 151 \\ 173 \end{bmatrix}$$

The solutions are $a = 142$, $b = 147$, $c = 151$, $d = 173$. These can be checked by adding each possible pair and comparing to the six totals given in the problem.

We conclude the section by investigating how matrices can be used to create specific polynomial models.

That's it for now. Hope it was fun!

STUDENT DISCOURSE

Pete: Let me see if I understand this. I know how to solve the system $\begin{aligned} 3x - 5y &= 2 \\ 4x + 7y &= -1 \end{aligned}$ using elimination, but I could also find the solution using matrices?

Sandy: That's right. What would you do?

Pete: I'd write the coefficient matrix and the constant matrices first.

Sandy: What are they in this case?

Pete: Let's see. The coefficient matrix is $\begin{bmatrix} 3 & -5 \\ 4 & 7 \end{bmatrix}$ and the constant matrix is $\begin{bmatrix} 2 \\ -1 \end{bmatrix}$.

Sandy: Very good. Then what?

Pete: I'd find the inverse of the coefficient matrix.

Sandy: Does it have an inverse?

Pete: If the determinant is not 0. Let's see . . . the determinant is $3(7) - 4(-5) = 41$. Yes, there is an inverse.

Sandy: Then what?

Pete: Just multiply the inverse by the constant matrix. The numbers in the product matrix are my solutions for x and y.

Sandy: That's the way I understand it. One last question. What if your system look like $\begin{matrix} 4x = 2y - 8 \\ 3y - 2 = 5x \end{matrix}$? Could you use matrices to find the solution?

Pete: I think so. But I'd have to write both equations in standard form first, wouldn't I?

Sandy: Yes, they must be in $Ax + By = C$ form so that you can write the coefficient matrix and the constant matrix.

1. In the Glossary, define the words in this section that appear in ***italic boldface*** type.

2. A concert hall has 1345 total seats divided into three types: main floor, mezzanine, and balcony. Main floor tickets cost \$45, mezzanine tickets cost \$35, and balcony tickets cost \$20. Find the total number of tickets of each type sold if the concert hall is full and the total ticket receipts are \$38,230.

3. Find the solution to each systems using matrices. If there is no solution, state why.

a. $3.7x - 5.2y = -1.6$
$4.9 + 7.2x = 5.7$

b. $15x - 10y = -5$
$-12x + 8y = 4$

c. $x - y + 2z = -5$
$4x + 3y - 7z = 1$
$3x - 8y + 5z = 11$

d. $4x - 3y + z - 8w = 4$
$x + y - 3z + w = -2$
$5x + 11y + 5z + 7w = 13$
$-4x + 3y - 6z - 3w = -5$

e. $y = 2x - 4z + 7$
$x = z - 5y + 2$
$z = 3x + 2y - 8$

f. $2x - 3y + 11z = 4$
$x + y - 5z = -1$

4. You have probably seen a *magic square* before. For example, the task might be to use the digits 1–9, without repetition, to complete Figure 1 below so that the sum of all rows, columns, and diagonals is 15.

FIGURE 1 **Magic Square**

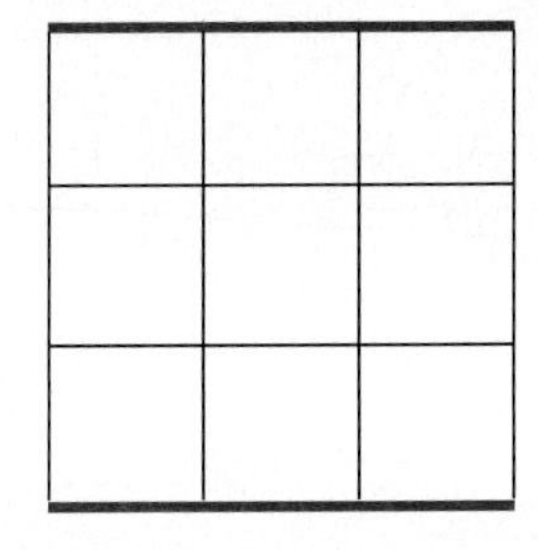

a. Fill in the magic square using any technique you wish.

b. This problem can be expressed as a system of equations. Do so.

c. Can you solve the system uniquely as written? Why or why not?

d. Write the digit 5 in the middle square. Write a system of equations for this situation.

e. Can you solve the system in part **d** uniquely? Why or why not?

f. Solve the problem using matrices on an appropriate system.

5. Handshakes: There is a group of people in a room. Each person shakes hands once with everyone else.

a. Complete Table 2 to look at the relationship between the number of people and the number of handshakes.

TABLE 2 **Handshakes**

Number of People	Number of Handshakes
1	
2	
3	

b. These data are modeled by a polynomial function. What is the degree of the function?

c. Use any technique you wish to find the specific polynomial function that models the data.

6. Let's review some system-solving techniques. Use either substitution or elimination to solve the following systems.

a. $3a - 5b = 17$

$11 - 7b = 4a$

b. $3r - 2t = -41$

$7t = 5 + r$

7. A quartic function contains the points (2, 1), (7, –3), (5, 2), (–3, –7), and (–1, 3). Find the quartic function using a system of equations. Check your answer using quartic regression on your calculator.

8. World Cigarette Production per Person: The data in Table 3 display the world cigarette production per person by year.

TABLE 3 World Cigarette Production per Person

Year	Cigarettes per Person
1983	968
1984	982
1985	1000
1986	1009
1987	1020
1988	1029
1989	1017
1990	1027
1991	999
1992	984
1993	982

Source: Lester R. Brown, Hal Kane, David Malin Roodman, *Vital Signs, 1994*, (New York: W. W. Norton, 1994), p. 101.

a. Construct a scatter plot for the data.

b. Find the best-fit regression model for the data.

c. Use the model to predict the production this year. Do you think this is an accurate prediction? Why or why not?

9. **A Buddy's Job:** Your buddy accepts a job that pays $27,000 initially with a raise of $2500 each year.

a. Create a table displaying the relationship between years worked and annual salary.

b. Analyze this table to determine if the data are linear, quadratic, cubic, quartic, or exponential. Justify your answer.

c. Write a specific function that models the situation.

d. Determine how long it will take for the annual salary to exceed $100,000. Explain what you did.

10. Refer to the preceding exploration. Let the input be the year and let the output be the total salary earned since beginning the job.

a. Create a table displaying the relationship between years worked and the total salary earned through that year.

b. Analyze this table to determine if the data are linear, quadratic, cubic, quartic, or exponential. Justify your answer.

c. Write a specific function that models the situation.

d. Determine how long it will take for the total salary to exceed $200,000. Explain what you did.

11. There is a nice relationship between the degree of the polynomial model and the minimum number of data points required to create the model.

a. Create a table. The left column corresponds to the degree of a polynomial function. The right column corresponds to the minimum number of data points required to create the model.

b. If you wish to create a specific polynomial model, what is the minimum number of data points you must know? Justify.

Figure 2 consists of circles, a number of points on the circles, and the maximum number of line segments (***chords***) joining these points. The line segments divide the circles into a certain number of regions. For example, the first circle is divided into two regions. Two points were chosen on this circle and one chord was formed. The second circle is divided into four regions. Three points were chosen on this circle and three chords were formed.

FIGURE 2

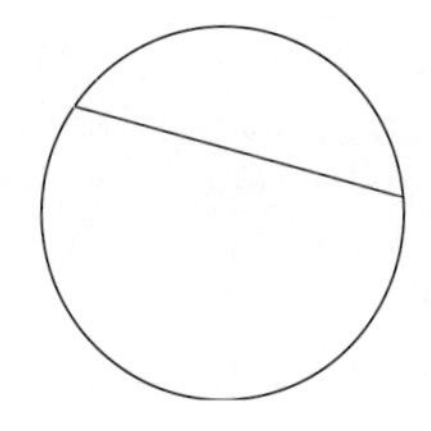
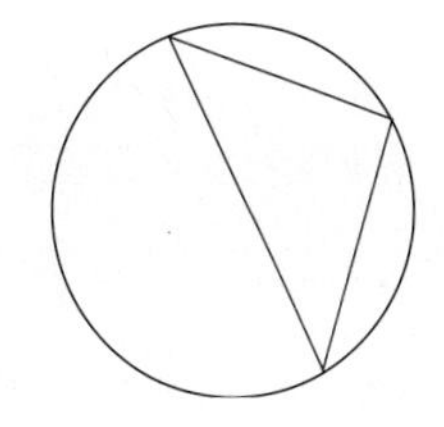
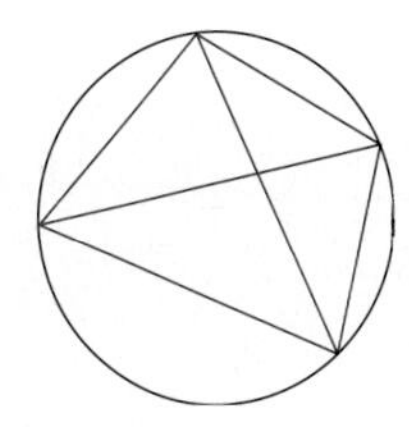

12. Let's determine the relationship between the number of points chosen and the number of chords formed.

a. Complete Table 4.

TABLE 4 **Points and Chords**

Number of Points	Number of Chords
1	
2	
3	
4	
5	
6	

b. Use finite differences to determine the general model for the data.

c. Use the minimum number of points to write a specific function that models the situation.

d. Use your model to determine how many chords there will be if 50 points are selected.

13. Let's determine the relationship between the number of points chosen and the number of regions formed.

a. Complete Table 5.

TABLE 5 Points and Regions

Number of Points	Number of Regions
1	
2	
3	
4	
5	
6	

b. Use finite differences to determine the general model for the data.

c. Use the minimum number of points to write a specific function that models the situation.

d. Use your model to determine how many regions there will be if 50 points are selected.

REFLECTION

Discuss the various techniques you have learned to solve systems of equations. Include advantages and limitations of each technique in your discussion.

SECTION 5.6

Considering Several Models Simultaneously: Review

Purpose

- Reflect on the key concepts of Chapter 5.

In this section you will reflect upon the mathematics in Chapter 5—what you've done and how you've done it.

1. State the most important ideas in this chapter. Why did you select each?

2. Identify all the mathematical concepts, processes, and skills you used to investigate the problems in Chapter 5?

There are a number of important ideas that you might have listed, including systems of equations, substitution method, elimination method, matrices, coefficient matrix, constant matrix, determinant, matrix inverse, matrix identity, operations on matrices, and matrix equations.

1. Identify the slope and y-intercept of the following equations by inspection.

a. $7x - 9y = 2$

b. $x - 2y = -17$

c. $-5x + 2y = 19$

d. $9x + 6y = -1$

2. Use inspection to determine if the following systems have no solution, exactly one solution, or an infinite number of solutions.

a. $7x + 2y = 0$
$7x + 2y = 2$

b. $9x - 5y = -2$
$5x - 9y = 2$

c. $y = \frac{2}{3}x - 1$
$y = \frac{2}{3}x + 15$

d. $x = 2y - 1$
$-x + 2y = 1$

3. Solve the following systems algebraically. Check results using both a table and a graph. If there is no solution, explain why. If there is a solution, defend your choice of solution technique.

a. $4x - y = 16$
$3y + 5x = -9$

b. $11x + 3y = -1$
$x + 7y = -3$

c. $y = 4x - 7$
$y = 5x + 11$

d. $y = 11 - 9x$
$x = 4y + 7$

e. $4x - 3y = 0$
$5x = 2y - 1$

f. $4x + 3y = 2$
$4x + 3y = 11$

g. $5x - 6y = 1$
$5x + 6y = 1$

h. $7x + 11y = -3$
$11x - 7y = 4$

i. $2a + 3b = 4$
$3a + 2b = -4$

j. $p = q + 4$
$3p - 3q = 12$

k. $5t + 6s = -1$
$3t - 4s = 8$

l. $9p = 2t$
$3t - 4p = 11$

4. Clothing Sale: In honor of its 35 years in business, a store marks all of its clothing down 35%. The savings on a pair of shoes is $23.80. We wish to find both the original price and the sale price for the pair of shoes.

a. Write a system of equations that represents this problem situation.

b. Solve the system. Write a sentence or two describing your answers in terms of the problem situation.

5. Sue's Fencing Company: Sue will use 238 feet of fencing to enclose a rectangular yard. The length of the yard is 10 feet more than the width. We wish to find the length and width of the yard.

a. Write a system of equations that represents this problem situation.

b. Solve the system. Write a sentence or two describing your answers in terms of the problem situation.

6. Cigarette Smoking: Table 1 displays the percent of males and percent of females in the United States who indicated they were smokers by year.

TABLE 1 Cigarette Smoking

Year	Male Smokers (%)	Female Smokers (%)
1965	51.9	33.9
1974	43.1	32.1
1979	37.5	29.9
1983	35.1	29.5
1985	32.6	27.9
1987	31.2	26.5
1990	28.4	22.8
1992	28.6	24.6
1993	27.7	22.5

Source: U.S. Bureau of the Census, *Statistical Abstract of the United States: 1996*, 116th ed. (Washington, D.C., 1996), p. 145.

a. Create a regression line for the percent of male smokers by year.

b. Create a regression line for the percent of female smokers by year.

c. Use the regression lines to estimate the year when the percent of male and percent of female smokers will be approximately the same. Briefly describe and defend your answer.

7. Use graphical or numeric techniques to estimate the solutions to the following systems.

a. $y = 4x + 1$

$y = -2x^2 + x + 5$

b. $y = 3x^2 - x - 4$

$y = -x^2 = 2x + 7$

c. $y = 2x - 3$

$y = x^3 + x - 2x + 4$

d. $y = 3x^2 + 2x - 1$

$y = -3x^2 + 2x - 1$

8. Write the coefficient matrix, the constant matrix, and the matrix equation for each of the following linear systems.

a. $2x = 3y + 9$

$4x - y = -11$

b. $3x - 2y - 7z = 12$

$4x + 7z = 15$

c. $y = 3x + 2$

$x = 9x - 5$

d. $x + y = 0$

$3x - 2y = 9$

$3x + 4y = -8$

9. Given the following graphs of two functions, estimate the solutions to the system.

a.

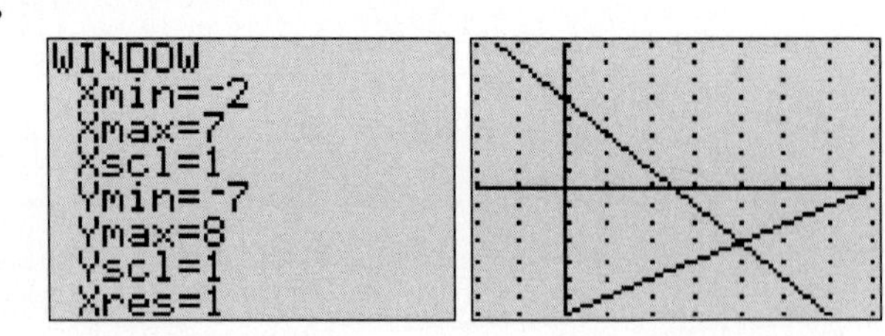

b.

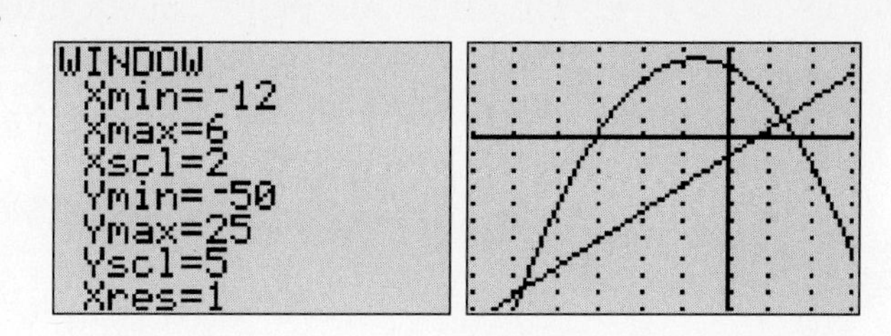

10. Perform the following matrix operations, if possible. If not possible, explain why.

a. $\begin{bmatrix} 3 & -5 & 2 \\ 0 & 7 & -8 \end{bmatrix} + \begin{bmatrix} -8 & 4 & 0 \\ -3 & 9 & -2 \end{bmatrix}$

b. $\begin{bmatrix} 5 & -1 \\ 9 & 0 \\ -8 & -7 \end{bmatrix} - \begin{bmatrix} -8 & 2 \\ 0 & 5 \\ -3 & 7 \end{bmatrix}$

c. $-5\begin{bmatrix} 6 \\ 2 \\ -7 \end{bmatrix}$

d. $4\begin{bmatrix} 3 & 0 & -6 \\ 5 & -1 & 2 \end{bmatrix}$

e. $\begin{bmatrix} 6 & -5 \\ 8 & 0 \end{bmatrix} + \begin{bmatrix} 8 \\ -1 \end{bmatrix}$

f. $\begin{bmatrix} 4 \\ -5 \\ 7 \end{bmatrix}\begin{bmatrix} -8 & 7 \end{bmatrix}$

g. $\begin{bmatrix} 4 & -5 & 6 \\ 3 & -9 & 0 \end{bmatrix}\begin{bmatrix} 7 & -9 & 2 \\ 6 & 5 & -1 \end{bmatrix}$

h. $\begin{bmatrix} 5 & 7 \\ 0 & -8 \\ -1 & -3 \end{bmatrix}\begin{bmatrix} 4 & -1 & 6 \\ -5 & 7 & 2 \end{bmatrix}$

i. $\begin{bmatrix} 4 & -1 & 6 \\ -5 & 7 & 2 \end{bmatrix}\begin{bmatrix} 5 & 7 \\ 0 & -8 \\ -1 & -3 \end{bmatrix}$

j. $det\left(\begin{bmatrix} 4 & -7 \\ 2 & 8 \end{bmatrix}\right)$

k. $\begin{bmatrix} 4 & -7 \\ 2 & 8 \end{bmatrix}^{-1}$

l. $det\left(\begin{bmatrix} -5 & 2 & 8 \\ -7 & 8 & 0 \\ 6 & -3 & 2 \end{bmatrix}\right)$

m. $\begin{bmatrix} -5 & 2 & 8 \\ -7 & 8 & 0 \\ 6 & -3 & 2 \end{bmatrix}^{-1}$

n. $\begin{bmatrix} 4 & -8 & 2 \\ 9 & -1 & 0 \\ 2 & -4 & 5 \end{bmatrix}^{-1}\begin{bmatrix} 4 \\ -6 \\ 2 \end{bmatrix}$

11. Use matrices to find the solution to each of the following systems.

a. $4.9x - 7.6y = -2.7$
$-2.7x + 5.9y = 1.1$

b. $3x - 4y + 2z = 17$
$5x + y - z = -2$
$-x + 3y + 7z = -6$

c. $3x + 2y + z - 4w = 15$
$5x + y + 3z - 5w = -2$
$x - 3y - 7z - w = 12$
$4x - 7y + 2z + 3w = -9$

d. $x = 2y - z + 5$
$y + 3x = 4 - 2z$
$z = x + y$

CONCEPT MAP

Construct a Concept Map centered on one of the following:

a. system of equations

b. matrix

c. solving systems of equations

REFLECTION

Select one of the important ideas you listed in Investigation 1. Write a paragraph to Izzy explaining your understanding of this idea. How has your thinking changed as a result of studying this idea?

6 Manipulating Quadratic Models

SECTION 6.1

Key Features: Factors, Zeros, and the Vertex

Purpose

- Investigate the relationship between the zeros of a quadratic function, the factors of a quadratic function, and the x-intercepts of the graph of a quadratic function.
- Solve quadratic equations and inequalities by finding zeros.

We investigated quadratic models in some detail in Chapters 3 and 4. In particular, given data in which the second finite differences in successive outputs are constant, but the first differences vary, the data are modeled by a quadratic function.

The general form of a quadratic function is $y(x) = ax^2 + bx + c$, where a, b, and c are parameters.

In this chapter we focus solely on the various features of a quadratic function. In particular, we look at the effect of the parameters on the graph. We talk about different forms of quadratic functions and look at algebraic techniques for solving equations that involve quadratics.

In this section we will focus on how to use specific features of a quadratic function to express the function in factored form, if possible. In addition, we discuss a method of solving equations that involves quadratics.

Recall from Chapter 4 that we can use quadratic regression to write the equation of the quadratic function through three given points. We begin the chapter with just such an investigation.

1. Figure 1 contains the graph of a quadratic function.

FIGURE 1

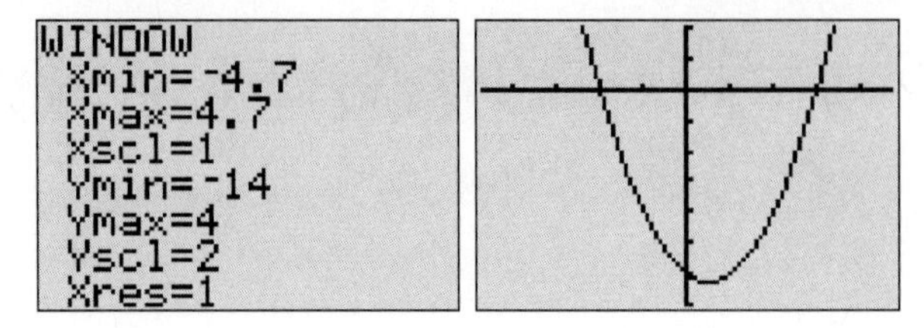

a. Write ordered pairs for three points from the graph. Which points did you choose and why?

b. Using the three points from part **a,** use quadratic regression to write the equation for the specific quadratic function. Check your answer by matching your graph with the one given in Figure 1.

2. The graph of any quadratic function is a parabola. How many times can a parabola intersect the:

a. x-axis? Why?

b. y-axis? Why?

In Figure 1, the easiest points to identify are the intercepts—that is, the points where the graph intersects the x-axis and the y-axis.

An ***x*-intercept** is a point where the graph of a function intersects the x-axis. Ordered pairs are used when writing x-intercepts. The second, or y components, of these ordered pairs must be 0. The graph in Figure 1 has x-intercepts $(-2, 0)$ and $(3, 0)$.

A ***y*-intercept** is a point where the graph of a function intersects the y-axis. Ordered pairs are used when writing y-intercepts. The first, or x components, of these ordered pairs must be 0. The graph in Figure 1 has y-intercept $(0, -12)$.

A parabola can intersect the x-axis 0, 1, or 2 times (Figure 2). A parabola can intersect the y-axis only once. A y-intercept has an input of 0. If the graph had more than one y-intercept, there would be more than one output for the input

FIGURE 2

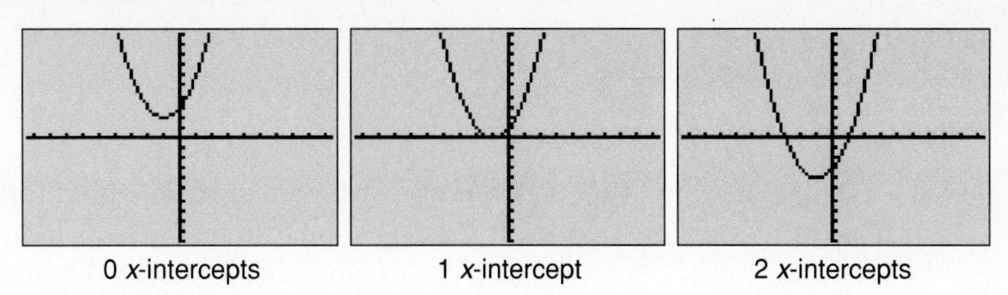

0. This violates the definition of function, which says that there is only one output for each input.

Given the intercepts of the graph in Figure 1, we use quadratic regression to find the equation for the function (Figure 3).

FIGURE 3

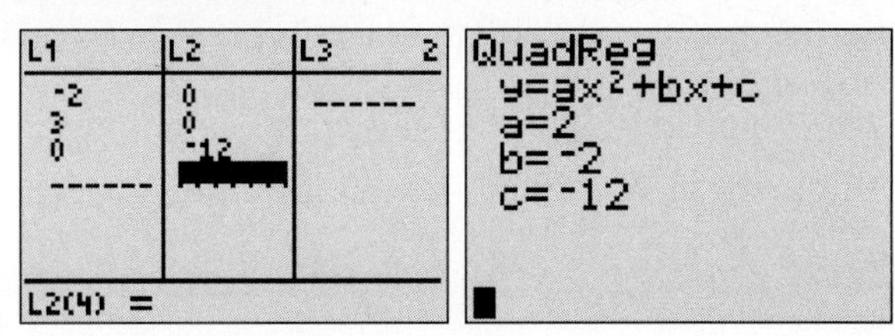

So the equation of the quadratic function for the graph in Figure 1 is $y = 2x^2 - 2x - 12$. We will investigate this function as written and in its factored form.

3. Consider the functions $y = 2x^2 - 2x - 12$ and $y = 2(x + 2)(x - 3)$.

a. Use a table to compare outputs of the two functions. What do you notice?

b. Using the table, write each input that has an output of 0. What is the relationships between your answers and the intercepts?

c. What are the factors of the function $y = 2(x + 2)(x - 3)$?

d. Set each factor of $y = 2(x + 2)(x - 3)$ that contains the variable x equal to zero and solve for x. Compare your answers to the answers in part **b.**

e. Graph $y = 2x^2 - 2x - 12$ and $y = 2(x + 2)(x - 3)$ in an appropriate window. What do you notice?

Figure 4 displays a table and graph for both functions. Notice that the outputs of the two functions are the same for each input and that the graphs coincide. This suggests the two functions are the same function. But why? One way to see this is to expand $y = 2(x + 2)(x - 3)$ by multiplying the factors.

FIGURE 4

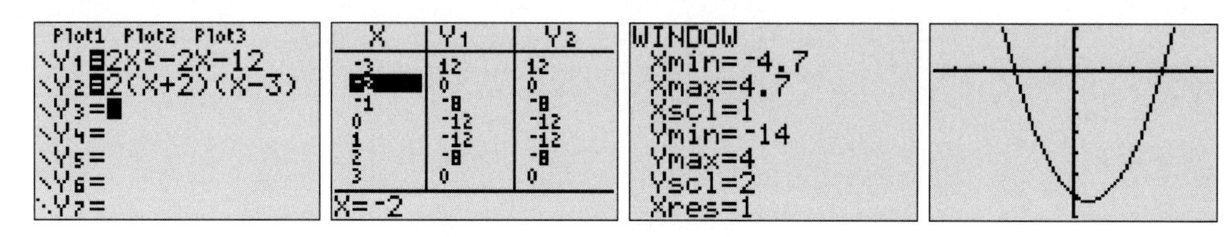

Multiply $(x + 2)$ and $(x - 3)$ and then distribute the 2:

$$\begin{aligned} 2(x+2)(x-3) &= 2(x^2 - 3x + 2x - 6) \\ &= 2(x^2 - x - 6) \\ &= 2x^2 - 2x - 12. \end{aligned}$$

This shows that the two functions are the same algebraically. They are just in different forms.

Expanded form:	$y = 2x^2 - 2x - 12$
Factored form:	$y = 2(x + 2)(x - 3)$

A function is in ***factored form*** if the last operation when evaluating the function is multiplication.

In factored form, the quadratic function $y = 2x^2 - 2x - 12$ is expressed as a product of three factors: a constant 2 and two linear expressions $(x + 2)$ and $(x - 3)$.

The output of $y = 2x^2 - 2x - 12$ is zero when the input is either –2 or 3. Recall that the x-intercepts are $(-2, 0)$ and $(3, 0)$.

Any input at an x-intercept is called a ***zero of the function***. So the zeros of $y = 2x^2 - 2x - 12$ are –2 and 3. A zero of a function is an input that produces 0 as output. Notice in Figure 5 the connection between the zeros in the table, the x-intercepts on the graph, the ordered pairs that represent the x-intercepts, the zeros of the function, and the factors of the function (the viewing window matches the one used in Figure 4).

FIGURE 5

Zeros in the table		
x-intercepts on graph		
Ordered pairs for x-intercepts on graph	$(-2, 0)$	$(3, 0)$
Stating the zeros	A zero of the function is -2.	A zero of the function is 3.
Stating the factors	A factor of the function is $(x + 2)$.	A factor of the function is $(x - 3)$.

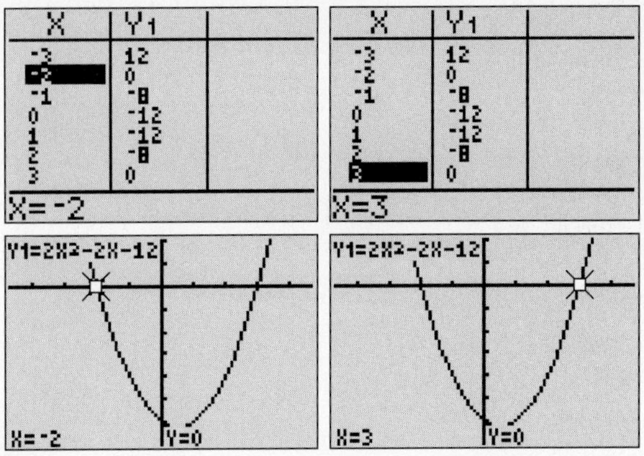

The relationship between zeros and factors depends on an important property.

Zero Product Property: If $ab = 0$, then $a = 0$ or $b = 0$. In words, if the product of two factors is zero, then one factor is zero or the other factor is zero.

We know $y = 2x^2 - 2x - 12$ can be written in factored form as $y = 2(x + 2)(x - 3)$.

If we set the linear factors $(x + 2)$ and $(x - 3)$ equal to 0, using the Zero Product Property, and solve for x, we get:

$$x + 2 = 0 \qquad x - 3 = 0$$

$$x = -2 \qquad x = 3$$

These are the zeros of the quadratic function. Notice that we can create the zeros if we know the factors of a quadratic function. To find the factors when we know the zeros, we reverse this process—that is, if we know that c is a zero, we write

$$x = c$$

Subtract c from both sides to make one side 0.

$$x - c = 0$$

Then $(x - c)$ is a factor of the quadratic.

Given the expanded form $y = 2x^2 - 2x - 12$, notice that 2 is a factor of every term. Using the distributive property of multiplication over addition/subtraction, we can factor out the 2 and get $y = 2(x^2 - x - 6)$.

This is an example of a technique called ***factoring out the common monomial factor.***

A ***monomial*** is a constant, a variable, or a product of constants and variables. For instance, 2 is a monomial since it is a constant. We still need to know how to write $x^2 - x - 6$ in factored form.

Let's spend some time investigating the relationship among the x-intercepts, the zeros, and the factors of a quadratic function. We begin by looking at quadratics $y = ax^2 + bx + c$ where $a = 1$.

One final note: We will only consider factoring over the rational numbers. That means that only rational numbers will appear in the factors and as zeros. If a polynomial does not have a rational number factorization, we will say the polynomial is ***prime*** over the rational numbers.

4. Complete Table 1. Use a graph to identify the x-intercepts. Use the x-intercepts or a table to identify the zeros. Use the zeros to predict the factorization. Use a symbol manipulator or table to check your factorization. Modify your factorization if necessary.

5. Refer to Table 1.

a. What is the connection between the signs of the zeros and the signs on b and c in the quadratic $y = x^2 + bx + c$?

b. How could you use the zeros to determine the value of parameter c in $y = x^2 + bx + c$?

c. How could you use the zeros to determine the value of parameter b in $y = x^2 + bx + c$?

6. Consider the quadratic trinomial $x^2 + 9x + 14$.

a. How can you use the parameter $c = 14$ to identify the possible zeros?

TABLE 1 **Quadratic Trinomials 1**

Quadratic	x-intercepts	Zeros	Product of Factors
$x^2 + 9x + 8$			
$x^2 + 10x + 21$			
$x^2 + 8x + 15$			
$x^2 - 9x + 8$			
$x^2 - 10x + 21$			
$x^2 - 8x + 15$			
$x^2 + 2x - 15$			
$x^2 + 9x - 22$			
$x^2 - 6x - 7$			
$x^2 - 3x - 28$			
$x^2 + 6x + 9$			

b. How can you use the parameter $b = 9$ to identify the possible zeros?

c. What are the zeros of the trinomial? Verify these both in a table and on a graph.

d. Use the zeros to factor the trinomial. Check your answer by multiplying the two factors together.

7. Consider the quadratic trinomial $x^2 - 9x + 14$.

a. How can you use the parameter $c = 14$ to identify the possible zeros?

b. How can you use the parameter $b = -9$ to identify the possible zeros?

c. What are the zeros of the trinomial? Verify these both in a table and on a graph.

d. Use the zeros to factor the trinomial. Check your answer by multiplying the two factors together.

8. A quadratic trinomial has zeros –5 and –2 and parameter $a = 1$.

a. Write the trinomial in factored form.

b. Multiply the factors to write the trinomial in the form $ax^2 + bx + c$.

9. Consider the quadratic trinomial $x^2 + 2x - 24$.

a. How can you use the parameter $c = -24$ to identify the possible zeros?

b. How can you use the parameter $b = 2$ to identify the possible zeros?

c. What are the zeros of the trinomial? Verify these both in a table and on a graph.

d. Use the zeros to factor the trinomial. Check your answer by multiplying the two factors together.

10. Consider the quadratic trinomial $x^2 + 2x - 4$.

a. What are the x-intercepts?

b. Can this polynomial be factored within the rational numbers? Why or why not?

11. Consider the quadratic trinomial $x^2 + 2x + 15$.

a. What are the x-intercepts?

b. Can this polynomial be factored within the rational numbers? Why or why not?

We wish to factor a quadratic of the form $x^2 + bx + c$. Here are some generalizations we can draw from Table 1.

- The zeros both have the opposite sign of parameter b if the parameter c is positive. The zeros have different signs if the parameter c is negative. For example, the zeros of $x^2 - 9x + 8$ are 1 and 8. They have the same sign since $c = 8$ is positive. They are positive since $b = -9$ is negative. On the other hand, the zeros of $x^2 + 2x - 15$ are –5 and 3. They have different signs since $c = -15$ is negative.

- The product of the two zeros is the constant term. For example, the product of the zeros of $x^2 + 2x - 15$ is $(-5)(3) = -15$, which is the value of parameter c.

- The sum of the two zeros is the opposite of the parameter b. For example, the sum of the zeros of $x^2 + 2x - 15$ is $-5 + 3 = -2$, which is the opposite of the parameter b.

Let's look at some examples.

Consider $x^2 + 9x + 14$. The parameters are $b = 9$ and $c = 14$. Both zeros must be negative. The product of the zeros is 14 and the sum is the opposite of 9.

A table and a graph of $y(x) = x^2 + 9x + 14$ appear in Figure 6.

FIGURE 6

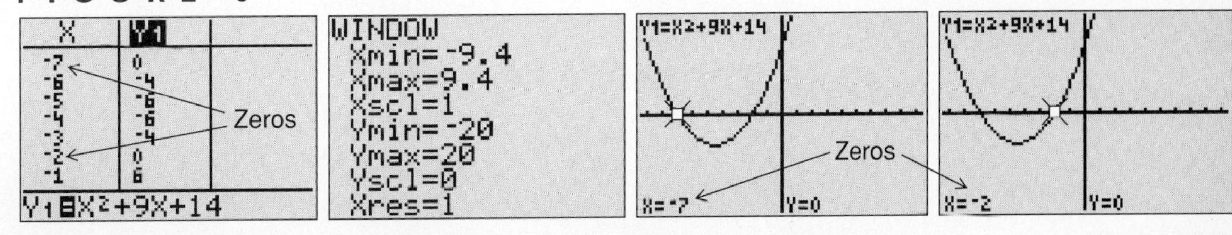

The x-intercepts are $(-2, 0)$ and $(-7, 0)$. The numbers -2 and -7, the x values of the x-intercepts, are the zeros. To create the factors, set x equal to each zero and rewrite so that one side of the equation is zero.

$$x = -2 \qquad x = -7$$

$$x + 2 = 0 \qquad x + 7 = 0$$

The factors are $(x + 2)$ and $(x + 7)$.

The quadratic trinomial $x^2 + 9x + 14$ factors as $(x + 2)(x + 7)$.

Consider $x^2 - 9x + 14$. The parameters are $b = -9$ and $c = 14$. Both zeros must be positive. The product of the zeros is 14 and their sum is 9.

A table and a graph of $y(x) = x^2 - 9x + 14$ appear in Figure 7.

FIGURE 7

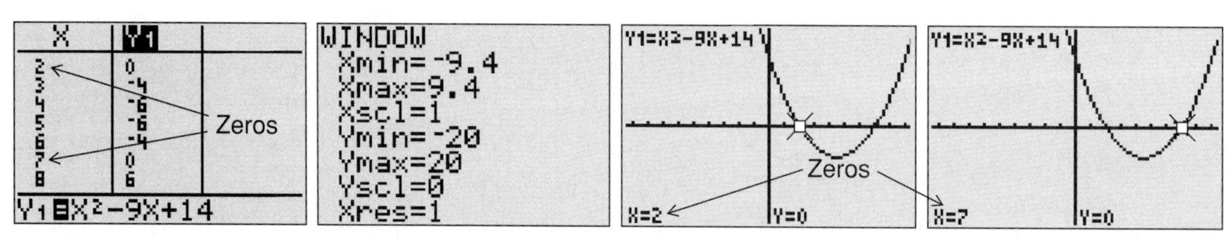

The x-intercepts are $(2, 0)$ and $(7, 0)$. The numbers 2 and 7 are the zeros. To create the factors, set x equal to each zero and rewrite so that one side of the equation is zero.

$$x = 2 \qquad x = 7$$

$$x - 2 = 0 \qquad x - 7 = 0$$

The factors are $(x - 2)$ and $(x - 7)$.

The quadratic trinomial $x^2 - 9x + 14$ factors as $(x - 2)(x - 7)$.

Consider $x^2 + 2x - 24$. The parameters are $b = 2$ and $c = -24$. The zeros have different signs. The product of the zeros is -24 and the sum of the zeros is -2.

Just a graph of $y(x) = x^2 + 2x - 24$ appears in Figure 8. A table could also be used to identify the zeros, as demonstrated previously.

FIGURE 8

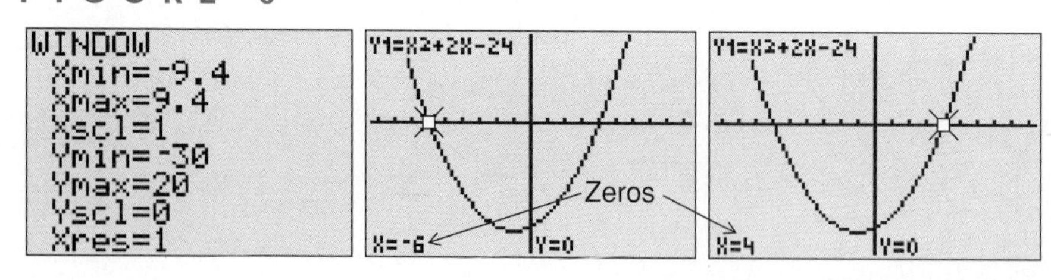

The x-intercepts are $(-6, 0)$ and $(4, 0)$. The numbers -6 and 4, the x values of the x-intercepts, are the zeros.

To create the factors, set x equal to each zero and rewrite so that one side of the equation is zero.

$$x = -6 \qquad x = 4$$

$$x + 6 = 0 \qquad x - 4 = 0$$

The factors are $(x + 6)$ and $(x - 4)$.

The quadratic trinomial $x^2 + 2x - 24$ factors as $(x + 6)(x - 4)$.

The quadratic trinomials $x^2 + 2x - 4$ and $x^2 + 2x + 15$ do not factor over the rational numbers. In each case there are no numbers whose product is c and whose sum is the opposite of b.

The function $y(x) = x^2 + 2x - 4$ has irrational zeros (Figure 9). The table and the graph suggest a zero between -4 and -3 and a zero between 1 and 2, but these zeros are not rational.

FIGURE 9

X	Y1
-4	4
-3	-1
-2	-4
-1	-5
0	-4
1	-1
2	4

Y1=X²+2X-4

No output of 0 for integer inputs

WINDOW
Xmin=-9.4
Xmax=9.4
Xscl=1
Ymin=-10
Ymax=10
Yscl=0
Xres=1

Y1=X²+2X-4
X=1 Y=-1

The graph of $y(x) = x^2 + 2x + 15$ has no x-intercepts (Figure 10).

FIGURE 10

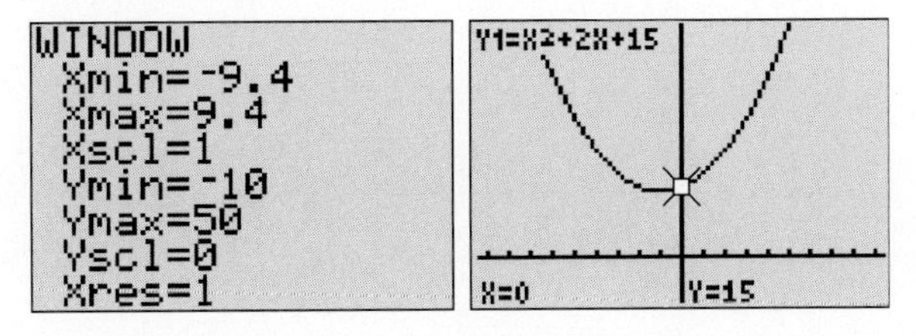

The last quadratic $x^2 + 6x + 9$ in Table 1 deserves special mention. Since the graph intersects the x-axis at only one point, $(-3, 0)$, the zero of the function, -3 is called a double zero (Figure 11). The corresponding factor, $(x + 3)$, occurs twice. So $x^2 + 6x + 9$ factors as $(x + 3)(x + 3)$ or $(x + 3)^2$. Quadratics with exactly one distinct zero factor into a linear expression squared and are called ***square trinomials***.

FIGURE 11

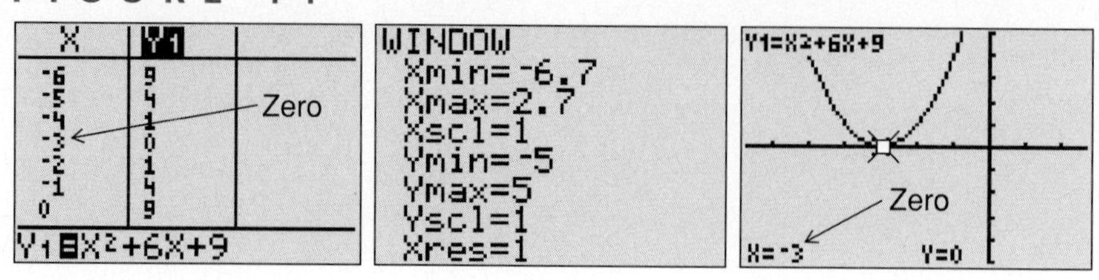

Finally, we consider quadratic trinomials $ax^2 + bx + c$, in which the parameter a is not 1.

12. Complete Table 2. Use a table, a graph, or a symbol manipulator to find the exact x-intercepts and zeros of each quadratic. Write the zeros in either integer or fractional form—do not write as decimals. Predict the factorization. Use a symbol manipulator, if available, to factor and compare to your factorization. If a symbol manipulator is not available, check your factorization by multiplying the factors.

TABLE 2 Quadratic Trinomials 2

Quadratic	x-intercepts	Zeros	Factored Form
$3x^2 - 17x + 24$			
$2x^2 + 13x + 15$			
$4x^2 - 12x + 9$			
$4x^2 - 4x - 3$			
$15x^2 - x - 2$			
$16x^2 - 24x + 9$			

13. Refer to Table 2.

a. How can the signs of the parameters a, b, and c help you determine the signs of the zeros?

b. Compare the numerators of the zeros with the value of parameter c in $ax^2 + bx + c$. What relationship do you see?

c. Compare the denominators of the zeros with the value of parameter a in $ax^2 + bx + c$. What relationship do you see?

d. Multiply the zeros. How does the product compare to the ratio $\frac{c}{a}$ where a and c are the parameters in $ax^2 + bx + c$?

14. Consider the quadratic trinomial $6x^2 + 31x + 35$.

a. How can you use the parameter $c = 35$ to identify the numerators of the possible zeros?

b. How can you use the coefficient of the parameter $a = 6$ to identify the denominators of the possible zeros?

c. Write the zeros of the trinomial as fractions. Verify these both in a table and on a graph.

d. Use the zeros to factor the trinomial. Check your answer by multiplying the two factors together.

15. Consider the quadratic trinomial $2x^2 + 3x - 35$.

a. How can you use the parameter $c = -35$ to identify the numerators of the possible zeros?

b. How can you use the coefficient of the parameter $a = 2$ to identify the denominators of the possible zeros?

c. Write the zeros of the trinomial as fractions. Verify these both in a table and on a graph.

d. Use the zeros to factor the trinomial. Check your answer by multiplying the two factors together.

We wish to factor general quadratic trinomials $ax^2 + bx + c$ over the rational numbers. Assume that a, b, and c are integers and a is positive. Here are some generalizations we can draw from Table 2.

- If parameter c is positive, the zeros will have the same sign. In this case, the sign of the zeros is opposite of the sign of parameter b. For example, the zeros of $3x^2 - 17x + 24$ are 3 and $\frac{8}{3}$. They are both positive since $b = -17$ is negative. If parameter c is negative, the zeros will have different signs. For example, the zeros of $15x^2 - x - 2$ are $\frac{2}{5}$ and $-\frac{1}{3}$. They have different signs since $c = -2$ is negative.

- The product of the zeros equals the ratio $\frac{c}{a}$ where a and c are the parameters in $ax^2 + bx + c$. This means the numerators of the zeros must be factors of parameter c and the denominators of the zeros must be factors of parameter a. For example, the product of the zeros of

  $$15x^2 - x - 2 \text{ is } \left(\frac{2}{5}\right)\left(-\frac{1}{3}\right) = -\frac{2}{15},$$

  which is the ratio of $c = -2$ to $a = 15$.

> **In general, rational number zeros have the form**
>
> $$\frac{\textbf{factor of } c\textbf{, the constant term}}{\textbf{factor of a, the numerical coefficient of squared term}}.$$

If we find a zero like $\frac{2}{5}$, we can form the factor as follows.

Let x equal the zero:	$x = \frac{2}{5}$
Multiply both sides by denominator 5 to eliminate fractions:	$5x = 2$
Make one side zero by subtracting 2 from each side:	$5x - 2 = 0$

So the factor $(5x - 2)$ may be derived from the zero $\frac{2}{5}$.

Following are some examples factored by this method.

Consider $6x^2 + 31x + 35$. The parameters $a = 6$ and $c = 35$ suggest that the product of the zeros is $\frac{35}{6}$.

Figure 12 displays a table, a graph, the zero –3.5, and the conversion of the zero to a fractional form.

FIGURE 12

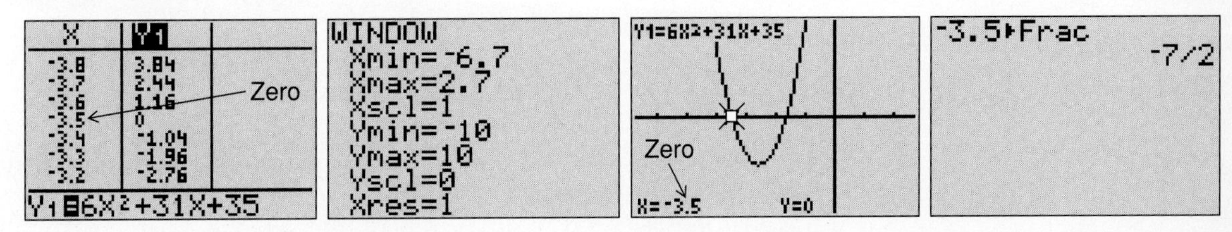

So one zero is $-\frac{7}{2}$. The product of the zeros is $\frac{35}{6}$. The second zero must be negative, have a numerator of 5 since $7(5) = 35$, and have a denominator of 3 since $2(3) = 6$.

The second zero is $-\frac{5}{3}$. Let's create the factors.

$$x = -\frac{7}{2} \qquad x = -\frac{5}{3}$$

$$2x = -7 \qquad 3x = -5$$

$$2x + 7 = 0 \qquad 3x + 5 = 0$$

So $6x^2 + 31x + 35$ factors as $(2x + 7)(3x + 5)$.

Consider $2x^2 + 3x - 35$. The parameters $a = 2$ and $c = -35$ suggest that the product of the zeros is $-\frac{35}{2}$.

Figure 13 displays a table and a graph indicating that –5 is a zero.

FIGURE 13

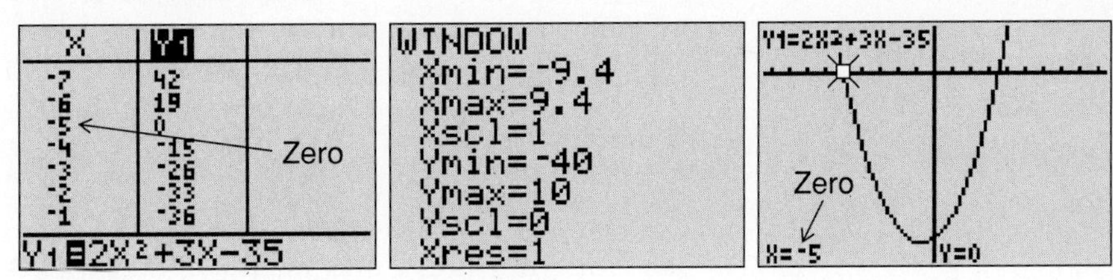

So one zero is –5. The product of the zeros is $-\frac{35}{2}$. The second zero must be positive, have a numerator of 7 since $7(5) = 35$, and have a denominator of 2 since $2(1) = 1$.

The second zero is $\frac{7}{2}$. Let's create the factors.

$$x = \frac{7}{2} \qquad\qquad x = -5$$

$$2x = 7 \qquad\qquad x + 5 = 0$$

$$2x - 7 = 0$$

So $2x^2 + 3x - 35$ factors as $(2x - 7)(x + 5)$.

The quadratic $16x^2 - 24x + 9$ is another example of a square trinomial. The graph has one x-intercept at $\left(\frac{3}{4}, 0\right)$ (Figure 14).

FIGURE 14

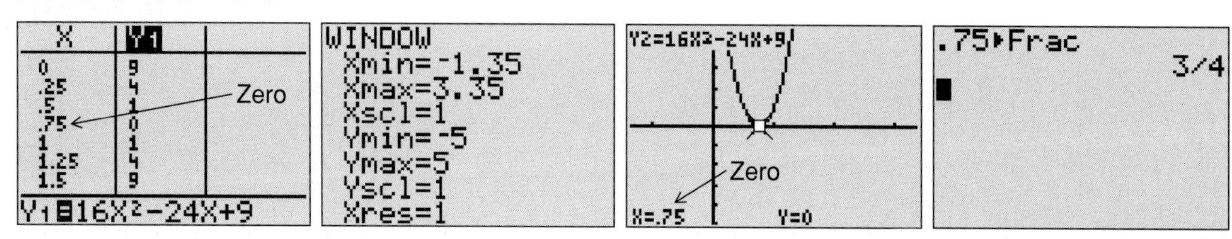

The zero is $\frac{3}{4}$ and the corresponding factor is $(4x - 3)$. So $16x^2 - 24x + 9 = (4x - 3)^2$.

Next, we look at factoring quadratic binomials over the rational numbers. In particular, we consider quadratics of the form $ax^2 + bx + c$ where the parameter $b = 0$.

16. Complete Table 3. Use a table, a graph, or a symbol manipulator to find the exact x-intercepts and zeros of each quadratic. Write the zeros in either integer or fractional form—do not write as decimals. Predict the factorization. Use a symbol manipulator, if available, to factor and compare to your factorization. If a symbol manipulator is not available, check your factorization.

17. Refer to Table 3.

a. Compare and contrast the quadratics that factored with those that did not factor. What was the key difference?

b. Consider the zeros of the functions that factored. What is the relationship between the two zeros?

TABLE 3 **Quadratic Binomials**

Quadratic	x-intercepts	Zeros	Factorization
$x^2 - 25$			
$x^2 - 16$			
$x^2 + 16$			
$4x^2 - 9$			
$4x^2 + 9$			
$25x^2 - 4$			

c. Several functions in Table 3 were the sum of two squares. Did they factor? Did their graphs have x-intercepts?

d. Several functions in Table 3 were the difference of two squares. Did they factor?

18. Do the following quadratics factor over the rational numbers? Why or why not?

a. $x^2 - 5$

b. $2x^2 - 9$

When factoring, we always look for a common monomial factor first. Next we count terms. We have factored both binomials and trinomials in this section.

There are several cases to consider when factoring $ax^2 + bx + c$ if $\boldsymbol{b = 0}$.

Case 1: If parameters a and c have the same sign, the graph does not intersect the x-axis, which means there are no real number zeros. The quadratic does not factor within the real numbers. An example is $x^2 + 16$.

Case 2: If parameters a and c have different signs, then the quadratic has two zeros that are opposites of each other. The quadratic factors only if the zeros are rational numbers. The zeros are rational only if the quadratic has general form $(\ )^2 - (\ )^2$, which is called the difference of two squares. So $4x^2 - 9$ will factor, but $2x^2 - 9$ will not factor since the first term, $2x^2$, is not a square.

Thus, with the exception of factoring out a common monomial factor, the only quadratic binomials that factor over the rational numbers have the form difference of two squares.

Given $4x^2 - 9$, the zeros are $-\frac{3}{2}$ and $\frac{3}{2}$ (Figure 15).

FIGURE 15

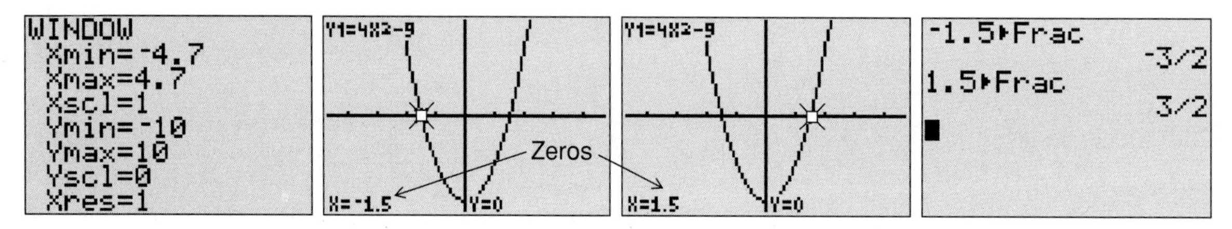

Let's form the factors.

$$x = -\frac{3}{2} \qquad x = \frac{3}{2}$$

$$2x = -3 \qquad 2x = 3$$

$$2x + 3 = 0 \qquad 2x - 3 = 0$$

So the factored form of $4x^2 - 9$ is $(2x + 3)(2x - 3)$.

Notice the form of the factorization.

$$\square^2 - \triangle^2 = (\square + \triangle)(\square - \triangle)$$

The difference of two squares factors as the product of the sum and the difference of the terms that were squared. Factors, such as $(2x + 3)$ and $(2x - 3)$, are called ***conjugate factors***.

Quadratic equations arise when we are given the output of a quadratic function and asked to find the input. The next investigation suggests a way to use the zeros of a quadratic function to solve quadratic equations.

19. Consider the function $y(x) = 2x^2 - 3x - 14$.

a. Write the equation that results if the output is 13 and you wish to find the input.

b. Rewrite the equation in the form $ax^2 + bx + c = 0$. Identify the values of the parameters a, b, and c.

c. Find the exact zeros of the quadratic $ax^2 + bx + c$ that you wrote in part **b.**

d. Check that the zeros when used as input to $y(x)$ produce an output of 13. If they do not, correct your work.

e. Write the inequality that results if the output is less than 13 and you wish to find the input.

f. Use the answer to the equation from part **a** along with the graph to solve the inequality.

20. Explain to your friend Izzy how to find the input to a quadratic equation given the output.

If we want $y(x) = 2x^2 - 3x - 14$ to have an output of 13, the resulting equation is $2x^2 - 3x - 14 = 13$. We could use a table or a graph to estimate where the function has an output of 13, but a common procedure when solving quadratic equations is to rewrite the equation so that one side is zero. If we subtract 13 from both sides of $2x^2 - 3x - 14 = 13$, we get

$$2x^2 - 3x - 27 = 0.$$

This is a quadratic equation of form $ax^2 + bx + c = 0$, where $a = 2$, $b = -3$, and $c = -27$. Since we want to know when $2x^2 - 3x - 27$ is zero, we look for the x-intercepts of the graph of $2x^2 - 3x - 27$ (Figure 16). The zeros of $2x^2 - 3x - 27$ are -3 and $4.5 = \frac{9}{2}$.

FIGURE 16

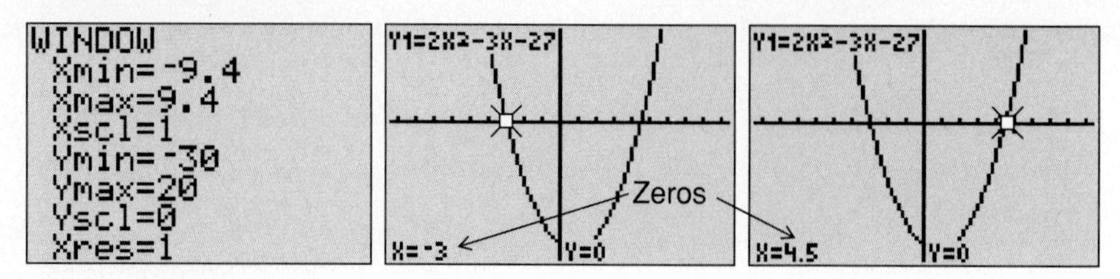

Substituting each of these zeros into $y(x) = 2x^2 - 3x - 14$, we get

$$y(-3) = 13 \text{ and } y\left(\frac{9}{2}\right) = 13, \text{ which are the desired outputs.}$$

Note that $2x^2 - 3x - 27$ has factored form $(x + 3)(2x - 9)$. The equation $2x^2 - 3x - 27 = 0$ can be written as $(x + 3)(2x - 9) = 0$. If we use the Zero Product Property, we could write $x + 3 = 0$ or $2x + 9 = 0$. Solving these linear equations results in the zeros that are also the solutions to the equation.

If we want the output of $y(x) = 2x^2 - 3x - 14$ to be less than 13, the resulting inequality is $2x^2 - 3x - 14 < 13$. Adding 13 to both sides, we get

$$2x^2 - 3x - 27 < 0.$$

This states that we want the output of $y = 2x^2 - 3x - 27$ to be negative. We know the output of $y = 2x^2 - 3x - 27$ is zero at –3 and 4.5. Output is negative when the graph is below the x-axis. This occurs for the shaded x values in Figure 17.

FIGURE 17

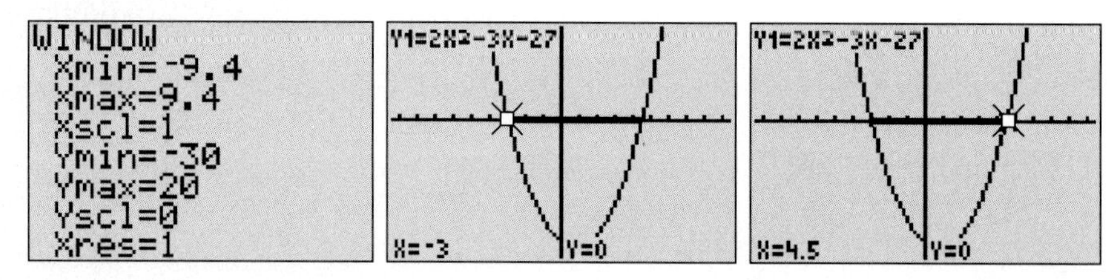

So $2x^2 - 3x - 27 < 0$ when x is between the two zeros, –3 and 4.5.

Summarizing, given a quadratic equation, one method of solution is to write the equation in the form $ax^2 + bx + c = 0$ and then to locate the zeros of $ax^2 + bx + c$. If you can easily factor $ax^2 + bx + c$, writing the quadratic in factored form, using the Zero Product Property, and solving the two linear equations is a second method of solution.

Given a quadratic inequality, find the zeros of the related equation $ax^2 + bx + c = 0$ first. Graph $y = ax^2 + bx + c$. If you wish to solve $ax^2 + bx + c < 0$, find intervals on x-axis where graph is below the x-axis. If you wish to solve $ax^2 + bx + c > 0$, find intervals on x-axis where graph is above the x-axis.

STUDENT DISCOURSE

Pete: I have a question about the last investigation we did.

Sandy: What's that?

Pete: We used zeros to solve quadratic equations.

Sandy: Right.

Pete: What if the zeros aren't "nice" numbers, like integers or fractions? Or what if the graph doesn't intersect the x-axis? Then there aren't any zeros, right?

Sandy: I wondered the same thing myself. If the zeros aren't "nice" numbers, as you call them, then they may be irrational numbers. Right now, the best we can do is approximate them. But later in the chapter, you will learn a formula that will help you identify the irrational zeros exactly.

Pete: Okay, but what about the other case?

Sandy: You mean when the graph doesn't intersect the x-axis?

Pete: Yes.

Sandy: There aren't any real number zeros then. However, you will learn about some numbers called the complex numbers. These numbers can be used to identify zeros in this case.

Pete: Sounds like there is a lot more to learn about quadratics. Thanks for the information.

1. In the Glossary, define the words in this section that appear in ***italic boldface*** type.

2. A quadratic function has zeros at $-\frac{2}{3}$ and $\frac{2}{3}$. Write the factors of the function.

3. The graph of a quadratic function appears in Figure 18.

FIGURE 18

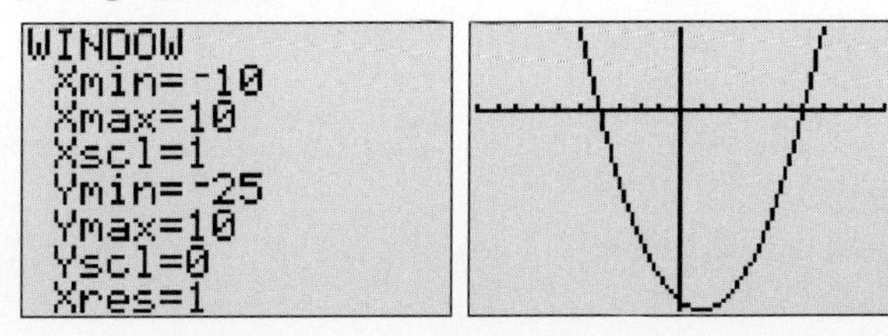

a. What are the zeros of this function?

b. Write the factors of this function.

c. For what x values is the output positive?

d. For what x values is the output negative?

4. A quadratic function intersects the x-axis at 4 only. Write a factor of this function.

5. Given the quadratic function $y(x) = x^2 - 5x - 3$. Find the inputs if the output is 21. Explain what you did.

6. Consider the factored quadratic $(5x - 2)(3x + 7)$.

a. Write the zeros as fractions.

b. Write the product of the factors.

7. a. Multiply and simplify each of the following:

$(x - 2)(x^2 + 2x + 4)$ $\quad$ $(x + 3)(x^2 - 3x + 9)$

$(2x - 5)(4x^2 + 10x + 25)$ $\quad$ $(4x + 1)(16x^2 - 4x + 1)$

b. Use your results to write factoring formulas for $a^3 - b^3$ and for $a^3 + b^3$.

8. The graph of a quadratic function appears in Figure 19.

a. What are the zeros of this function?

FIGURE 19

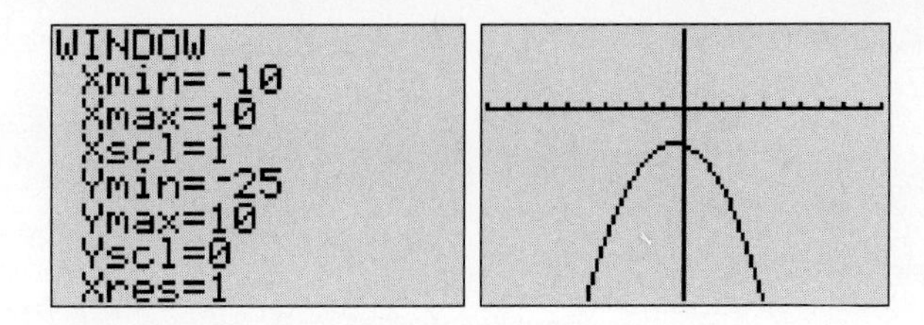

b. Does this function factor over the rational numbers?

c. For what x values is the output positive?

d. For what x values is the output negative?

9. Find the solutions to each of the following equations, inequalities, or systems of equations.

a. $4x - 7 = 2 - 8x$

b. $x^2 - 5x + 6 = 0$

c. $4x - 7y = 2$
$3x + y = -9$

d. $x^2 + 9x - 2 = -16$

e. $5 - 3x = 17x + 5$

f. $4x - 7y + 2z = -5$
$3x + y - z = 1$
$x + 4y + 7z = 9$

g. $6x^2 + x - 3 = 2$

h. $6x^2 + 7x = 5$

i. $3x - 2 < 7x + 8$

j. $x^2 - 5x + 6 > 0$

k. $2x^2 + 5x > -2$

l. $x^2 + x + 7 < 0$.

10. Find the zeros and factor completely. If the polynomial is prime, say so and identify whether the zeros are irrational or nonreal.

a. $5x^2 + 4x + 1$

b. $6x^2 - x - 1$

c. $x^2 - 12x + 36$

d. $2x^2 + 30x + 48$

e. $x^2 - 9x + 14$

f. $8x^2 + 18x - 5$

g. $16y^2 - 121$

h. $2y^2 + 5y + 2$

i. $t^2 - 21t$

j. $x^2 + 6x - 27$

k. $6x^2 + x - 35$

l. $4x^2 + 12x + 9$

m. $9y^2 + 27$

n. $4x^2 + 11x - 45$

o. $x^2 + 16x + 64$

p. $x^2 + 3x - 2$

11. Use a graph to estimate the solutions to the following systems of equations involving quadratics.

a. $y = 2x^2 + x - 1$
$y = 3$

b. $y = x^2 - 5x - 7$
$3x + 2y = 1$

c. $y = -3x^2 + x + 5$
$y = 2x - 1$

d. $y = -x^2 - x + 2$
$y = x^2 + 2x - 11$

REFLECTION

Write a paragraph discussing what you know about factoring. Clearly describe the relationships between the zeros of a function, the x-intercepts of a function, and the factors of a function. Indicate what representation (table, graph, equation) you use to find the zeros of a function, the x-intercepts of a function, and the factors of a function.

SECTION 6.2

Graphical Analysis of Quadratic Models

Purpose

- Investigate transformations on the basic quadratic model.
- Illustrate important relationships between quadratic functions and parabolas in the plane.
- Explore the role of the parameters of a quadratic function in the graph of the function.

In this section, we continue to analyze quadratic functions. Aside from the intercepts, another important feature of the graph of a quadratic function is the vertex, which is the lowest or highest point on the graph depending on whether the parabola opens upward or downward. We will see how to use the vertex and one other point to write the equation of a quadratic function. In the process, we introduce some new parameters and discuss their effect on the positioning of the graph of a quadratic function.

We begin the section by considering the graph of a quadratic function in which we don't have sufficient information to write the equation of the function.

1. Figure 1 contains a portion of the graph of a quadratic function.

   a. Two points are easily identified. What are they?

FIGURE 1

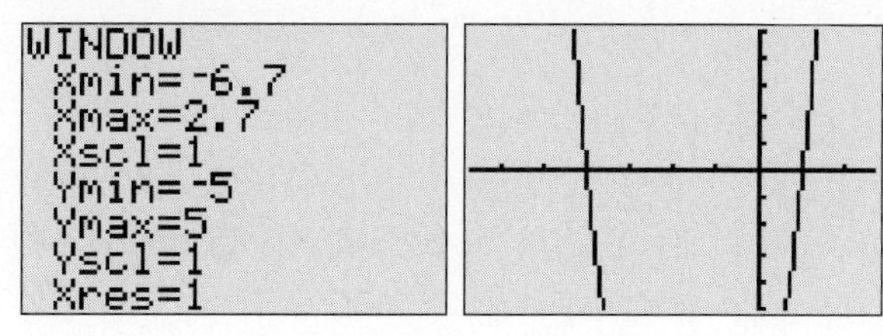

b. Write two factors of the quadratic pictured in Figure 1.

c. Multiply the factors from part **b** together and write the resulting expression.

d. Using the same viewing window as in Figure 1, graph the quadratic you created in part **b**. Does your graph match the one in Figure 1?

2. Consider the following quadratic functions:

$y(x) = x^2 + 3x - 4$ $\quad$ $y(x) = 2x^2 + 3x - 4$

$y(x) = 3x^2 + 3x - 4$ $\quad$ $y(x) = 4x^2 + 3x - 4$

$y(x) = 0.8x^2 + 3x - 4$ $\quad$ $y(x) = 0.6x^2 + 3x - 4$

a. How are the six functions different?

b. Graph all six functions. Explain how the difference in the six functions affects their graphs.

3. Consider the following quadratic functions.

$y(x) = 2x^2 + 3x - 4$ $\quad$ $y(x) = -2x^2 + 3x - 4$

a. How are the two functions different?

b. Graph the two functions. Explain how the difference in the two functions affects their graphs.

4. Given any quadratic function $y(x) = ax^2 + bx + c$, use the results of Investigations 2 and 3 to describe to your confused friend, Izzy, the effect of parameter a on the graph of the function.

The graph in Figure 1 has x-intercepts $(-4, 0)$ and $(1, 0)$. The zeros of the function are -4 and 1 so the factors are $(x+4)$ and $(x-1)$. Multiplying the factors together, we get

$$(x+4)(x-1) = x^2 - x + 4x - 4 = x^2 + 3x - 4$$

If we graph $y(x) = x^2 + 3x - 4$ over the graph in Figure 1, we see that they have the same x-intercepts, but the graphs don't quite match (Figure 2).

FIGURE 2

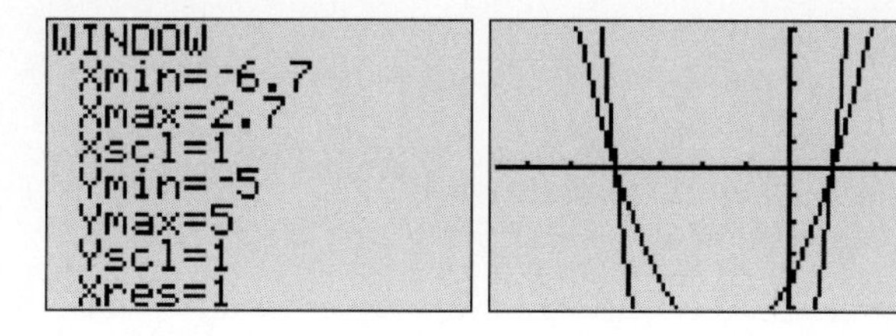

The original graph seems to be "steeper" than the graph of $y(x) = x^2 + 3x - 4$. We need to talk about the graphs of quadratic function a little to understand what is happening.

The graph of $y(x) = ax^2 + bx + c$ is called a ***parabola***. The sign of parameter a determines if the parabola opens upward or downward. The graph looks like one of the two graphs in Figure 3.

FIGURE 3

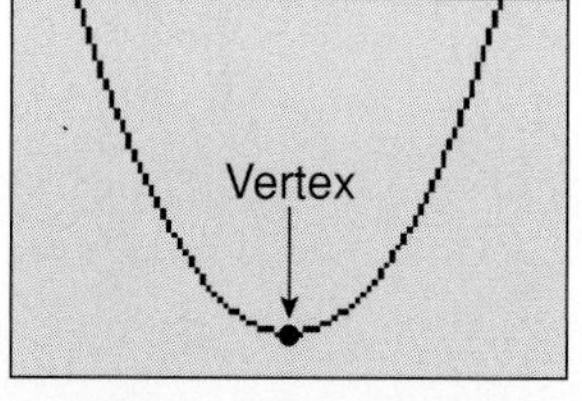

Coefficient of squared term is positive. Parameter a is positive.

or

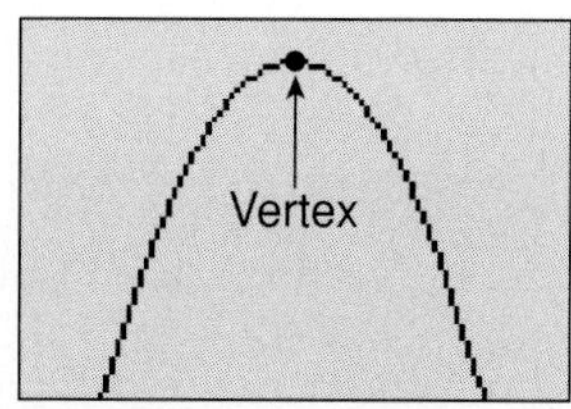

Coefficient of squared term is negative. Parameter a is negative.

Given $y(x) = ax^2 + bx + c$,

- if a is positive, the parabola opens upward and the graph has a lowest point called the ***vertex***.
- if a is negative, the parabola opens downward and has a highest point called the ***vertex***.

If we draw a vertical line through the vertex and fold the parabola along this line, one branch of the parabola will lie on top of the other branch. This demonstrates the ***symmetry*** of the parabola with respect to the vertical line through the vertex.

The parameter a also controls the "steepness" of the parabola (Figure 4).

FIGURE 4

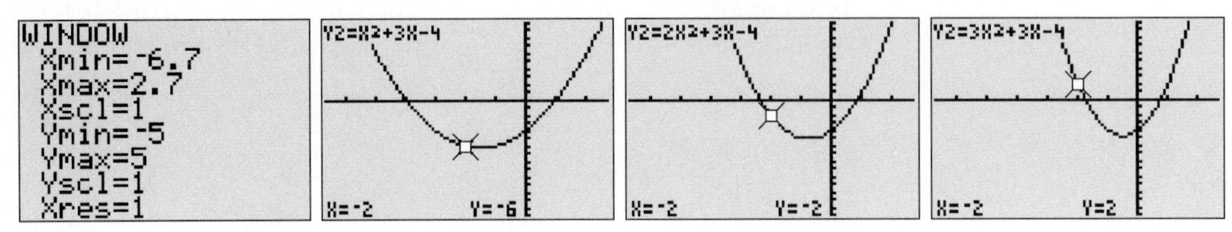

As you saw in Investigation 2,

- the larger the value of $|a|$, the "narrower" the parabola.
- the smaller the value of $|a|$, the "wider" the parabola.

So, if we return to the graph in Figure 1, we discovered that the function $y(x) = x^2 + 3x - 4$ did not quite fit the graph even though they had the same x-intercepts. We need to multiply the expression $x^2 + 3x - 4$ by some number so we can match the "width" of the function graphed in Figure 1.

In other words, if a function has zeros -4 and 1, then multiplying the corresponding factors $(x + 4)$ and $(x - 1)$ does not necessarily create the function. We may need to multiply by a constant a to obtain the correct "width."

The equation for the function in Figure 1 is $y(x) = a(x + 4)(x - 1)$ or $y(x) = a(x^2 + 3x - 4)$. To find a we need one other piece of information. Recall that we need three pieces of information to determine a quadratic and we've only used two so far.

In the next investigation, we see how to find the value of a. Also, we look for a relationship between the location of the zeros and the location of the vertex.

5. Figure 5 contains the graph in Figure 1 but in a different viewing window.

FIGURE 5

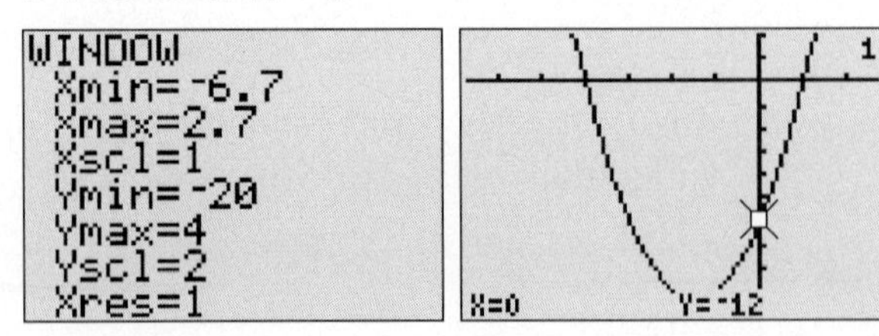

a. What additional point can you now identity?

b. Substitute the point you've identified into $y(x) = a(x^2 + 3x - 4)$ and solve for parameter a.

c. Write the equation for the graph in Figure 1. Check to make sure your function's graph matches the graph in Figure 1.

6. Complete Table 1. Given the zeros, predict the x value of the vertex. Use the graph in a "friendly" window to determine the x value and the y value of the vertex.

TABLE 1 Vertex of Quadratic Graphs

Function	Zeros	x Value of Vertex	y Value of Vertex
$x^2 - 10x + 9$			
$x^2 - 16$			
$x^2 - 4x - 21$			

7. Use the data collected in Table 1 to answer these questions. Assume that $y(x)$ is a quadratic function.

a. If the graph of $y(x)$ has x-intercepts (1, 0) and (5, 0), what is the x-coordinate of the vertex?

b. Describe in words how you find the x value of the vertex if you know the zeros of a quadratic function.

c. If you know the x value of the vertex, how can you find the y value of the vertex without graphing?

The graph in Figure 5 has a y-intercept of $(0, -12)$. If $x = 0$ and $y = -12$, the function $y(x) = a(x^2 + 3x - 4)$ with parameter a becomes

$$-12 = a((0)^2 + 3(0) - 4)$$

Simplify: $$-12 = a(-4)$$

Divide both sides by -4: $$3 = a$$

So $y(x) = 3(x^2 + 3x - 4)$ or, finally, $y(x) = 3x^2 + 9x - 12$. This is the function whose graph appears in Figures 1 and 5.

The graph of a quadratic function is symmetric with respect to a vertical line through the vertex. If we fold the parabola along the line of symmetry, one branch will fold onto the other branch. In particular, one zero of the function will fold onto the other zero. This suggests that the x value of the vertex is the midpoint of the line segment between the two zeros. The *mean* of the two zeros is the midpoint of that line segment.

For example, consider a quadratic function with zeros 1 and 5. The mean of 1 and 5 is 3. The x value of the vertex is 3.

In general, if a quadratic function has zeros r and s, then the x-coordinate of the vertex is $\frac{r+s}{2}$.

If we know the x value of the vertex, we can find the y value of the vertex by evaluating the function.

For example, the zeros of $y(x) = x^2 - 10x + 9$ are 1 and 9. The mean of 1 and 9 is 5. So 5 is the x value of the vertex.

Let's evaluate the function at 5:

$$y(5) = (5)^2 - 10(5) + 9 = -16$$

So the vertex of the graph of $y(x) = x^2 - 10x + 9$ is $(5, -16)$ (Figure 6).

FIGURE 6

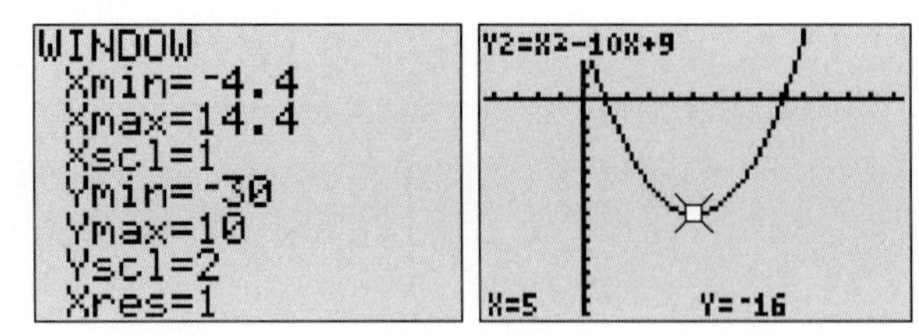

Now we investigate how to create the equation of a quadratic function if we know the vertex. We begin by looking at functions whose vertex is either on the x- or y-axes. For now we will consider functions of the form $y(x) = x^2 + bx + c$. In other words, the parameter a is 1.

Use a "friendly" viewing window for these investigations.

8. Consider functions of the form $y(x) = x^2 + k$. Select four different values of k (two positive and two negative). Record the value of parameter k, the new function, and the ordered pair for the vertex in Table 2.

TABLE 2 **Quadratic Transformations I**

Value of k	Expression	Vertex
3	$x^2 + 3$	

9. Given a quadratic function of the form $y(x) = x^2 + k$, where is the vertex located?

10. If a function has a vertex at $(0, -7)$ and the parameter $a = 1$, write the equation of the function.

11. Consider functions of the form $y(x) = (x - h)^2$. Select four different values of h (two positive and two negative). Record the value of parameter h, the new function, and the ordered pair for the vertex in Table 3.

TABLE 3 **Quadratic Transformations 2**

Value of k	Expression	Vertex
3	$(x-3)^2$	

12. Given a quadratic function of the form $y(x) = (x - h)^2$, where is the vertex located?

13. If a function has a vertex at $(6, 0)$ and the parameter $a = 1$, write the equation of the function.

14. If a function has a vertex at $(-5, 0)$ and the parameter $a = 1$, write the equation of the function.

DISCUSSION

As you saw in the preceding investigation, the graph of $y(x) = x^2 + k$ has a vertex of $(0, k)$ (Figure 7, for example).

FIGURE 7

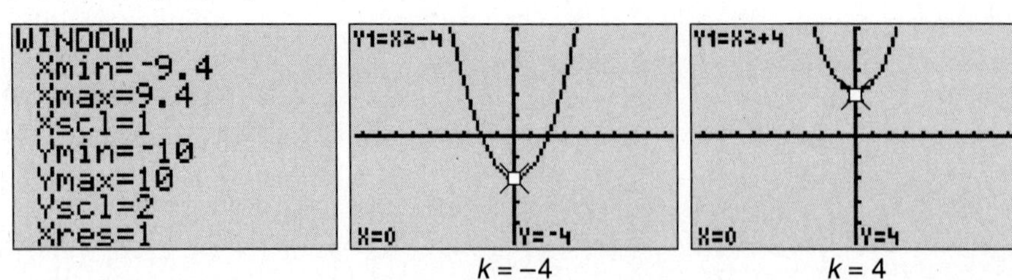

The graph of $y(x) = (x - h)^2$ has a horizontal intercept of $(h, 0)$. You'll notice that $(x - h)$ is acting as a factor of the function. The corresponding zero is h. Thus the x-intercept of $y(x) = (x - h)^2$ is $(h, 0)$. Since this is the only x-intercept, this point is also the vertex of the parabola (Figure 8, for example).

FIGURE 8

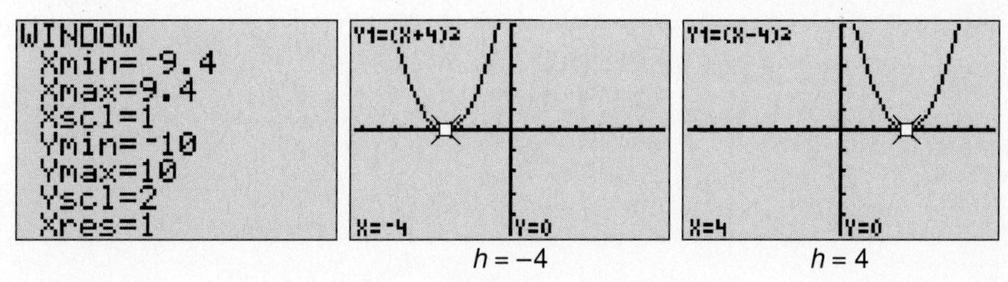

Now let's look at the case where the vertex is not on an axis.

15. Consider functions of the form $y(x) = (x - h)^2 + k$. Select four different values of h and k using both negative and positive numbers. Record the value of parameter h, the value of parameter k, the new function, and the ordered pair for the vertex in Table 4.

TABLE 4 **Quadratic Transformations 3**

Value of h	Value of k	Expression	Vertex
2	−3	$(x - 2)^2 - 3$	

16. Given a quadratic function of the form $y(x) = (x - h)^2 + k$, where is the vertex located?

17. If a function has a vertex at $(-2, 5)$ and the parameter $a = 1$, write the equation of the function.

The function $y(x) = (x - h)^2 + k$ will have a vertex at (h, k). For example, a function with vertex at $(-2, 5)$ and the parameter $a = 1$ has the equation $y(x) = (x + 2)^2 + 5$ (Figure 9).

FIGURE 9

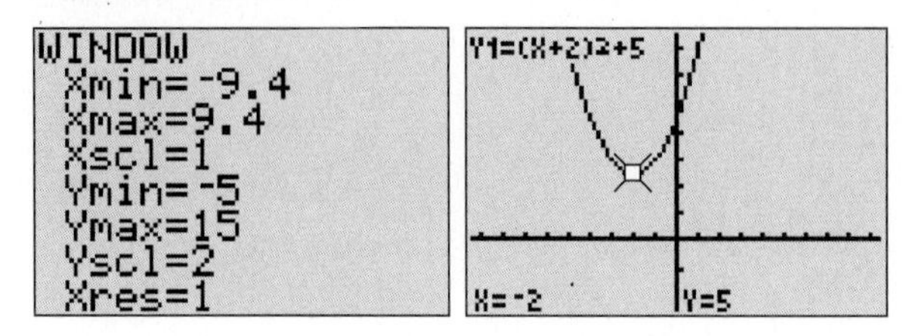

If we wish to make the graph "narrower" or "wider," we use a third parameter a. The more general form is $y(x) = a(x - h)^2 + k$. Thus we have another form of a quadratic function that has three parameters a, h, and k. The parameter a in this form is the same as the parameter a in the standard form $y(x) = ax^2 + bx + c$.

18. Consider the standard form $y(x) = ax^2 + bx + c$ of a quadratic function. What information can you obtain from the parameters?

19. Consider the "vertex" form $y(x) = a(x - h)^2 + k$ of a quadratic function.

a. What information can you obtain from the parameters?

b. What is the domain and the range of the function $y(x) = a(x - h)^2 + k$?

20. The graphs of a quadratic function appear in Figure 10.

FIGURE 10

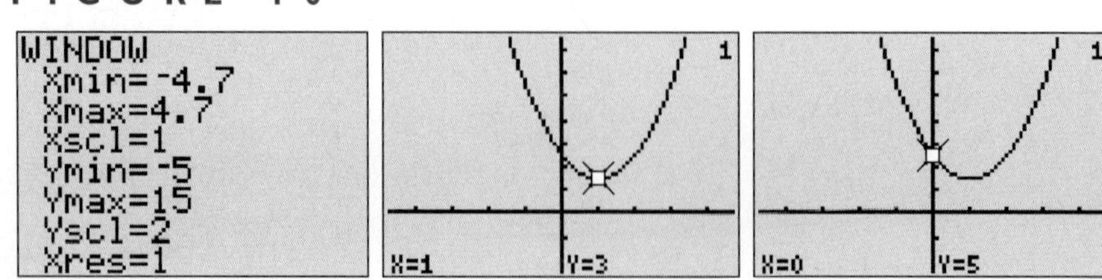

a. What is the vertex?

b. What are the intercepts?

c. Use the vertex and intercepts to create the equation of the function. Briefly state what you did.

Given $y(x) = ax^2 + bx + c$, we immediately know the value of the second finite differences in successive outputs using the value of a. We know that the y-intercept is at $(0, c)$. The value of a also determines the "width" of the parabola and whether the parabola opens upward or downward.

Given $y(x) = a(x - h)^2 + k$, the value of a provides the same information as stated above. The values of parameters h and k provide the location of the vertex.

The domain of any quadratic function is the set of all real numbers. The range, however, is limited by the location of the vertex.

If the parameter a is positive, the parabola opens upward and thus the vertex (h, k) is the lowest point (absolute minimum). The range is the set of all real numbers greater than or equal to k.

If the parameter a is negative, the parabola opens downward and thus the vertex (h, k) is the highest point (absolute maximum). The range is the set of all real numbers less than or equal to k.

In Figure 10, the vertex is at $(1, 3)$ and the y-intercept is $(0, 5)$. Since we know the vertex, the equation must have form $y(x) = a(x - 1)^2 + 3$. We must find the value of parameter a by using the y-intercept to substitute for x and y.

Let $x = 0$ and $y = 5$:	$5 = a(0 - 1)^2 + 3$
Simplify right side of "=":	$5 = a + 3$
Subtract a from both sides:	$2 = a$

So the equation of the function displayed in Figure 10 is $y(x) = 2(x - 1)^2 + 3$. The range of this function is the set of all real numbers greater than or equal to 3.

If we expand and simplify, we can write this in standard form.

Square $(x - 1)$:	$y(x) = 2(x^2 - 2x + 1) + 3$
Distribute the 2:	$y(x) = 2x^2 - 4x + 2 + 3$
Add like terms:	$y(x) = 2x^2 - 4x + 5$

So we see that given the equation of a quadratic in the "vertex" form $y(x) = a(x - h)^2 + k$, we can use algebraic simplification to write the equation in standard form $y(x) = ax^2 + bx + c$.

STUDENT DISCOURSE

Pete: How do I know which form of equation for a quadratic function to use?

Sandy: It really depends on the kind of information you have. For example, what would you do to find the equation if you knew three points?

Pete: I'd use quadratic regression. That seems the quickest.

Sandy: If you use quadratic regression, what form do you get for the equation?

Pete: The calculator gives me the values for parameters a, b, and c, so I guess I have the standard form $y(x) = ax^2 + bx + c$.

Sandy: Makes sense. What if you only know the zeros?

Pete: Then I would first write the function in factored form. But don't I need some other information too?

Sandy: Yes, you need some kind of information, such as another point, so you can determine the value of a.

Pete: The result when I multiplied out would still be the standard form, right?

Sandy: Yes. Finally, suppose you know the vertex and another point.

Pete: That's easy. I would use the form $y(x) = a(x - h)^2 + k$ since I know the vertex.

Sandy: Great.

Pete: I have one final question. I know how to convert the form $y(x) = a(x - h)^2 + k$ into the form $y(x) = ax^2 + bx + c$. That was demonstrated in the last discussion. How can I reverse this and go the other way?

Sandy: Well, you know the value of a already. You need to find the vertex. Do you remember how to do that?

Pete: Let's see. I remember that I can average the zeros to find the x value of the vertex. Oh, I see. Then I can find the y value by evaluating the function. Then I have everything I need to write the equation in the "vertex" form.

Sandy: Very good. I think you have it.

1. In the Glossary, define the words in this section that appear in ***italic boldface*** type.

2. Create two different quadratic trinomials that have –3 as the x value of the vertex.

3. Virtual Reality Experiences Revenue: In Chapter 3, we investigated the function $R = -2x^2 + 10{,}000x$, where the input x is the number of headsets and R is the total revenue if x headsets are sold.

a. What are the zeros of this function? What do the zeros mean in the context of number of headsets sold and revenue?

b. What is the vertex of the graph of this function? What does the vertex mean in the context of number of headsets sold and revenue?

c. Write the function in factored form.

4. Fencing In the Dog: In Chapter 3, we investigated the function $A = -l^2 + 100l$, where l is the length of a rectangular dog pen and A is the area of the pen.

a. What are the zeros of this function? What do the zeros mean in the context of the length and area of the dog pen?

b. What is the vertex of the graph of this function? What does the vertex mean in the context of the length and area of the dog pen?

c. Write the function in factored form.

5. In Chapter 3, we introduced the following model: If an earthly object is projected straight up into the air, then the height h (in feet) of the object above the Earth is given by a quadratic function of the time t (in seconds) elapsed since the object was shot or thrown into the air. The general model is $h(t) = -16t^2 + v_0 t + h_0$, where v_0 is the initial velocity (in feet per second) and h_0 is the initial height (in feet) above the Earth. Suppose the initial velocity is 64 feet per second and the initial height is 192 feet.

a. Write the specific model for the given information.

b. What are the zeros of this function? What do the zeros mean in the context of the time and height of the object?

c. What is the vertex of the graph of this function? What does the vertex mean in the context of the time and height of the object?

d. Write the function in factored form.

6. Write in the form $y(x) = ax^2 + bx + c$. Identify the values of the parameters a, b, and c.

a. $y(x) = -7(x+5)^2 + 2$ **b.** $y(x) = 6(x-9)^2 - 10$

7. Write in the form $y(x) = a(x-h)^2 + k$. Identify the vertex.

a. $y(x) = 5x^2 + 4x + 1$ **b.** $y(x) = 6x^2 - x - 1$

c. $y(x) = x^2 - 12x + 36$ **d.** $y(x) = 2x^2 + 30x + 48$

8. The numbers $\frac{2}{3}$ and $-\frac{5}{2}$ are the zeros of a quadratic function $y(x)$ with integer coefficients. The y-intercept is $(0, -2)$.

a. Write the factors of the function. Your answer should not contain fractions.

b. Write an equation for the function.

9. Explain the difference between the functions $y(x) = 2x$ and the function $Q(x) = x^2$.

10. Describe how you would visualize the graph of $y(x) = 0.3(x + 2)^2 + 7$.

11. Create a quadratic function with vertex at (5, 6) such that the second finite differences in successive outputs is 4.

12. A quadratic function has a line of symmetry at $x = 2$ and an absolute maximum of 3. The point (0, 1) is on the graph of the function. Find the algebraic representation (equation).

13. The graph of a quadratic function appears in Figure 11.

FIGURE 11

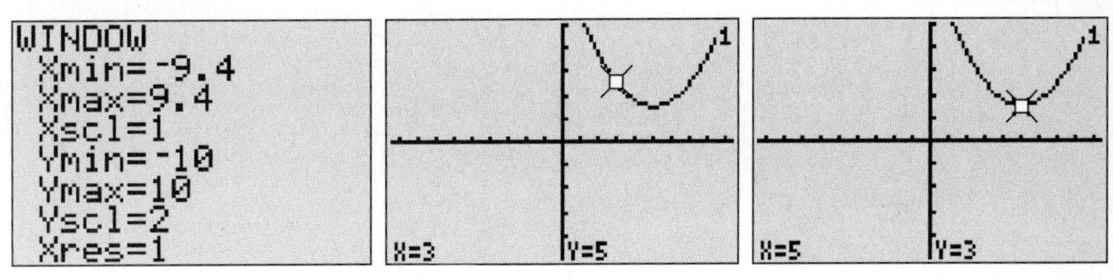

Write the algebraic representation (equation). Explain what you did.

14. Given the function $y(x) = x^2 + 2x - 3$:

a. Identify the values of the parameters a, b, and c from the general quadratic $y(x) = ax^2 + bx + c$.

b. Predict the effect on the graph if only the sign of c is changed.

c. Graph the given function along with the function in which the sign of c is changed. Describe the effect if only the sign of c is changed.

d. Predict the effect on the graph if only the sign of b is changed.

e. Graph the given function along with the function in which the sign of b is changed. Describe the effect if only the sign of b is changed.

f. Predict the effect on the graph if only the sign of a is changed.

g. Graph the given function along with the function in which the sign of a is changed. Describe the effect if only the sign of a is changed.

15. When just the sign of a is changed, the graphs are ***tangent*** to each other.

a. What is the point of tangency?

b. Find the equation of the tangent line to the curves at the point of tangency. You might do this using trial and error graphically.

c. Compare the equation of the line of tangency to the original function. Write a conjecture about the point of tangency and the equation of tangency when two quadratic functions differ only in the sign of a.

16. Create a quadratic function. Test your conjecture from the preceding exploration. Modify if necessary.

17. Given the quadratic function $y(x) = x^2 - 5x - 3$, find the inputs if the output is 21. Explain what you did.

18. Find the solutions to each of the following equations, inequalities, or systems of equations.

a. $4x - 7 = 2 - 8x$

b. $x^2 - 5x + 6 = 0$

c. $4x - 7y = 2$
$3x + y = -9$

d. $x^2 + 9x - 2 = -16$

e. $5 - 2x > 7x + 3$

f. $6x^2 + x - 3 > 2$

g. $5 - 3x = 17x + 5$

h. $4x - 7y + 2z = -5$
$3x + y - z = 1$
$x + 4y + 7z = 9$

i. $x^2 - 5x + 6 < 0$

j. $6x^2 + 7x = 5.$

19. Find the zeros and factor completely. If the polynomial is prime, say so.

a. $5x^2 + 4x + 1$

b. $6x^2 - x - 1$

c. $x^2 - 12x + 36$

d. $2x^2 + 30x + 48$

e. $x^2 - 9x + 14$

f. $x^2 + 13x + 40$

REFLECTION

Write a paragraph describing the critical features of the graph of a quadratic function.

SECTION 6.3

Finding Roots

Purpose

- Investigate the difference between a square and a square root.
- Investigate the square root in which the radicand is a square.
- Simplify square roots in which the radicand contains squares.

In the previous two sections of this chapter, we studied quadratic functions in which the zeros of the function are rational numbers. In fact, a quadratic function will only factor (with the exception of factoring a common monomial factor) over the rational numbers if the zeros of the function are rational numbers.

The rest of the chapter will look at the case in which the zeros of a quadratic function are not rational. There are two such possibilities. The zeros may be irrational and thus the graph intersects the x-axis at irrational numbers. However, if the parabola does not intersect the x-axis, then the zeros are not real numbers. Instead the zeros are nonreal complex numbers, which we investigate in Section 6.4.

In this section, we begin to investigate the case in which the zeros of the function are irrational. In particular, these zeros involve square roots. As a result, this section will develop some important ideas about square roots.

Falling Objects: If an object is dropped from height h feet, the distance (in feet) of the object from the ground (ignoring wind resistance) is given by the expression $h - 16t^2$, where t is the number of seconds since the object was released.

1. Suppose the object is dropped from a height of 448 feet.

a. Write the specific function in which the distance d (in feet) from the ground is the output and time t (in seconds) since the object was released is the input.

b. What is the domain of the function you wrote in part **a** considering the problem situation?

c. What is the range of the function you wrote in part **a** considering the problem situation?

d. Express the function you wrote in part **a** in factored form.

e. Approximate the t-intercepts of the function. Were you able to find the t-intercepts exactly? Why or why not?

f. Interpret the meaning of the zeros of this function in terms of the time and the distance the object is from the ground.

If the object is dropped from a height of 448 feet, the function is

$$d(t) = 448 - 16t^2$$

Since t is the input and represents time, the domain is the set of all nonnegative real numbers less than or equal to the time the object hits the ground (which we don't know right now).

Since d is the output and represents height, the range is the set of all nonnegative real numbers less than or equal to 448, the initial height.

We can write the function in factored form by factoring the common monomial factor, 16 from both terms.

$$d(t) = 16(28 - t^2)$$

We know that the difference of two squares factors, but 28 is not a square, so we cannot factor the function further within the rational numbers.

Figure 1 displays a graph of the function along with the approximate value of the positive t-intercept.

FIGURE 1

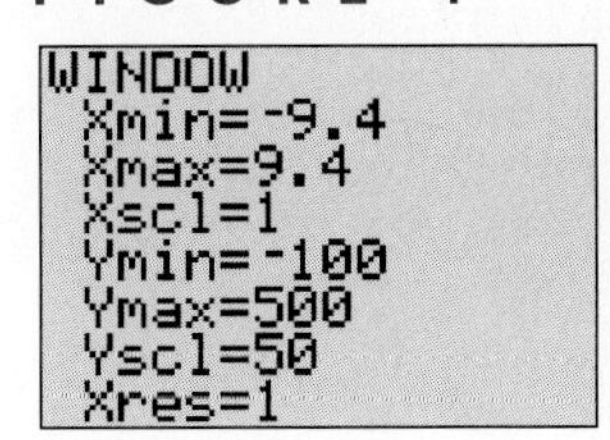

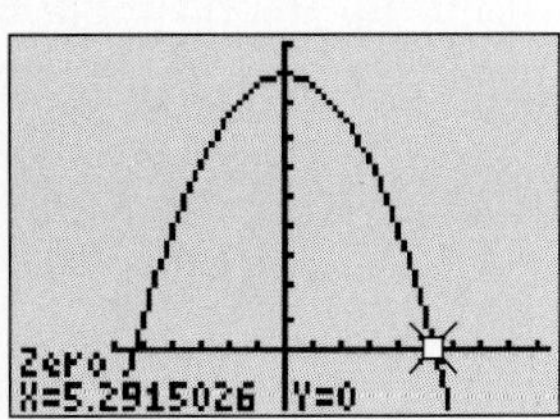

Both t-intercepts are irrational numbers and, thus, cannot be written exactly as decimals. Of course, the negative zero is not in the domain of the function considering the problem situation.

Interpreting the t-intercept displayed in Figure 1, we can say that the object will strike the ground after about 5.29 seconds.

What is going on algebraically? Let's take a look.

2. Write an equation that expresses the fact that the output of $d(t) = 16(28 - t^2)$ is 0.

3. Solve the equation for t^2.

4. In order to calculate t, what operation must we use? Perform this operation and write the value of t that results.

If we want to know when the output is zero, the related equation is

$$16(28 - t^2) = 0$$

Divide both sides by 16: $0 = 28 - t^2$

Add t^2 to both sides: $t^2 = 28$

Now we must reverse the process of squaring. Reversing the process requires that we compute a square root. When we take the square root of any positive real number, we get two answers: a positive value and a negative value. Since this process of taking the square root yields two outputs for all positive inputs, it is a relation, but not a function.

A relation machine appears in Figure 2.

FIGURE 2

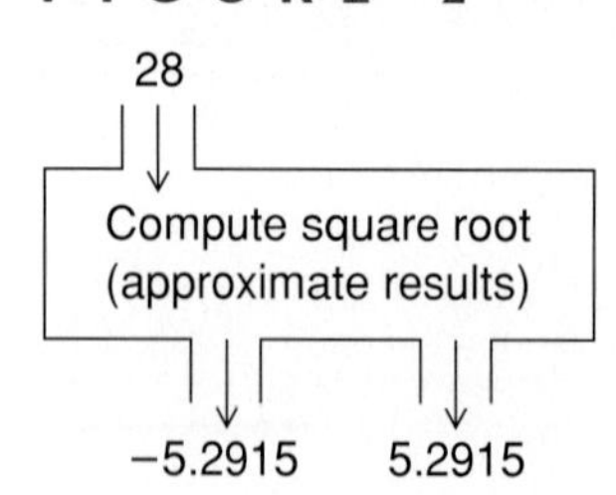

So, if $t^2 = 28$, then $t \approx -5.2915$ or $t \approx 5.2915$

Since the problem situation requires that t be nonnegative, we see that the output is zero when the input is about 5.2915. This matches the point we highlighted on the graph in Figure 1.

One purpose of this section is to develop some initial ideas about square roots. Let's investigate.

5. By definition, a number a is a ***square root*** of another number b if $a^2 = b$.

a. Which letter acts as the square root? Why?

b. What letter acts as the square? Why?

c. If $a^2 = b$, complete Table 1. Based on your results, modify your answers to parts a and b, if necessary.

TABLE 1 Squares and Square Roots I

b	a
25	
81	
100	
9	
12	
1	
5	
0	

6. Complete Table 2. Compare and contrast the outputs of the two functions $y = x^2$ and $y = \sqrt{x}$.

TABLE 2 Squares and Square Roots 2

x	$y = x^2$	$y = \sqrt{x}$
2		
4		
6		
8		
10		
12		

7. Compare and contrast the graphs of $y = x^2$ and $y = \sqrt{x}$.

8. Compare and contrast the graphs of $y = \sqrt{x}$ and $y = -\sqrt{x}$.

As stated in the investigation, a number a is a ***square root*** of another number b if $a^2 = b$.

In this definition, a is acting as the square root and b is acting as the square.

For example, 7 is a square root of 49 since $7^2 = 49$. In this case, $a = 7$ and $b = 49$.

Also, –7 is a square root of 49 since $(-7)^2 = 49$.

Every positive real number has two square roots that are opposites of each other. In the last example, both 7 and –7 are square roots of 49. This means that each row in the a column of Table 1 should contain two values—a positive value and a negative value.

The **positive** square root of a given real number is called the ***principal square root*** of the given number.

For example, 7 is the principal square root of 49.

The symbol $\sqrt{b}$ represents the principal square root of b. In this notation, the $\sqrt{\ }$ is called the ***radical*** and b is called the ***radicand***.

So $\sqrt{49}$ means the principal square root of 49, which is 7. Given $\sqrt{49}$, 49 is called the radicand.

To represent the square root of b that is negative, we write $-\sqrt{b}$.

For example, $-\sqrt{49} = -7$.

Figure 3 displays a table and a graph for the squaring and "square rooting" functions.

FIGURE 3

X	Y1	Y2
2	4	1.4142
4	16	2
6	36	2.4495
8	64	2.8284
10	100	3.1623
12	144	3.4641
14	196	3.7417

X=2

WINDOW
Xmin=-9.4
Xmax=9.4
Xscl=1
Ymin=-10
Ymax=10
Yscl=2
Xres=1

Y1=X²
X=0 Y=0

Y2=√(X)
X=0 Y=0

Notice that, for inputs larger than 1, $y = x^2$ produces outputs that are larger than the inputs while $y = \sqrt{x}$ produces outputs that are smaller than the inputs.

From the graph we see that $y = x^2$ is a parabola symmetric about the vertical axis. The graph of $y = \sqrt{x}$ can be interpreted as "half of a parabola" that opens to the right. The fact that we are graphing the principal square root requires that the outputs be positive so we only have the part of the parabola that is above (or on) the x-axis.

The graph of $y = -\sqrt{x}$ produces the other half of the parabola. The outputs are negative so the graph appears below (or on) the x-axis (Figure 4).

FIGURE 4

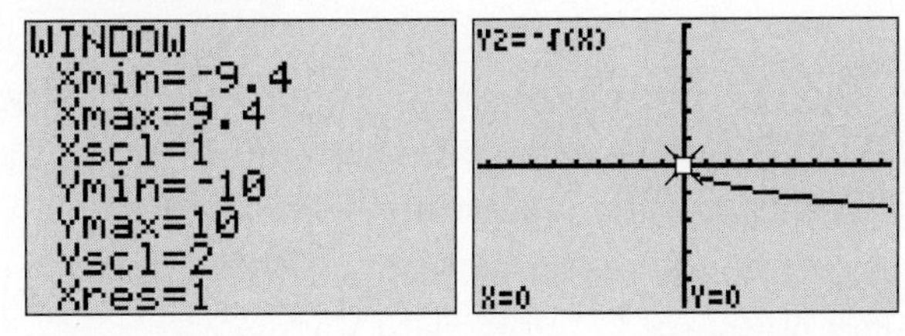

Now let's investigate some properties of square roots and squaring.

9. What do you think $\sqrt{(a)^2}$ represents for any number a?

10. Complete Table 3. Predict the result. Use a calculator to check the result. Modify your prediction if necessary.

TABLE 3 Square Roots of Squares

Expression	Simplification
$\sqrt{(5)^2}$	
$\sqrt{(63)^2}$	
$\sqrt{(25)^2}$	
$\sqrt{(-3)^2}$	
$\sqrt{(-2)^2}$	
$\sqrt{(-5)^2}$	
$\sqrt{(-7)^2}$	

11. Based on the results in Table 3, what does $\sqrt{x^2}$ equal?

In Table 3, we investigated the process of computing the square root of a square. Notice that the domain to this process is all real numbers. This occurs since the first step is squaring. The range is all nonnegative numbers since the last operation is principal square root. Figure 5 displays a function machine and a graph for the process $\sqrt{x^2}$.

This process is equivalent to finding the absolute value of a number. The output is the original input if the input was nonnegative. The output is the opposite of the input if the input is negative.

In general, $\sqrt{(x)^2} = |x|$.

This idea is helpful in solving equations involving squares algebraically.

FIGURE 5

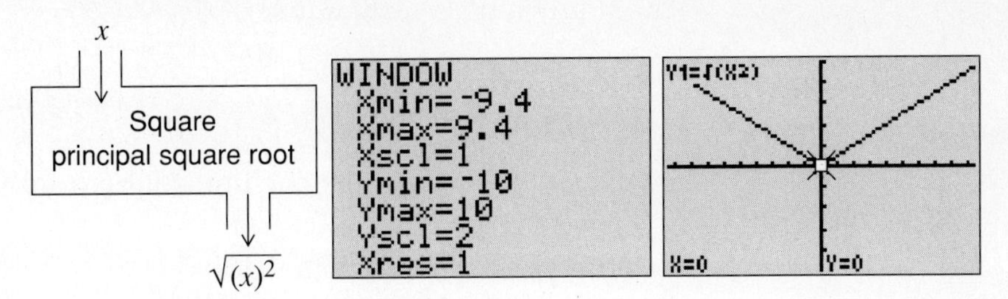

INVESTIGATION

12. Recall when we solved the equation $0 = 16(28 - t^2)$, we stopped at the point $t^2 = 28$ and stated that the answers for t involved the square roots of 28. Let's look at the algebraic steps that produce these answers. Begin with $t^2 = 28$.

a. Reverse the process on the variable by applying the square root operation to both sides of the equation. Do not simplify.

b. Replace any square roots of squares with the absolute value function.

c. Simplify the numerical term by calculating the approximate square root.

d. Identify the possible answers.

13. If s is the length of the side of a square, then the area of the square is s^2. Suppose that the area of the square is 6241 square inches if we increase the length of each side by 3.

a. Write an expression involving s that expresses the length of sides after they are increased by 3 inches.

b. Using the expression in part **a**, write an expression that represents the area of the square after the sides have been increased by 3 inches.

c. Write an equation that expresses the fact that the area of the square is 6241 square inches.

d. Solve the equation in part **c** using an approach similar to that of the previous Investigation.

We wish to look at an algebraic procedure to solve the equation $t^2 = 28$.

Apply square root to both sides:	$\sqrt{t^2} = \sqrt{28}$
Replace with absolute value:	$\lvert t \rvert = \sqrt{28}$
Compute square root:	$\lvert t \rvert \approx 5.2915$
Definition of absolute value:	$t \approx 5.2915$ or $-t \approx 5.2915$
Possible answers:	$t \approx 5.2915$ or $t \approx -5.2915$

If s is the length of the side of a square, then $(s + 3)$ is the length of each side if the sides are increased by 3 inches. The resulting area is $(s + 3)^2$. If the resulting area is 6241 square inches, the equation is

$$(s + 3)^2 = 6241.$$

A solution procedure follows.

Apply square root to both sides:	$\sqrt{(s + 3)^2} = \sqrt{6241}$
Replace with absolute value:	$\lvert s + 3 \rvert = \sqrt{6241}$
Compute square root:	$\lvert s + 3 \rvert = 79$
Definition of absolute value:	$(s + 3) = 79$ or $-(s + 3) = 79$
Possible values for absolute value:	$(s + 3) = 79$ or $s + 3 = -79$
Solve for x:	$s = 76$ or $s = -82$

Since s represents the length of the original side of the square, s must be positive, so we omit the negative solution and conclude that the original side had length 76 inches.

We should investigate one other issue. The equation $\lvert t \rvert = \sqrt{28}$ involves an irrational answer. We approximate the square root of 28 to finish the solution process. If we wanted an exact answer, we might leave the answer in the form $\sqrt{28}$. Is there a simpler exact form? Let's investigate.

14. Predict the *exact* simplified form of each given expression. Use a symbol manipulator, if available, to simplify the expressions without approximating. Modify your predictions if necessary.

TABLE 4 Simplifying Radicals

Expression	Simplification
$\sqrt{3^2}$	
$\sqrt{3^4}$	
$\sqrt{3^6}$	
$\sqrt{3^8}$	
$\sqrt{3^9}$	
$\sqrt{3^{15}}$	
$\sqrt{2^2 3^2}$	
$\sqrt{3^2 5^2 7^2}$	
$\sqrt{(2^2 \cdot 3 \cdot 5 \cdot 7^2)}$	
$\sqrt{(2^4 \cdot 3^5 \cdot 5^3 \cdot 7^2)}$	
$\sqrt{x^2 y^4}$	

15. Discuss in general how to simplify a radical whose radicand involves numbers to powers.

Let's analyze a few of the expressions from Table 4.

Simplify $\sqrt{3^6}$. The exponential expression 3^6 can be thought of as $3^2 3^2 3^2$. We rewrite in terms of squares to match the fact that we are computing a square root.

So $\sqrt{3^6} = \sqrt{3^2 3^2 3^2} = |3||3||3| = 3^3 = 27$

We have intuitively used an important property of radicals, sometimes referred to as the ***distributive property of roots over products***.

Written algebraically, $\sqrt{a \cdot b} = \sqrt{a} \cdot \sqrt{b}$.

Similarly, $\sqrt{3^9} = \sqrt{3^2 3^2 3^2 3^2 3} = |3||3||3||3|\sqrt{3} = 3^4\sqrt{3} = 81\sqrt{3}$.

Note that the form $a\sqrt{b}$ represents the product of a and the square root of b.

Simplify $\sqrt{2^2 3^2}$. The radicand is the product of two squares. The square root of each is the absolute value of the base.

$$\sqrt{2^2 3^2} = |2| \cdot |3| = 2(3) = 6$$

Let's look at a few more.

Simplify $\sqrt{(2^2 \cdot 3 \cdot 5 \cdot 7^2)}$. The radicand contains two factors that are squares and two that are not. This suggests that the nonsquares will remain in the radicand after simplification.

$$\begin{aligned} \sqrt{(2^2 \cdot 3 \cdot 5 \cdot 7^2)} &= \sqrt{2^2} \cdot \sqrt{3} \cdot \sqrt{5} \cdot \sqrt{7^2} = |2| \cdot \sqrt{3} \cdot \sqrt{5} \cdot |7| \\ &= 2 \cdot 7 \cdot \sqrt{3} \cdot \sqrt{5} = 14\sqrt{15} \end{aligned}$$

$\sqrt{15}$ cannot be simplified without approximating, since 15 does not contain a square factor.

Simplify $\sqrt{(2^4 \cdot 3^5 \cdot 5^3 \cdot 7^2)}$.

$$\begin{aligned} \sqrt{(2^4 \cdot 3^5 \cdot 5^3 \cdot 7^2)} &= \sqrt{2^4 \cdot 3^4 \cdot 3 \cdot 5^2 \cdot 5 \cdot 7^2} \\ &= |2^2| \cdot |3^2| \cdot |5| \cdot |7| \cdot \sqrt{3 \cdot 5} \\ &= 4(9)(5)(7) \cdot \sqrt{15} \\ &= 1260\sqrt{15}. \end{aligned}$$

Simplify $\sqrt{x^2 y^4}$.

$$\sqrt{x^2 y^4} = |x| \cdot |y^2| = |x| \cdot y^2$$

Since we do not know the sign on x, we cannot eliminate the absolute value from $|x|$. Since y^2 cannot be negative, we can eliminate the absolute value from $|y^2|$.

The simplification process just investigated requires that the radicand be written as a product of powers of primes. Let's look at how to do this.

Let's look at simplifying, without approximating, $\sqrt{3960}$.

16. Write the prime factorization of the radicand, 3960. You might use a factor tree such as

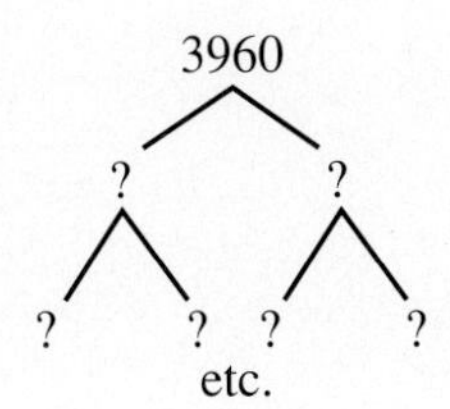

a. Find two numbers whose product is 3960. Write each on a branch coming from 3960.

b. For each branch, find two numbers whose product equals the number. Write these on a branch from the number.

c. Continue the process until the numbers at the ends of the branches are all prime numbers.

d. The prime factorization is the product of the numbers at the ends of the branches.

17. Use the prime factorization of 3960 to simplify, without approximating, $\sqrt{3960}$.

18. Use the prime factorization of 28 to simplify, without approximating, $\sqrt{28}$.

DISCUSSION

In Table 4, we simplified radicals when the radicand contained a ***prime factorization*** of a number. Recall that a ***prime number*** is a whole number with exactly two divisors, 1 and itself.

To create a prime factorization, we often use a ***factorization tree*** as shown in Figure 6.

Prime factor 3960.

FIGURE 6

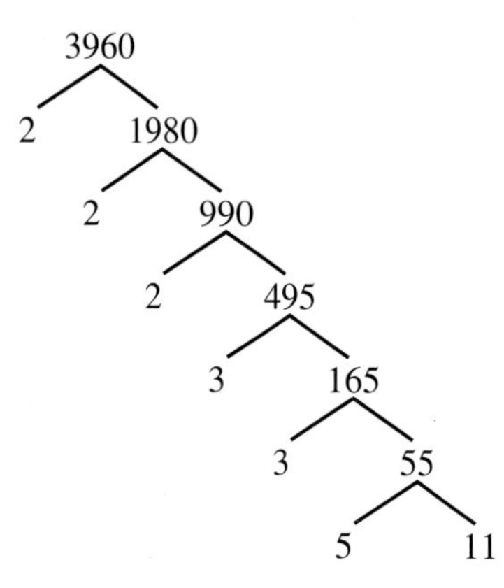

So the prime factorization of 3960 is $2^3 \cdot 3^2 \cdot 5 \cdot 11$.

Then $\sqrt{3960} = \sqrt{(2^3 \cdot 3^2 \cdot 5 \cdot 11)} = 2 \cdot (3) \cdot \sqrt{2 \cdot 5 \cdot 11} = 6\sqrt{110}$.

Similarly, the prime factorization of 28 is $2^2 \cdot 7$.

So $\sqrt{28} = \sqrt{2^2 \cdot 7} = 2\sqrt{7}$.

If we return to the equation $|t| = \sqrt{28}$, we can now write $|t| = 2\sqrt{7}$ and conclude that $t = 2\sqrt{7}$ or $t = -2\sqrt{7}$. This is the simplest form of the exact solutions to the equation.

We will investigate more general roots later in the text.

STUDENT DISCOURSE

Pete: I always thought that $\sqrt{4}$ was equal to 2 and –2. Am I wrong about that?

Sandy: It's true that the square roots of 4 are 2 and –2. But $\sqrt{4}$ doesn't mean all the square roots of 4.

Pete: Well, I read it meant the principal square root of 4, but I'm still a little shaky on that.

Sandy: The principal square root means the positive square root, not the negative square root. Try it on your calculator.

Pete: You're right. When I punch in $\sqrt{4}$, the calculator just gives me 2. But what do I write if I want the negative square root?

Sandy: Think about it. You can answer that yourself.

Pete: Probably $-\sqrt{4}$. Right?

Sandy: Exactly.

Pete: But then what does $\sqrt{-4}$ mean?

Sandy: What do you think?

Pete: Let me punch it in on my calculator. Oh, I get an error message. I guess square roots of negative numbers don't exist.

Sandy: Look what my calculator says, $2i$.

Pete: That's weird. What's that i there for and what does it mean?

Sandy: That's called an imaginary number and you get to investigate it in the next section.

Pete: Why did your calculator do that and mine didn't?

Sandy: I have my calculator in a complex number mode. That means it can compute with these new numbers. You need to check to see if your calculator has such a mode.

Pete: Thanks. I will.

1. In the Glossary, define the words in this section that appear in ***italic boldface*** type.

2. Compare and contrast the processes of squaring and square rooting.

3. Discuss the difference between $3\sqrt{2}$ and $2\sqrt{3}$.

4. Find the approximate value of $5 + 7\sqrt{11}$. State the processes you used to approximate the number.

5. There is one number with only one square root. What is it?

6. Compare and contrast $\sqrt{(x)^2}$ and $(\sqrt{x}\,)^2$. Include in your answer a discussion of domain and range for each expression. Also state under what conditions they are the same.

7. Compare and contrast $-\sqrt{x}$ and $\sqrt{-x}$. What is the domain and range of each?

8. Complete Table 5. Based on the results, what can you say about the relationship between $\sqrt{a+b}$ and $\sqrt{a}+\sqrt{b}$?

TABLE 5 Roots and Sums

a	b	$\sqrt{a+b}$	$\sqrt{a}+\sqrt{b}$
4	9		
25	36		
81	49		
2	3		
11	16		
9	27		

9. Complete Table 6. Based on the results, what can you say about the relationship between $\sqrt{x}$ and $x^{1/2}$?

TABLE 6 Base to One-half Power

x	$\sqrt{x}$	$x^{1/2}$
1		
2		
3		
4		
5		
6		

10. Use prime factorization to simplify the radicals in Table 7.

TABLE 7 Simplifying Radicals 2

Expression	Prime Factorization of Radicand	Simplification
$\sqrt{244}$		
$\sqrt{54}$		
$\sqrt{108}$		
$\sqrt{410}$		
$\sqrt{1573}$		
$\sqrt{3927}$		
$\sqrt{9800}$		
$\sqrt{71,825}$		

11. Square Dog Pen: We need a square dog pen. Let s represent the length of the side and A represent the area.

a. Write the algebraic representation for the function that expresses area in terms of length of a side.

b. Write the equation that must be solved to find the length of a side if the area is 200 square meters. Solve the equation.

c. The length of a side is decreased by 2 m. Write an algebraic expression for this in terms of s.

d. Refer to part **c**. Suppose the area is 162 square meters when the length of a side is decreased by 2 meters. Write an equation that will find s. Solve the equation.

12. Find the solutions to each of the following equations, inequalities, or systems of equations.

a. $x^2 = 48$

b. $3x - 7 = 4 - 9x$

c. $4x^2 = 16x - 15$

d. $(x - 5)^2 = 68$

e. $4 - x > 3 + x$

f. $5y - 2x = 15$
$3x + 2y = -9$

g. $x^2 - 2x - 24 < 0$

h. $y = 7x + 2$
$14x - 2y = 1$

i. $(2x + 9)^2 = 12$

j. $2x^2 - 3x = x^2 + 2x + 6$

13. Write the following functions in factored form. Identify the vertex.

a. $y(x) = x^2 - 8x + 12$

b. $y(x) = 4x^2 + 16x - 15$

14. The numbers –5 and $\frac{6}{5}$ are the zeros of a quadratic function $y(x)$ with integer coefficients. The y-intercept is $(0, -1)$.

a. Write the factors of the function. Your answer should not contain fractions.

b. Write an equation for the function.

15. Create the equation of the quadratic function through the points $(1, 2)$, $(-3, 5)$, and $(0, -4)$. Approximate the zeros and the vertex of the function.

16. Write the equation of the linear function parallel to the line $2x - 3y = 7$ if the linear function has an x-intercept of $(4, 0)$.

REFLECTION

Write a paragraph describing how to simplify radicals involving square roots.

SECTION 6.4

The Reality of Complex Numbers

Purpose

- Review relationships among number systems.
- Investigate the square root of a negative number.
- Investigate the complex number system.
- Investigate operations on complex numbers.

In the preceding section, we investigated both the meaning and the simplification of square roots. We noted that the square root of a negative number is not a real number. But does the square root of a negative number exist? The answer is yes, but we need a new number system—the complex number system—in order to define such a quantity. Defining the square root of a negative number may go against your intuition, but, without the mathematics of complex numbers, we wouldn't be able to analyze such quantities as electricity, the flow of blood in the bloodstream, the flow of air over the wings of a plane, or the flow of water through a pipe. When complex numbers were initially "discussed" by mathematicians, they were called "imaginary" numbers. The name is unfortunate since it suggests that the numbers have no useful purpose.

Seriously, we need to investigate complex numbers if we are to fully understand quadratic functions. The goal of this section is to introduce you to complex numbers and their operations.

1. Consider the function $y(x) = 3x + 5$. Write an example, if possible, of an output value for which the corresponding input value is

a. a whole number. Justify your answer.

b. an integer, but not a whole number. Justify.

c. a rational number, but not an integer. Justify.

d. a real number, but not a rational number. Justify.

2. Consider the basic quadratic function $Q(x) = x^2$. Write an example, if possible, of an output value for which the corresponding input value is:

a. a whole number. Justify your answer.

b. an integer, but not a whole number. Justify.

c. a rational number, but not an integer. Justify.

d. a real number, but not a rational number. Justify.

e. not a real number. Justify.

One justification for the hierarchy of number systems is that, with each new number system, we can solve more problems and answer more questions. You may wish to refresh your memory about this hierarchy by reviewing Section 2.4.

For example, suppose we want to know the input to function y that produces an output of 2.

Resulting equation: $3x + 5 = 2$

Subtract 5 from both sides: $3x = -3$

If we only know about whole numbers, this equation is not solvable—it has no solution within the whole numbers. However, the equation does have a solution within the integers.

Divide both sides by 3: $x = -1$

We need the integers, minimally, to be able to answer the question.

Let's change the desired output to 6.

Resulting equation: $3x + 5 = 6$

Subtract 5 from both sides: $3x = 1$

Now the integers are insufficient for solving this problem. This equation has no solution over the integers. The equation does have a solution if we move to the next superset, the rational numbers. ***Superset*** means a "larger" set that contains all the elements of a given set. For example, the integers are a superset of the whole numbers and the rational numbers are a superset of the integers.

Divide both sides by 3: $x = \frac{1}{3}$

Suppose we want to know if there is an input to function x that produces an output that is not in the set of rational numbers.

Equation: $x^2 = 5$

No rational number will serve as a solution to this equation. The number π is irrational. The arithmetic on π results in irrational numbers. We must advance to the next superset, the real numbers, to solve the equation.

Take the square root of each side: $\sqrt{x^2} = \sqrt{5}$

$x = \sqrt{5}$ or $x = -\sqrt{5}$

The solutions can be approximated by the decimals $x \approx 2.236$ and $x \approx -2.236$.

The exact solutions $\sqrt{5}$ and $-\sqrt{5}$ are irrational numbers that can be approximated by rational numbers. Provided that we choose an output that is a real number, the resulting equation always has a solution that is a real number.

This is not the case when we consider the function $Q(x) = x^2$. If we select a square number as output, then the inputs are two integers that are opposites. If we select a nonsquare whole number as output, then the inputs are two irrational numbers that are opposites.

However, suppose the output is a negative number, such as –1.

Resulting equation:	$x^2 = -1$
Take square root of both sides:	$\sqrt{x^2} = \sqrt{-1}$
State both answers for x:	$x = \sqrt{-1}$ or $x = -\sqrt{-1}$

We do not write $|x|$ for $\sqrt{x^2}$ when x is not a real number. Nonreal numbers cannot be classified as positive or negative. We still must list both answers for x.

Unfortunately, this simple-looking equation has no solution over the real numbers. The solution involves the square root of a negative number. This is not a real number. Thus we need a superset of the real numbers in order to handle such problems.

Such a superset is called the ***complex number system***. Let's investigate.

3. Use a computer or calculator capable of manipulating complex numbers. Enter the expression and record the simplification in Table 1.

TABLE 1 Roots of Negative Numbers I

Expression	Simplification
$\sqrt{-1}$	
$\sqrt{-4}$	
$\sqrt{-6}$	
$\sqrt{-18}$	

4. Based on the results of Table 1, how is $\sqrt{-1}$ defined? Describe how the simplifications are done.

5. For Table 2, predict the simplification of each expression. Use a computer or calculator capable of manipulating complex numbers to find the simplification. Modify your predictions as necessary.

TABLE 2 Roots of Negative Numbers 2

Expression	Simplification
$\sqrt{-9}$	
$\sqrt{-16}$	
$\sqrt{-5}$	
$\sqrt{-7}$	
$\sqrt{-12}$	
$\sqrt{-48}$	
$\sqrt{-72}$	

6. Your friend, Izzy, wants you to describe, in writing, how to simplify the square root of a negative integer. Do so and provide an example to illustrate your procedure.

The set of integers is created by defining some new numbers, the opposites of the whole numbers. The set of rational numbers is created by defining some new numbers, the ratios of integers.

In a similar fashion, we extend the real number system by defining a new number, the number ***i***. This number is used to represent the principal square root of –1. Symbolically,

$$i = \sqrt{-1}$$

The number i is called the ***complex unit***. **Leonhard Euler**, one of the best mathematicians of the eighteenth century, was the first to use i to represent the square root of negative one, though even he believed that complex numbers existed only in the imagination.

This one basic definition provides us with a representation for the square root of any negative number as follows. Assume x is positive.

$$\sqrt{-x} = \sqrt{-1 \cdot x} = \sqrt{-1} \cdot \sqrt{x} = i \cdot \sqrt{x}$$

The square root of a negative number is equal to the number i multiplied by the square root of the corresponding positive number. If possible, the square root of the positive number could then be simplified.

$$\sqrt{-4} = i \cdot \sqrt{4} = i \cdot 2 = 2i$$

and

$$\sqrt{-18} = i \cdot \sqrt{18} = i \cdot \sqrt{2 \cdot 3^2} = i \cdot 3 \cdot \sqrt{2} = 3i\sqrt{2}$$

When involved in a product such as $2i$, the i is usually the last factor. One common exception occurs when one of the factors is a radical. The i is usually written before the radical so that the i does not appear to be under the radical.

Before formally discussing the complex numbers, let's explore some operations involving this new number i.

7. Use a computer or calculator capable of manipulating complex numbers. Enter the expression and record the simplification in Table 3.

TABLE 3 Operations with Complex Numbers I

Expression	Simplification
$4i + 3i$	
$(2 + 3i) + (5 - 4i)$	
$i(i)$	
$(2i)(5i)$	
$(2 + i)(4 + 3i)$	

8. Based on the results of Table 3, explain to your friend, Izzy, how to:

a. add numbers involving the complex unit i.

b. multiply numbers involving the complex unit i.

9. For Table 4, perform the indicated operation and simplify each expression.

TABLE 4 Operations with Complex Numbers 2

Expression	Simplification
$7i - 11i$	
$(3 - 5i) - (6 + 2i)$	
$(5 - 8i) + (-6 - 4i)$	
$(7 + i) - (9 - i)$	
$(3i)(i)$	
$(-2i)(3i)$	
$(-9i)(-11i)$	
$3i(2 + 5i)$	
$(5 - 2i)(6 + 4i)$	
$(3 - 7i)(3 + 7i)$	
$(5 + 7i)^2$	

10. Explain to your friend, Izzy, why there is no i in the answer to $(3 - 7i)(3 + 7i)$. Make up another problem and demonstrate the multiplication for Izzy, who is quite confused.

11. a. Expand the polynomial $(2x+3)^2$.

b. Expand the complex number $(2+3i)^2$.

c. Compare and contrast your answers to parts **a** and **b**.

Numbers such as $2i$ (which is $0+2i$), $4+3i$, $9-2i$, 7 (which is $7+0i$), etc., are examples of ***complex numbers***.

Any number of the form $a+bi$ where a and b are real numbers is called a complex number.

If b is zero, the number is a real number. Thus the complex number system contains the real number system.

One use of complex numbers is in the simplification of square roots of negative numbers as you saw in Tables 1 and 2.

Since this is a new number system, we need to look at the operations on complex numbers. This was the purpose of Tables 3 and 4. An important fact to use when operating on complex numbers is that $i^2=-1$. When i^2 arises in a computation, replace it with -1.

The idea behind complex numbers, being able to define the square root of a negative number, has concerned mathematicians since the thirteenth century. The mathematician ***Jerome Cardan*** published a famous algebra book, ***Ars Magna***, in 1545. The book contains descriptions of how to solve third- and fourth-degree polynomial equations. In this book, he multiplies complex numbers, in spite of having to, in his own words, "put aside the mental tortures involved."[1]

In the general complex number form $a+bi$, the number is called ***pure imaginary*** if $a=0$. Thus numbers like $3i$ and $-7i$ are complex numbers that are referred to as "pure imaginary."

The famous French mathematician ***René Descartes*** coined the word *imaginary* for complex numbers in 1620. This was unfortunate since these numbers are no more "imaginary" than real numbers are "real."

The great German mathematician ***Gottfried von Leibniz*** studied complex numbers extensively in 1673, using them to derive viable algebraic results. He thought of them as "a fine and wonderful refuge of the divine spirit—almost an amphibian between being and nonbeing."[2]

1. David M. Brown, *The History of Mathematics* (Boston: Allyn & Bacon, 1985).
2. Ely Fordham, *Leapfrogs: Complex Numbers* (Cambridge, England: E. G. M. Mann & Son, 1980).

By the eighteenth century, complex numbers were widely accepted and used. If you have some doubts about their uses, ask a friend who is knowledgeable in engineering or electricity to convince you.

One special note about multiplication of complex numbers. You noticed from Tables 3 and 4 that multiplying two numbers of the form $a + bi$ resulted in an answer of the same form. However, when you multiply $(3 - 7i)(3 + 7i)$, you obtain the real number 58. Notice the relationship between the two factors. They are identical except that the sign on the term containing i is different. Two numbers are called ***complex conjugates*** when they are related this way. Multiplying complex conjugates results in a real number.

Let's conclude with an equation to solve.

12. Find all numbers such that the square of the sum of the number and 5 is -20.

a. Write an equation that models the problem.

b. Solve the equation. Simplify completely.

The problem stated in Investigation 12 is modeled by the equation

$(x + 5)^2 = -20$ where x is the number we wish to find.

Let's solve the equation for x.

Original equation:	$(x + 5)^2 = -20$
Take square root of both sides:	$\sqrt{(x + 5)^2} = \sqrt{-20}$
Rewrite radical using i:	$\sqrt{(x + 5)^2} = i\sqrt{20}$
Prime factor radicand:	$\sqrt{(x + 5)^2} = i\sqrt{2^2 \cdot 5}$
Simplify radical:	$\sqrt{(x + 5)^2} = 2i\sqrt{5}$

Remove square root by stating both cases:

$$x + 5 = 2i\sqrt{5} \quad \text{or} \quad x + 5 = -2i\sqrt{5}$$

Subtract 5 from both sides:

$$x = -5 + 2i\sqrt{5} \quad \text{or} \quad x = -5 - 2i\sqrt{5}$$

STUDENT DISCOURSE

Pete: These complex numbers seem to act a lot like polynomials, except when I have to replace i^2 with -1.

Sandy: What do you mean?

Pete: Well, when I add them, I add the terms that don't have i as a factor and I add the terms that do have i as a factor. It's similar to like terms.

Sandy: And...

Pete: When I multiply them, it's like multiplying binomials.

Sandy: But with one big difference.

Pete: Actually, I see several differences. When I multiply binomials like $(x+2)(x-3)$, I end up with three terms after combining like terms. In fact, sometimes there are even four terms if different letters are used in the binomials.

Sandy: So far, so good.

Pete: But when I multiply two complex numbers, I only get two terms—one with i as a factor and one without.

Sandy: Why do you think that happens?

Pete: As I said before, that goes back to the i^2 being equal to -1. If I multiply $(2+3i)(4+5i)$, I get $8+10i+12i+35i^2$.

Sandy: I agree.

Pete: But $35i^2$ becomes -35 and can be combined with the 8. The $10i$ and the $12i$ add to $22i$.

Sandy: That's right. It seems as if you have it. Now try to find $\frac{3+5i}{2-7i}$.

Pete: I'll work on that and let you know.

1. In the Glossary, define the words in this section that appear in ***italic boldface*** type.

2. Write two complex numbers whose product is a real number.

3. Use the patterns from Table 1 to simplify the following without technology.

a. $\sqrt{-81}$

b. $\sqrt{-64}$

c. $\sqrt{-54}$

d. $\sqrt{-40}$

4. Find all values of x so that the square of 2 less than x is equal to –90.

5. Investigate the effects of multiplying any complex number $a + bi$ by i.

6. Use a computer or calculator capable of manipulating complex numbers to complete Table 5 on page 474.

7. Use the pattern in Table 5 to simplify the following.

a. i^{53}

b. i^{241}

c. i^{164}

d. i^{122}

8. Poor Pete. Sandy asked him to calculate $\frac{3+5i}{2-7i}$. He did it on his calculator, which gave him an answer of $-\frac{29}{53}+\frac{31}{53}i$, but he doesn't understand. Help him by multiplying the numerator and the denominator by the conjugate of the denominator. Simplify and see what you get.

9. Use the patterns from Tables 3 and 4 to perform the following operations.

a. $9i - 5i$

b. $(3 - 11i) - (6 + 4i)$

c. $(5i)(-13i)$

d. $4i(7 - 2i)$

e. $(5 - 6i)(4 + 2i)$

f. $(4 - 3i)(4 + 3i)$

TABLE 5 Raising the Number i to a Power

Expression	Simplification
i^0	
i	
i^2	
i^3	
i^4	
i^5	
i^6	
i^7	
i^8	
i^9	
i^{10}	
i^{11}	
i^{12}	
i^{13}	

g. $(6-4i)^2$

h. $(2+i)(2-i)$

10. Consider the sequence generated by the function machine shown in Figure 1 on the next page.

The process begins by selecting any complex number as the first input. Then each output is used as the next input.

FIGURE 1

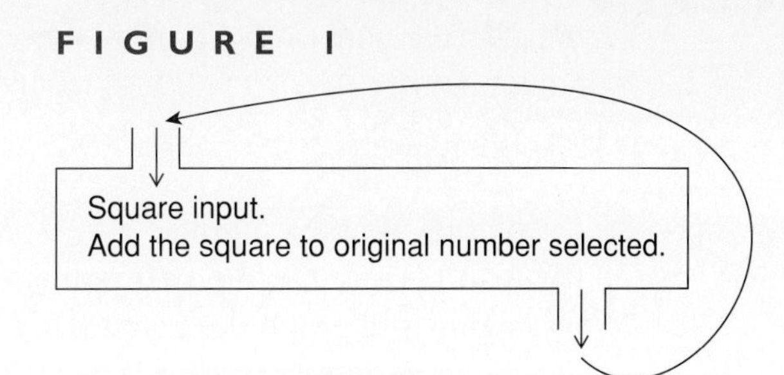

a. Generate the first five elements if the starting number is –2. What happens to the numbers in the sequence?

b. Generate the first five elements if the starting number is 1. What happens to the numbers in the sequence?

c. Generate the first five elements if the starting number is i. What happens to the numbers in the sequence?

d. If the sequence of numbers does not continuously grow larger and larger, the sequence is ***bounded***. If not, it is ***unbounded***. For which of the starting numbers in parts **a–c** is the sequence bounded?

e. The set of all complex numbers for which the sequence is bounded is called the ***Mandelbrot Set***. This is a very important set in the area of ***chaos theory*** and ***fractals***. Find three more complex numbers that are in the Mandelbrot Set. Justify your answer.

11. Geometric interpretations of the Mandelbrot Set lead to intriguing and interesting pictures. To understand the pictures, we need to be able to represent complex numbers graphically. To do so, we define a new plane in which the horizontal axis is called the ***real axis*** and the vertical axis is called the ***imaginary axis***. This is the ***complex plane***. Complex numbers of the form $a + bi$ are graphed by thinking of them as the ordered pair (a, b).

a. Draw the complex plane and mark points corresponding to $3 - 5i$, $7 + i$, -4, and $-6i$.

b. What is the general form of all numbers whose graph is on the vertical axis?

c. What is the general form of all numbers whose graph is on the horizontal axis?

d. Graph the complex number $3 + 2i$. Multiply the number by i. Graph the result. How are the two points related geometrically? Multiply by i again and graph the result. How is this point related to the original point geometrically?

e. Find a picture of the Mandelbrot Set in another resource. Interpret the general structure of the set geometrically.

12. Construct a complete diagram of all the major number systems: natural, whole, integers, rational, real, complex. (*Hint*: See Section 2.4.)

13. Find the solutions to each of the following equations, inequalities, or systems of equations.

a. $3x^2 + 11x - 4 = 0$

b. $5 - 3x > 7$

c. $(4x - 1)^2 = 18$

d. $3x + 11y - 2 = 0$
$x - 4y - 3 = 0$

e. $9x^2 - 25 = 18x^2 - 50$

f. $x^2 + 10x + 16 < 0$

g. $9 - 2x = 4x + 15$

h. $6x^2 = 13x - 6$

14. Write the following functions in factored form. Identify the vertex.

a. $y(x) = 2x^2 - 10x - 12$

b. $y(x) = 12x^2 - 13x + 3$

15. Simplify the following radicals completely without approximating.

a. $\sqrt{2520}$

b. $\sqrt{17{,}150}$

16. Write the equation for the function given the following graphs. Briefly describe what you did.

a.

FIGURE 2

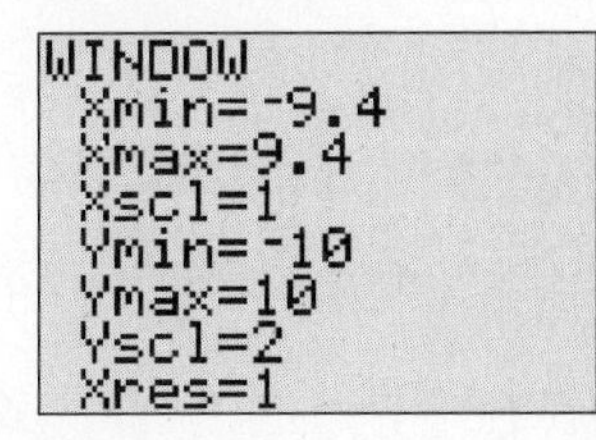

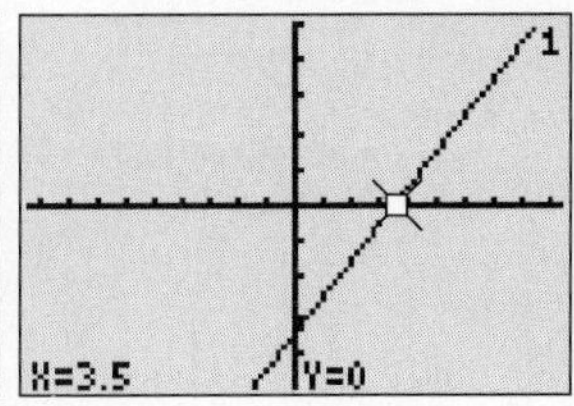

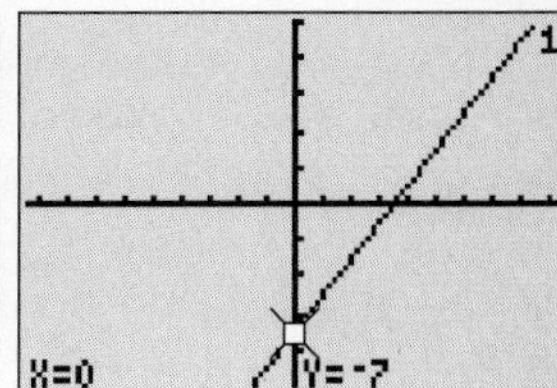

b.

FIGURE 3

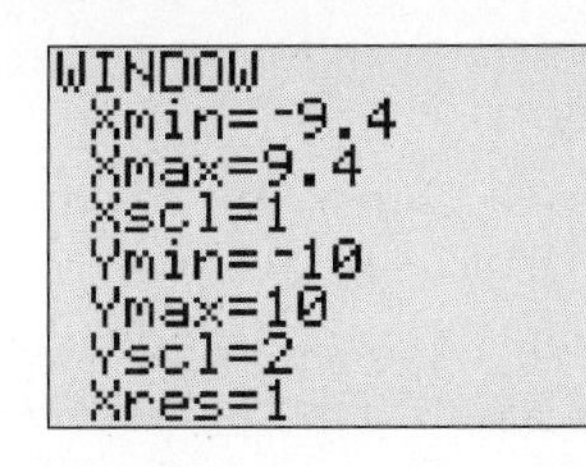

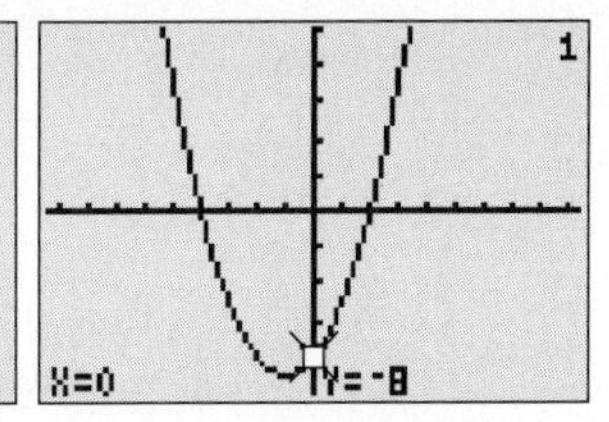

c.

FIGURE 4

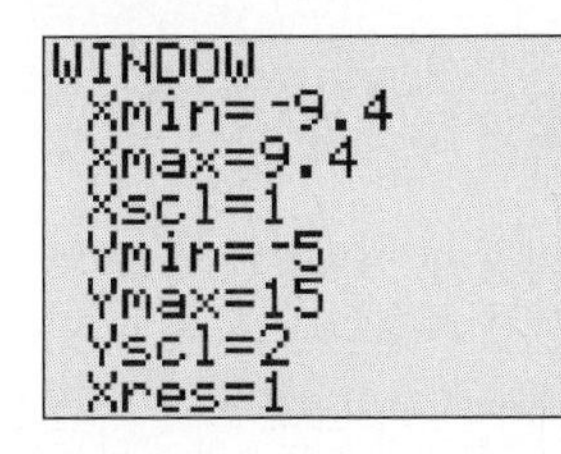

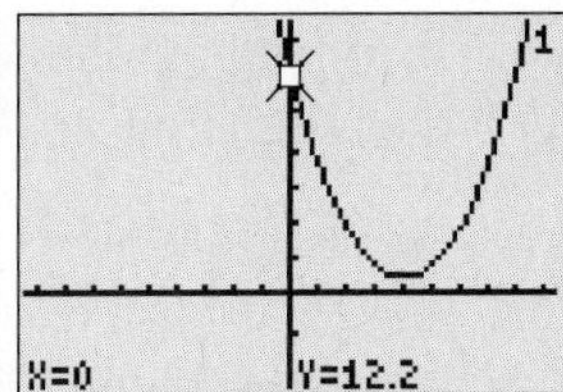

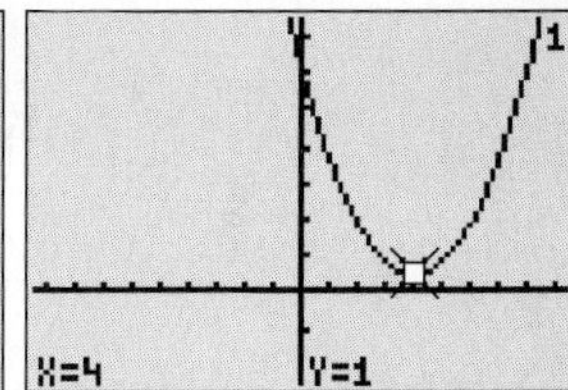

REFLECTION

Write a paragraph describing complex numbers and operations on complex numbers.

SECTION 6.5

General Solution Procedures for Quadratic Equations

Purpose

- Investigate the ***completing the square*** process.
- Use the ***completing the square*** process to derive the quadratic formula.
- Solve quadratic equations using the quadratic formula.

As we've noted several times, the standard form of a quadratic function is $y(x) = ax^2 + bx + c$. We can locate the zeros exactly if the zeros are rational numbers. In fact, if the zeros are rational numbers, we can write the quadratic function in factored form.

However, many quadratic functions have zeros that are either irrational or nonreal complex numbers. In this section, we develop a general formula, called the quadratic formula, that will produce the exact zeros of a quadratic function regardless of whether the zeros are rational, irrational, or nonreal complex numbers. In addition, we find that the formula only relies on the values of the parameters a, b, and c.

We begin by looking at the role of the parameters in finding the zero of a linear function. Next, we'll introduce a technique, called completing the square, that is used to develop the quadratic formula.

1. The general form of a linear function is $y(x) = ax + b$, where a and b are parameters.

a. What does a represent in this function? Give an example.

b. What does b represent in this function? Give an example.

c. Find the zero of this function in terms of the parameters a and b.

2. Consider the points $(-3, -6)$, $(1, 26)$, and $(2, 44)$.

a. Determine the equation of the quadratic function through the points.

b. Graph your function and approximate the zeros.

c. Locate the exact coordinates of the vertex.

d. Write the equation of the function in the form $y(x) = a(x-h)^2 + k$.

e. Use the equation in part d to find the exact values of the zeros.

In the general linear function $y(x) = ax + b$, the parameter a is the slope, or rate of change of the output with respect to the input, of the function. The parameter b is the y coordinate of the y-intercept, which is $(0, b)$.

We find the zero of the function by letting the output be zero and solving for the input.

Output equal to zero:	$ax + b = 0$
Subtract b from both sides:	$ax = -b$
Divide both sides by a assuming that a is not zero:	$x = -\frac{b}{a}$

Thus, the zero of $y(x) = ax + b$ is $-\frac{b}{a}$ assuming a is not zero.

This results in a general formula for finding the zero that depends solely on the value of the parameters.

$$\text{If } ax + b = 0 \text{ then } x = -\frac{b}{a} \text{ assuming } a \text{ is not zero.}$$

We'd like to find a similar formula to solve the quadratic equation $ax^2 + bx + c = 0$ that uses only the parameters a, b, and c.

The quadratic function through the points (–3, –6), (1, 26), and (2, 44) is $y(x) = 2x^2 + 12x + 12$ (Figure 1).

FIGURE 1

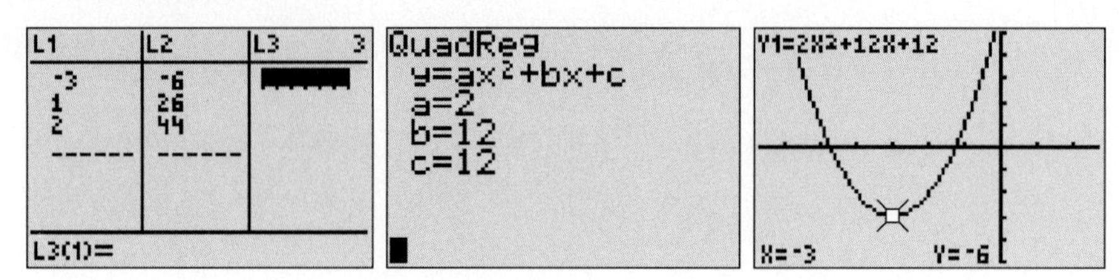

The zeros are irrational numbers—one is between –2 and –1; the other is between –5 and –4. The vertex is (–3, –6).

Since the parameter $a = 2$ and the vertex is (–3, –6), the "vertex" form of the function is $y(x) = 2(x+3)^2 - 6$.

Although we have no method yet to solve $2x^2 + 12x + 12 = 0$ for x, we have solved equations of the form $2(x+3)^2 - 6 = 0$ before.

Add 6 to both sides:	$2(x+3)^2 = 6$
Divide both sides by 2:	$(x+3)^2 = 3$
Apply square root to both sides:	$\sqrt{(x+3)^2} = \sqrt{3}$
Simplify roots:	$x + 3 = \sqrt{3}$ or $x + 3 = -\sqrt{3}$
Subtract 3 from both sides:	$x = -3 + \sqrt{3}$ or $x = -3 - \sqrt{3}$

The exact zeros of $y(x) = 2x^2 + 12x + 12$ are $-3 + \sqrt{3}$ and $-3 - \sqrt{3}$. Figure 2 supports this by displaying the approximations and the graph.

FIGURE 2

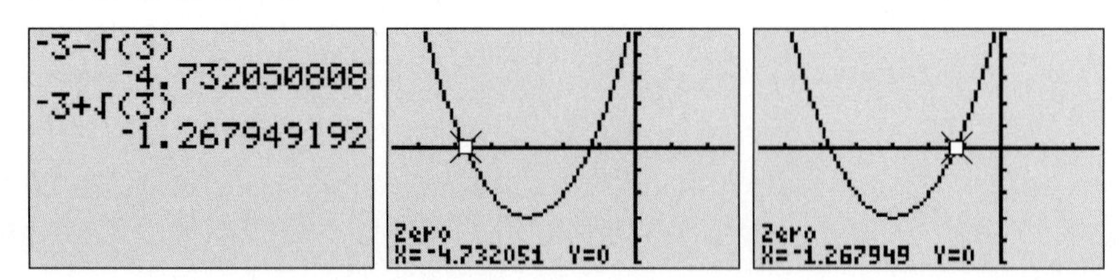

So, by transforming the equation $ax^2 + bx + c = 0$ into the form $a(x-h)^2 + k = 0$, we can locate the exact zeros of a quadratic function. This transformation depends on our ability to locate the vertex exactly.

There are several methods for locating the vertex algebraically. One method begins with creating the term $(x - h)^2$. To understand how, let's take a step back into history. The ancient Greeks thought of mathematics in terms of geometry.

For example, the term x^2 can be thought of geometrically as the area of a square with side length x.

The term $6x$ can be thought of as a rectangle with length 6 and width x.

Thus $x^2 + 6x$ can be thought of as a square of area x^2 and a rectangle of area $6x$. For our purposes, it will be easier to think of $6x$ as the area of two congruent rectangles, each with length 3 and width x (Figure 3).

FIGURE 3

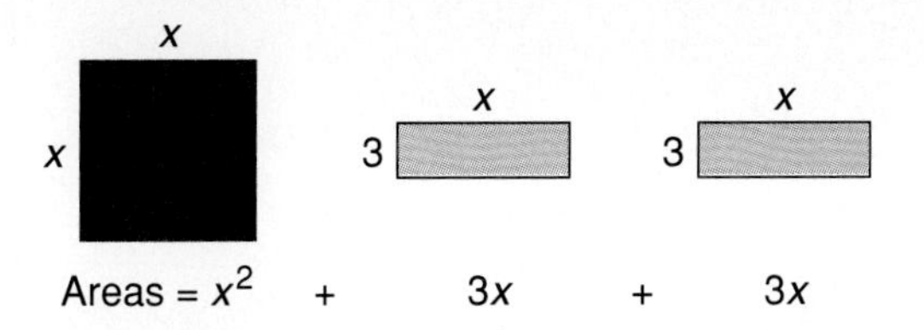

If we put these three figures together, we see a result in Figure 4.

FIGURE 4

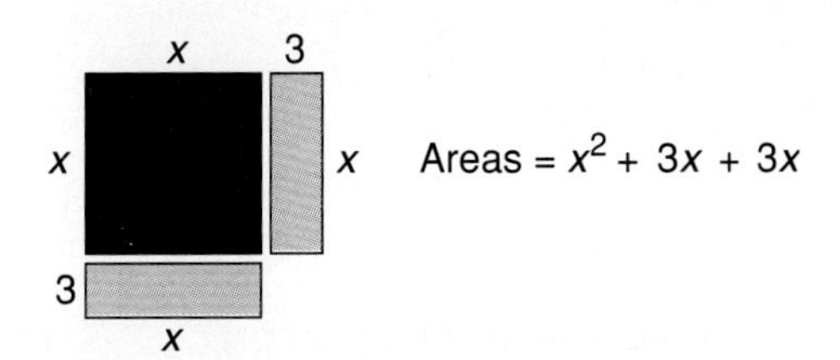

A goal is to add the correct amount in the bottom right corner so that the new figure is a square. This is called ***completing the square***.

3. Draw a geometric object in Figure 4 so that the four objects together create a square.

4. What is the area of the object you drew? What algebraic expression is represented by the sum of the areas of the four objects?

5. What is the total length of a side for the new square created in Figure 4? Using this length, write the area of the large square.

6. a. Represent $x^2 + 10x$ geometrically.

b. What must be added to this expression to complete the square?

c. Write the new expression for total area of the square in two ways—one as a sum and the other in factored form.

Figure 5 contains an additional square required to complete the large square.

FIGURE 5

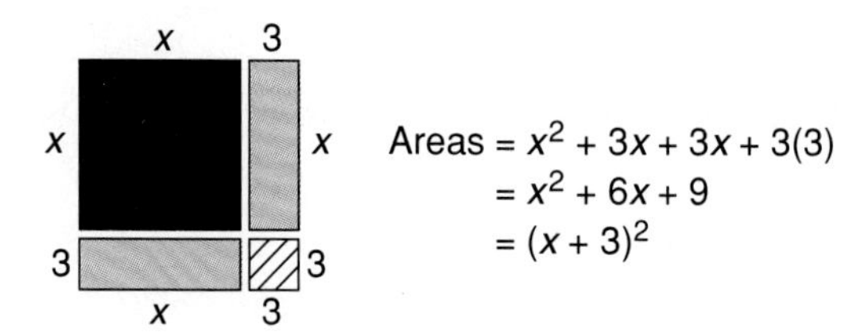

The length of each side of this square is 3 units and the area is 9 square units. Thus, $x^2 + 6x + 9$ is the sum of the areas of the four pieces.

But the large square has a side of length $x + 3$. The area of the large square is $(x + 3)^2$. The expressions $x^2 + 6x + 9$ and $(x + 3)^2$ are equivalent. The second is the factored form of the first.

To do a similar thing with $x^2 + 10x$, think of $10x$ as the sum of the areas of two rectangles, each with sides of length 5 and x. The required square that must be added has length 5 units. Therefore, this additional square has area 25 square units. The total area may be represented as $x^2 + 10x + 25$ or $(x + 5)^2$ (Figure 6).

FIGURE 6

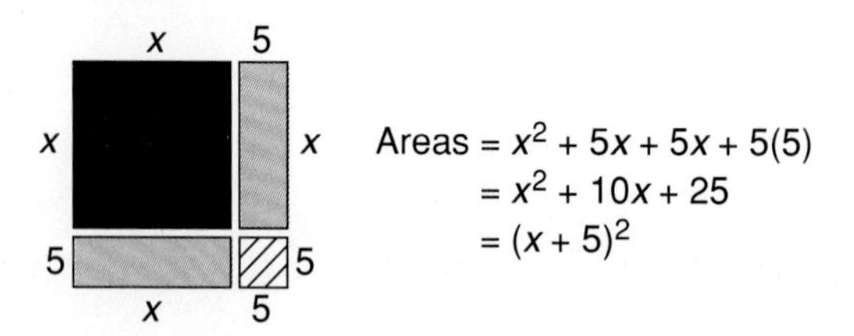

One final point is needed. Suppose we wish to represent $x^2 - 6x$. Now we are taking two rectangles of area $3x$ away from the original square with area x^2 (Figure 7).

FIGURE 7

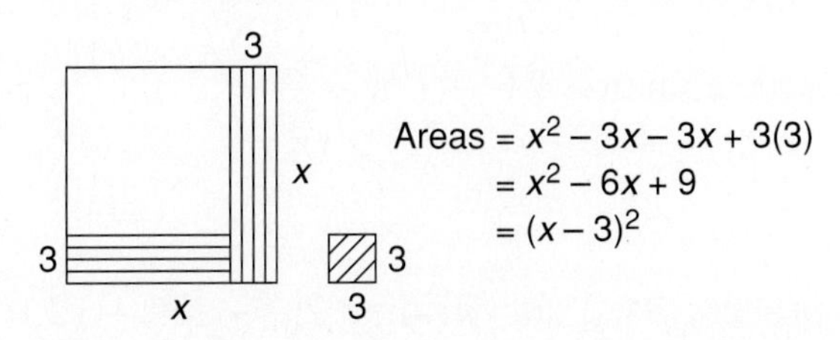

We want to represent the area of the square with the white region. When the $3x$ was subtracted from the square twice, the 3 by 3 region with area 9 square units was subtracted twice instead of just once. Thus to represent the area of the square we must add the 9 back on once, getting $x^2 - 6x + 9$ as the area of the square.

The white square has a side of length $x - 3$ and thus an area of $(x - 3)^2$. The expressions $x^2 - 6x + 9$ and $(x - 3)^2$ are equivalent.

This is the geometric approach to completing a square. The algebraic equivalent allows us to solve quadratic equations. Let's try it on the equation we have already solved—$2x^2 + 12x + 12 = 0$.

7. Given the equation $2x^2 + 12x + 12 = 0$, what are the values of the parameters a, b, and c?

8. Now you can complete the square on the terms containing x and x^2. To make life easier, rewrite the equation $2x^2 + 12x + 12 = 0$ so that the parameter c is no longer on the left side of the "=".

9. It is usually simpler if the coefficient of x^2 is 1. This is accomplished by factoring the parameter a out of the terms on the left side of the "=". Do so and write the new equation.

10. What must be added to both sides of the equation to complete the square? Add this to both sides and rewrite the equation.

11. Write the left side of the equation in factored form. Rewrite the equation so that one side is 0. Your equation should have the form $a(x-h)^2+k=0$. Compare this to the form we wrote when we identified the vertex from the graph.

Given the equation $2x^2+12x+12=0$, the values of the parameters are:

$$a=2, \qquad b=12, \qquad \text{and} \qquad c=12.$$

The process used for writing $2x^2+12x+12=0$ in the form $a(x-h)^2+k=0$ follows.

Subtract $c=12$ from both sides: $\quad 2x^2+12x=-12$

Factor out 2 on left side of "=": $\quad 2(x^2+6x)=-12$

Add 9 to x^2+6x to complete square. Add $2(9)$ on right side of "=" to maintain the equality:

$$2(x^2+6x+9)=-12+2(9)$$

Factor x^2+6x+9: $\quad 2(x+3)^2=6$

Subtract 6 from both sides: $\quad 2(x+3)^2-6=0$

This checks with the equation we wrote earlier in the section.

How does this get us to the formula we promised? We will use the completing the square process on the general quadratic equation $ax^2+bx+c=0$ to develop the desired formula. The next investigation leads you through the derivation.

Here's a generic quadratic equation $ax^2+bx+c=0$ where a, b, and c are real number parameters.

12. Subtract the parameter c from both sides. Write the new equation.

13. Unlike the previous investigation, the parameter a does not seem to be a factor of ax^2+bx. Instead of factoring, divide both sides (every term) of the equation by a. To do this, what assumption are we making about a? Write the equation.

We have done the following so far.

Original equation: $ax^2 + bx + c = 0$

Subtract c from both sides: $ax^2 + bx = -c$

Divide both sides by a, assuming that a is not zero:

$$x^2 + \frac{b}{a}x = -\frac{c}{a}$$

We need to complete the square.

14. Label the sides of each square-rectangle in Figure 8 to represent the left side of your equation.

FIGURE 8

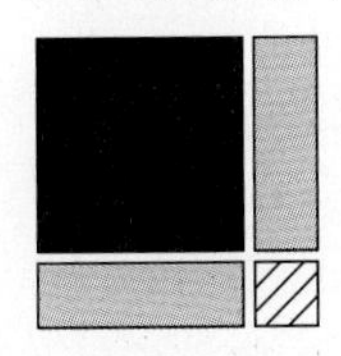

15. What must be added to both sides of the equation to complete the square? Add this to both sides and rewrite the equation.

16. Rewrite the equation so that the left side of the equation is in factored form.

17. Add the fractions on the right side of the equation by finding a common denominator. Write the resulting equation below.

To complete the square we concentrate on the coefficient of x, which is $\frac{b}{a}$. Since there will be two rectangles whose total area is $\frac{b}{a}x$, each must have a side of $\frac{1}{2}\left(\frac{b}{a}\right) = \frac{b}{2a}$. Figure 9 displays the geometric representation.

The area of the small square is $\left(\frac{b}{2a}\right)\left(\frac{b}{2a}\right) = \frac{b^2}{4a^2}$. This quantity must be added to both sides of the equation.

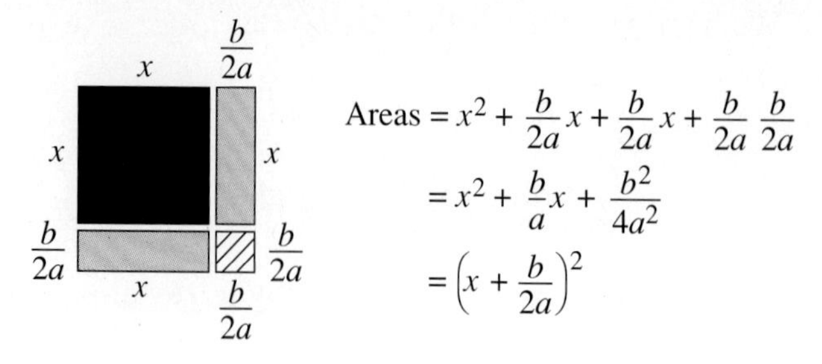

$$x^2 + \frac{b}{a}x + \frac{b^2}{4a} = -\frac{c}{a} + \frac{b^2}{4a^2}$$

The large square in Figure 9 has a side of length $x + \frac{b}{2a}$. Its area is $\left(x + \frac{b}{2a}\right)^2$. This is the factorization of the left side of the equation.

This leaves us with

$$\left(x + \frac{b}{2a}\right)^2 = -\frac{c}{a} + \frac{b^2}{4a^2}.$$

We need to add the expressions on the right side of the "=". This means we need a common denominator. Since a is a factor of $4a^2$, the least common denominator is $4a^2$. The rest of the work follows.

Original expression: $-\frac{c}{a} + \frac{b^2}{4a^2}$

Multiply $\frac{c}{a}$ by $\frac{4a}{4a}$: $-\frac{c}{a}\left(\frac{4a}{4a}\right) + \frac{b^2}{4a^2}$

Simplify: $\frac{-4ac}{4a^2} + \frac{b^2}{4a^2}$

Add fractions: $\frac{b^2 - 4ac}{4a^2}$

So the equation has form $\left(x + \frac{b}{2a}\right)^2 = \frac{b^2 - 4ac}{4a^2}$.

To solve the equation for x, we need to take square roots. Let's do it.

18. Take square roots of both sides. Then rewrite the left side using absolute value.

19. Simplify the radical in the denominator on the right side. Write the result below.

20. Remove the absolute value by writing two statements.

21. Isolate x in both statements. Write the result below.

Let's continue solving the general equation.

Equation: $$\left(x+\frac{b}{2a}\right)^2 = \frac{b^2-4ac}{4a^2}$$

Take square roots of both sides: $$\sqrt{\left(x+\frac{b}{2a}\right)^2} = \sqrt{\frac{b^2-4ac}{4a^2}}$$

Absolute value: $$\left|x+\frac{b}{2a}\right| = \sqrt{\frac{b^2-4ac}{4a^2}}$$

Simplify denominator: $$\left|x+\frac{b}{2a}\right| = \frac{\sqrt{b^2-4ac}}{2|a|}$$

Remove absolute value:

$$x+\frac{b}{2a} = \frac{\sqrt{b^2-4ac}}{2|a|} \quad \text{or} \quad x+\frac{b}{2a} = -\frac{\sqrt{b^2-4ac}}{2|a|}$$

Solve for x:

$$x = -\frac{b}{2a}+\frac{\sqrt{b^2-4ac}}{2|a|} \quad \text{or} \quad x = -\frac{b}{2a}-\frac{\sqrt{b^2-4ac}}{2|a|}$$

One last simplification is in order. We prefer not to leave the absolute value on the parameter a. One way to omit it is to guarantee that a is always positive when using this formula. We can do this by multiplying the quadratic equation by -1 if the original value of a is negative. Alternatively, simply dropping the absolute value on a reverses the two solutions.

Thus we have one of the most important results in all of mathematics.

Quadratic Formula: Given the quadratic equation $ax^2 + bx + c = 0$ where a, b, and c are real numbers, we can find the values of x by using the quadratic formula

$$x = -\frac{b}{2a} \pm \frac{\sqrt{b^2 - 4ac}}{2a}.$$

The plus or minus between the two terms is a shorthand way to write the two separate solutions.

Another way of looking at this relates to the zeros of $y(x) = ax^2 + bx + c$.

The exact zeros of $y(x) = ax^2 + bx + c$ can be found by substituting the parameters a, b, and c into $-\frac{b}{2a} \pm \frac{\sqrt{b^2 - 4ac}}{2a}$.

22. Let's use the quadratic formula to find the exact zeros of $y(x) = 2x^2 + 12x + 12$.

a. Write the equation associated with the problem.

b. Identify the values of the parameters a, b, and c.

c. Substitute the values of the parameters into the quadratic formula. Do not simplify yet. Write the result below.

d. Calculate the value under the radical. This is called the *discriminant*. Write this value below.

e. Simplify the radical, if possible. This involves prime factoring the discriminant and calculating the square root of any square factors. Write the result below.

f. If possible, reduce the fractions in each term of the solution. Write the result below. This is the final answer to the problem. Compare your answers to those obtained earlier in the section.

23. Consider the function $y(x) = 3x^2 + x + 7$. Use the quadratic formula to determine the exact value of the inputs that produce an output of 5.

To find the zeros of $y(x) = 2x^2 + 12x + 12$, we solve the equation $2x^2 + 12x + 12 = 0$. The parameters are $a = 2$, $b = 12$, and $c = 12$.

Identify parameters a, b, and c:	$a = 2,\ b = 12$, and $c = 12$
Quadratic formula:	$x = -\dfrac{12}{2(2)} \pm \dfrac{\sqrt{(12)^2 - 4(2)(12)}}{2(2)}$
Simplify:	$x = -3 \pm \dfrac{\sqrt{144 - 96}}{4}$
Simplify radical further:	$x = -3 \pm \dfrac{\sqrt{48}}{4}$
Prime factor 48:	$x = -3 \pm \dfrac{\sqrt{2^4 \cdot 3}}{4}$
Simplify radical:	$x = -3 \pm \dfrac{2^2\sqrt{3}}{4}$
Simplify:	$x = -3 \pm \dfrac{4\sqrt{3}}{4}$
Reduce fraction:	$x = -3 \pm \sqrt{3}$

Whew! Now that's some algebra.

To find the inputs to $y(x) = 3x^2 + x + 7$ that produce 5 as an output, we must solve the equation $3x^2 + x + 7 = 5$. To use the quadratic formula, we need the form $ax^2 + bx + c = 0$.

Subtract 5 from both sides:	$3x^2 + x + 2 = 0$
Identify parameters:	$a = 3, \quad b = 1, \quad \text{and} \quad c = 2$
Quadratic formula:	$x = -\dfrac{1}{2(3)} \pm \dfrac{\sqrt{(1)^2 - 4(3)(2)}}{2(3)}$
Simplify:	$x = -\dfrac{1}{6} \pm \dfrac{\sqrt{1 - 24}}{6}$
	$x = -\dfrac{1}{6} \pm \dfrac{\sqrt{-23}}{6}$
	$x = -\dfrac{1}{6} \pm i\dfrac{\sqrt{23}}{6}$

We see that the solution is a nonreal complex number. Let's check using a calculator that can manipulate complex numbers (Figure 10). In both cases, the output is 5 when the input is one of the solutions we just found.

FIGURE 10

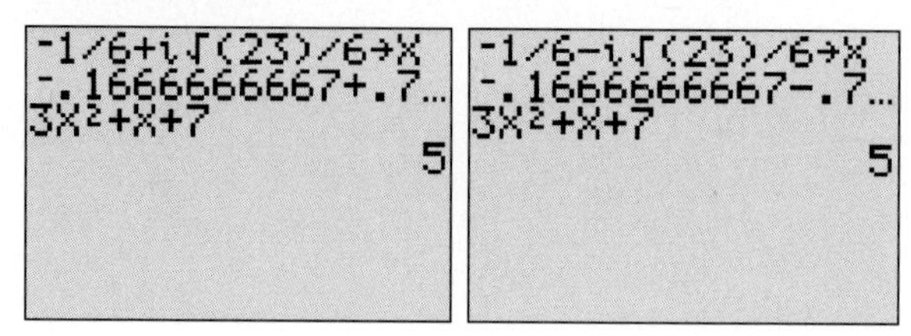

That completes our survey of quadratic functions. In summary, to use the quadratic formula:

- Write the equation in the form $ax^2 + bx + c = 0$.
- Identify the values of parameters a, b, and c.
- Substitute these values into the formula.
- Simplify the radical, if possible.
- Reduce the fractions, if possible.

Make this formula your friend. You'll be seeing it again in the next chapter.

STUDENT DISCOURSE

Pete: So the book is telling me that I can solve any quadratic equation by using the quadratic formula, is that right?

Sandy: That's the way I see it.

Pete: Can I use the formula even if the solutions are "nice" numbers like integers?

Sandy: Sure, though a table, a graph, or factoring can help you determine these also.

Pete: How do I know when the solution will involve the complex unit i?

Sandy: I think you can answer that yourself. Give it a try.

Pete: Well, I know that i arises when I take the square root of a negative number. So I guess I have to look at the radicand $b^2 - 4ac$.

Sandy: Good idea. Remember that $b^2 - 4ac$ is called the discriminant. What information does the discriminant tell you?

Pete: I guess I would get nonreal number answers if the discriminant is negative, right?

Sandy: Yes. What if the discriminant is positive.

Pete: I know that *i* won't arise so the solutions will be real. Is there a way to distinguish between rational and irrational?

Sandy: I'll give you a hint. Rational solutions do not contain a radical.

Pete: Oh, I think I see. The only way the radical drops out is if the discriminant is a square.

Sandy: Exactly. Why don't you summarize what you've just said.

Pete: Let's see. If the discriminant is negative, the solutions are nonreal complex numbers. If the discriminant is a square, the solutions are rational numbers, and if the discriminant is positive, but not a square, the solutions are irrational.

Sandy: You said a mouthful. I think you have a good handle on this. See you later.

1. In the Glossary, define the words in this section that appear in ***italic boldface*** type.

2. In Section 3.4 we developed the function $y(x) = 1000x^2 + 34{,}000x$ to output the accumulated salary at CDs-R-Us after x years.

a. Write an equation that represents how many years it will take for the accumulated salary to be $500,000.

b. Solve the equation algebraically. Interpret the answer in terms of the problem situation.

3. Virtual Reality Experiences Revenue: In Chapter 3, we investigated the function $R = -2x^2 + 10{,}000x$ where the input x is the number of headsets and R is the total revenue if x headsets are sold.

a. Write an equation that represents the number of headsets that must be sold if the desired revenue is $100,000.

b. Solve the equation. Analyze the solutions in terms of this problem.

4. Fencing In the Dog: In Chapter 3, we investigated the function $A = -l^2 + 100l$ where l is the length of a rectangular dog pen and A is the area of the pen.

a. Write an equation that represents situation in which the area of the pen is 1000 square feet.

b. Solve the equation. Analyze the solutions in terms of this problem.

5. In Chapter 3, we introduced the following model: If an earthly object is projected straight up into the air, then the height h (in feet) of the object above the Earth is given by a quadratic function of the time t (in seconds) elapsed since the object was shot or thrown into the air. The general model is $h(t) = -16t^2 + v_0 t + h_0$ where v_0 is the initial velocity (in feet per second) and h_0 is the initial height (in feet) above the Earth. Suppose the initial velocity is 125 feet per second and the initial height is 8 feet.

a. Write the specific model for the given information.

b. Write an equation that represents the time when the ball will hit the ground.

c. Solve the equation and interpret your solution in terms of the given problem.

6. Given the quadratic function $f(x) = 2x^2 + 6x + 3$, let's find the input so that the output is 4.

a. Write the equation associated with the problem.

b. Can the function be factored? Why or why not?

c. Solve by completing the square.

d. Solve by quadratic formula.

7. Given the quadratic function $f(x) = 3x^2 + 5x + 3$.

a. Use the quadratic formula to find the zeros of $f(x)$.

b. Does the graph of $f(x)$ intersect the x-axis? How could you tell based on your solution to part **a**?

c. How could you use the discriminant to determine if the graph of a quadratic function intersects the horizontal axis?

8. When using the completing the square process, what could happen to cause the solutions to be nonreal complex numbers?

9. In the book *Ars Magna*, published in 1545, Jerome Cardan discusses quadratics using the following problem: "If someone says to you, divide 10 into two parts, one of which multiplied into the other shall produce 40, it is evident that this case or problem is impossible. Nevertheless we shall solve it."

a. Write a system of equations that represents Cardan's problem.

b. Use substitution to solve the system.

c. Why did Cardan say that the problem is impossible?

10. Given the general quadratic function $f(x) = ax^2 + bx + c$.

a. Write a statement involving the parameters a, b, and c that represents the zeros of the function. What is a famous name for this statement?

b. Recall that the input value at the vertex is half the distance between the zeros. Use your answer to part **a** to identify the value of the input at the vertex.

11. Complete Table 1 by computing the discriminant and then using the discriminant to determine if the zeros are rational, irrational, or nonreal.

TABLE 1 **Discriminant and Zeros**

Function	Discriminant	Zeros	Rational, Irrational, or Nonreal?
$f(x) = x^2 + 9x + 20$			
$g(x) = x^2 - 2x + 5$			
$h(x) = x^2 - 4x + 2$			
$k(x) = 9x^2 - 6x - 4$			
$m(x) = 8x^2 - 2x - 15$			
$q(x) = x^2 - 4x + 7$			

12. To create the quadratic formula, we assumed that $\sqrt{\frac{s}{t}} = \frac{\sqrt{s}}{\sqrt{t}}$. Complete Table 2 to help in determining if this statement is true in general.

Based on the data, what is your conclusion?

13. Use the quadratic formula to answer the following questions.

a. What are the zeros of $y(x) = 6x^2 + 7x - 1$?

b. What inputs will produce an output of –8 for the function $y(x) = 2x^2 + 9x - 5$?

c. Solve the equation $x + 3 = (x - 2)^2$.

TABLE 2 Roots and Quotients

s	t	$\sqrt{\frac{s}{t}}$	$\frac{\sqrt{s}}{\sqrt{t}}$
4	1		
25	9		
625	400		
17	36		
81	11		

14. Find the solution(s) to the following equations.

a. $3x^2 - 5x - 11 = 0$

b. $5y^2 + y + 4 = 3$

c. $3x - 4 = 7 - 8x$

d. $3x^2 - x = 5x + 7$

REFLECTION Describe various techniques for solving quadratic equations.

SECTION 6.6

Manipulating Quadratic Models: Review

Purpose

- Reflect on the key concepts of Chapter 6.

In this section you will reflect upon the mathematics in Chapter 6—what you've done and how you've done it.

1. State the most important ideas in this chapter. Why did you select each?

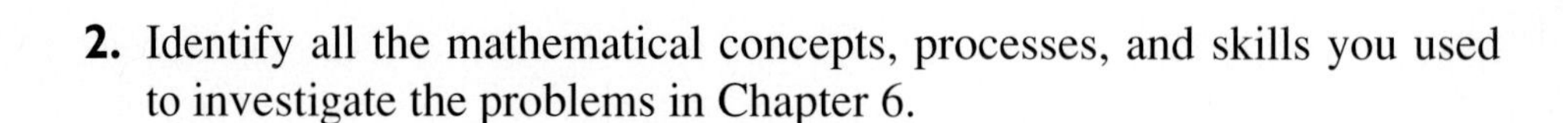

2. Identify all the mathematical concepts, processes, and skills you used to investigate the problems in Chapter 6.

DISCUSSION

There are a number of important ideas that you might have listed, including parabolas, vertex, zeros, quadratic functions, factoring, radicals, complex numbers, imaginary numbers, completing the square, discriminant, and quadratic formula.

REVIEW

1. Given the quadratic function $y(x) = 2x^2 - 3x + 7$, find the inputs if the output is 10. Explain what you did.

2. A quadratic function has zeros -4 and $\frac{3}{5}$ and a y-intercept of $(0, 1)$. Find the equation of the function. Explain what you did.

3. A quadratic function has a vertex at $(-5, 2)$ and a y-intercept at $(0, -3)$. Find the equation of the function. Explain what you did.

4. The graph of a quadratic function appears in Figure 1.

FIGURE 1

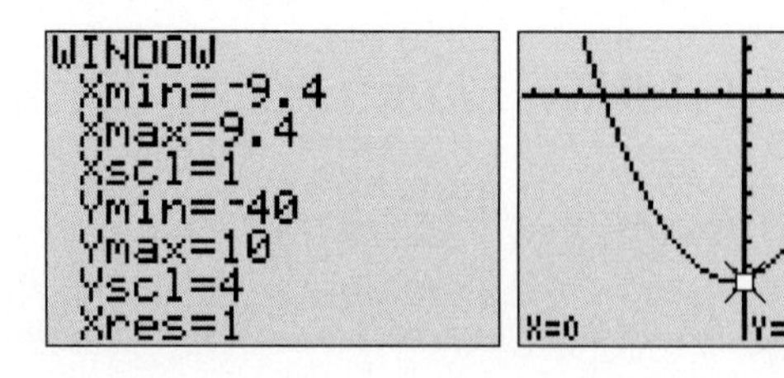

a. What are the zeros of this function?

b. Write the factors of this function.

c. Write the equation of the function. Explain what you did.

5. The graph of a quadratic function appears in Figure 2.

FIGURE 2

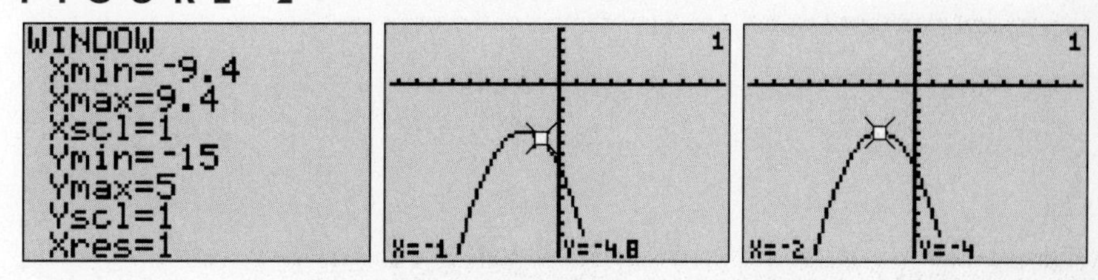

a. What is the vertex?

b. Write the algebraic representation (equation). Explain what you did.

6. Sketch the graph of a function that does not have any real number zeros.

7. Find the zeros and factor completely. If the polynomial is prime, say so.

a. $3x^2 - 11x - 4$

b. $x^2 + 14x + 49$

c. $4x^2 - 1$

d. $4y^2 + 18y + 8$

e. $9x^2 - 24x + 16$

f. $4x^2 - 15x - 4$

g. $x^3 - 27$

h. $8y^3 + 125$

8. Find the solutions to each of the following equations, inequalities, or systems of equations.

a. $5y + 9 = 2 - 7y$

b. $x^2 - 2x - 15 = 0$

c. $3x = 4 - 7y$
$2x + 9y = -2$

d. $3x^2 - 5x - 2 = 4$

e. $2.7x < 3.48x + 9.2$

f. $x - 2y - 3z = 4$
$2x + y + 7z = -1$
$3x - 2y + 4z = 17$

g. $5x - 2 > 7 + 3x$

h. $3x^2 - 5x + 2 = x^2 - 9x = 7$

i. $4x^2 = 3x - 9$

j. $(3x + 1)^2 = 2205$

k. $4x - y = 12$
$8x + 2y = 5$

l. $x^2 + 7x - 44 < 0$

m. $x^2 = 45$

n. $y = 2x^2 - x + 2$
$y = -5$

o. $3x = 9x^2 + 11$

p. $x^2 + x + 1 = 0$

q. $2x + 3y = -1$
$y = x^2 - x + 2$

r. $x^2 + 2x = 1$

s. $3x^2 - 2x = 7x + 9$

t. $4x^2 + 9x - 3 = 2 - 3x + 4x^2$

9. Rewrite the equation in the form $y(x) = ax^2 + bx + c$. Identify the values of the parameters a, b, and c.

a. $y(x) = 3(x - 5)^2 - 7$

b. $y(x) = -4(x + 2)^2 + 11$

10. Rewrite the equation in the form $y(x) = a(x - h)^2 + k$. Identify the vertex.

a. $y(x) = x^2 - 9x + 14$

b. $y(x) = x^2 + 13x + 40$

11. If an earthly object is projected straight up into the air, then the height h (in feet) of the object above the Earth is given by a quadratic function of the time t (in seconds) elapsed since the object was shot or thrown into the air. The general model is $h(t) = -16t^2 + v_0 t + h_0$ where v_0 is the initial velocity (in feet per second) and h_0 is the initial height (in feet) above the Earth. Suppose the initial velocity is 40 feet per second and the initial height is 56 feet.

a. Write the specific model for the given information.

b. What are the zeros of this function? What do the zeros mean in the context of the time and height of the object?

c. What is the vertex of the graph of this function? What does the vertex mean in the context of the time and height of the object?

d. Write the function in factored form.

e. Find the exact times when the height will be 60 feet.

12. Simplify the following radicals completely without approximating.

a. $\sqrt{173{,}901}$ **b.** $\sqrt{106{,}480}$

c. $\sqrt{\frac{25}{9}}$ **d.** $\sqrt{\frac{84}{49}}$

13. Write the equation of the function satisfying the given conditions.

a. A linear function through $(3, 1)$ and $(-2, 7)$.

b. A quadratic function with vertex $(3, -7)$ and y-intercept $(0, 6)$.

c. An exponential function with base 0.7 and y-intercept $(0, 4)$.

d. A quadratic function with zeros -9 and 6 with y-intercept $(0, -2)$.

e. A linear function perpendicular to the function $3x + 5y = -7$ and through the point $(1, 4)$.

f. A quadratic function through $(4, 6)$ and $(2, -1)$ and with vertex at $(-1, -12)$.

g. A linear function with intercepts $(0, -4)$ and $(7, 0)$.

h. An exponential function with base 1.9 and containing the point (2, 17).

14. Perform the following complex number operations.

a. $(2 + 5i) + (7 - 2i)$

b. $(4 - 9i) - (8 + 2i)$

c. $(3i)(-11i)$

d. $3i(6 - 4i)$

e. $(3 + 8i)(5 - 2i)$

f. $(3 - 10i)(3 + 10i)$

g. $(5 + 2i)^2$

h. $(1 + i)^2$

15. Write the equation for the function given the following graphs. Briefly describe what you did.

a.

FIGURE 3

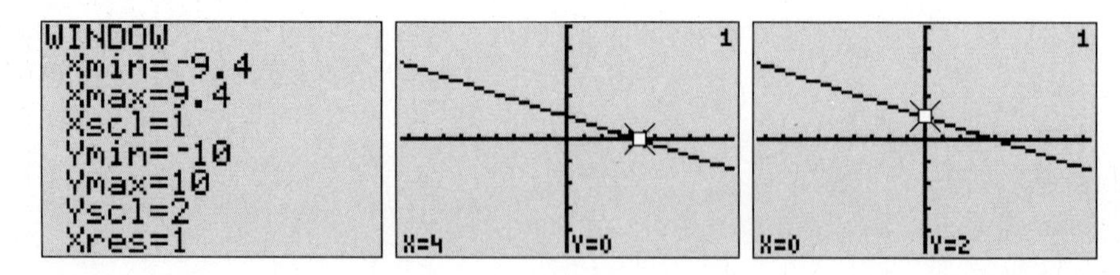

b.

FIGURE 4

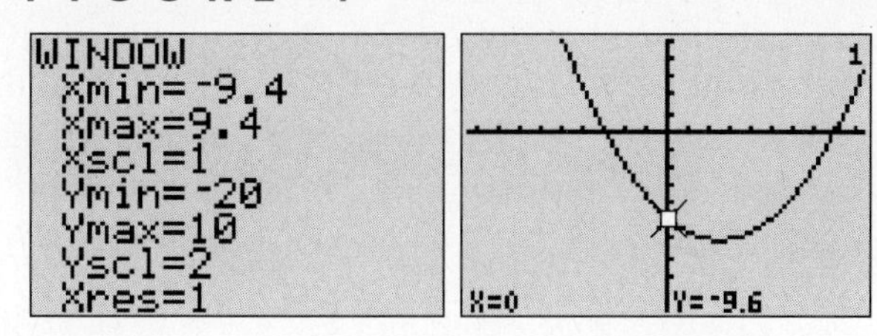

c.

FIGURE 5

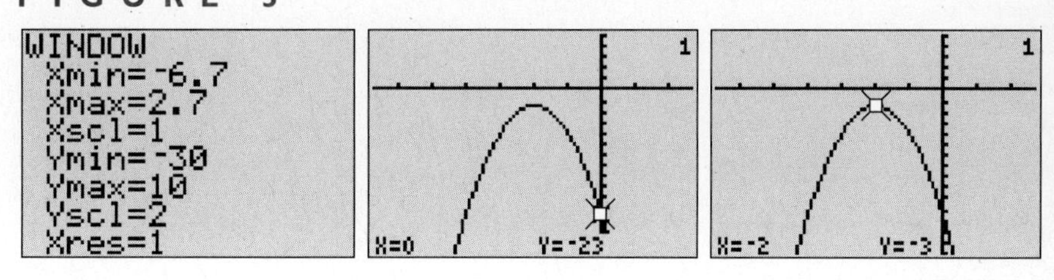

16. Use the quadratic formula to answer the following questions.

a. What are the zeros of $y(x) = 2x^2 - x - 5$?

b. What inputs will produce an output of -3 for the function $y(x) = 5x^2 + 2x + 1$?

c. Solve the equation $4x = x^2 - 9$.

CONCEPT MAP

Construct a Concept Map centered on one of the following:

a. graphs of quadratic functions

b. factoring

c. radical

d. complex number system

e. solving quadratic equations

REFLECTION

Select one of the important ideas you listed in Investigation 1. Write a paragraph to Izzy explaining your understanding of this idea. How has your thinking changed as a result of studying this idea?

7 Nonpolynomial Algebraic Functions

SECTION 7.1

Rational Functions

Purpose

- Define rational functions using the operation of division on polynomial functions.
- Explore domain, zeros, vertical intercepts, vertical asymptotes, end behavior, and discontinuities of rational functions.

Throughout the book so far, we have investigated linear, quadratic, and exponential functions in detail. Linear and quadratic functions belong to the class of functions called polynomial functions, which were briefly discussed in Section 4.6.

A ***polynomial function*** is a sum of terms in which the input variable is raised to some whole number power. Basically, every term of a polynomial function is either a constant, a variable, or a product of constants and variables. An example of a fifth-degree polynomial function in one variable is

$$f(x) = 2x^5 - 3x^4 + x^3 + 7x^2 - 5x - 4.$$

Recall that since subtracting is the same as adding the opposite, we usually just rewrite polynomials with subtraction when the coefficient is a negative number.

Polynomial functions belong to a larger class of functions called ***algebraic functions***. They are constructed by using the six basic operations of addition, subtraction, multiplication, division, taking a power, and/or taking a root on polynomial functions. It is this set of functions that have a lot of similarities to polynomials, but some major differences too, which we will study in this chapter. In particular, we investigate two ***nonpolynomial algebraic functions***, rational functions and radical functions.

Exponential functions belong to another class of functions, which we look at in Chapter 8. This class is called transcendental.

Rational functions are formed by dividing one polynomial function by another polynomial function. The divisor cannot be a constant function. This means the input variable will be in the denominator.

For example, $f(x) = \dfrac{2x + 5}{x^2 - x + 2}$ is a rational function formed by dividing the linear function $y = 2x + 5$ by the quadratic function $y = x^2 - x + 2$.

Radical functions are formed by including a term that contains the root (second, third, fourth, etc.) of a variable expression.

For example, $f(x) = 2 + \sqrt{3x + 1}$ is an example of a radical function.

In Sections 7.1 and 7.2, we investigate rational functions.

1. Driving to School: Your friend Izzy lives about 15 miles from school. For one week, Izzy keeps track of the time it takes him to get to school. He'd like to be able to figure out his average speed based on the time. Can you help him out?

TABLE 1 Izzy's Trips to School

Day	Time (hours)	Speed (mph)
Monday	0.5	
Tuesday	0.75	
Wednesday		
Thursday		
Friday		
General	t	

a. Write a formula for the relationship between speed, time, and distance.

b. Use your formula to calculate Izzy's average speed each day and record results in Table 1.

c. If t is the time and s is the average speed, write the equation for a function that computes Izzy's average speed when the input is time.

d. What is the domain of the function you wrote in part **c**? Why?

e. Graph the function you wrote in part **c** in a "friendly" window so that the inputs increment by some multiple of 0.1. (If you don't know how to do this, ask a classmate, your instructor, or consult the manual.) Sketch the graph and describe its behavior.

f. What is happening to the graph when the input is very close to zero? Interpret this behavior in terms of Izzy's time and average speed.

g. What is happening to the graph when the input is very large? Interpret this behavior in terms of Izzy's time and average speed.

In general, the product of speed and time represents distance. If s is speed, t is time, and d is distance, then

$$s \cdot t = d.$$

If we divide both sides by t, we get $s = \frac{d}{t}$.

Since we know the distance that Izzy travels to school is 15 miles, we can write the function $s = \frac{15}{t}$.

For example, if it takes Izzy 45 minutes to get to school, then $t = \frac{3}{4}$ hours.

$s = \frac{15}{\left(\frac{3}{4}\right)} = 15\left(\frac{4}{3}\right) = 20$. So Izzy's average speed would be 20 mph if it takes him 45 minutes to get to school.

The function $s = \frac{15}{t}$ is an example of a rational function because we are dividing the constant function $y = 15$ by the linear function $y = t$. The fact that the input variable appears as a term in the denominator is sufficient to create a rational function.

Since division by 0 is undefined, the domain of this algebraic function is all real numbers except 0. If we consider the context (Izzy's trip to school), the input cannot be negative either.

In general, the domain of any rational function is the set of all real numbers except those that make the denominator 0.

Figure 1 displays a graph and tables of the function $s = \frac{15}{t}$.

FIGURE 1

X	Y1
0	ERROR
.1	150
.2	75
.3	50
.4	37.5
.5	30
.6	25

Y1=15/X

X	Y1
5	3
6	2.5
7	2.1429
8	1.875
9	1.6667
10	1.5
11	1.3636

Y1=15/X

WINDOW
Xmin=-9.4
Xmax=9.4
Xscl=1
Ymin=-10
Ymax=30
Yscl=10
Xres=1

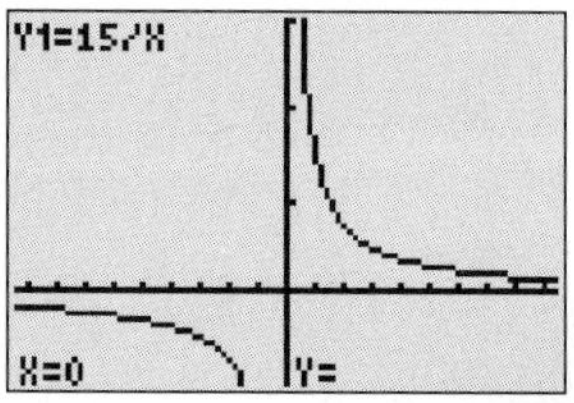

Y1=15/X
X=0 Y=

Unlike other graphs we have encountered, the graph of $s = \frac{15}{t}$ consists of two disconnected branches. When a graph has a break, the function is said to be ***discontinuous*** at the value of the input where the break occurs.

An invisible vertical line separates the two branches. This vertical line passes through the point (0, 0), whose input coordinate is exactly where the function is undefined. The line is called a vertical asymptote. It has the equation $x = 0$ and coincides with the y-axis. Such a line is not part of the graph. A ***vertical asymptote*** is an invisible line that separates the branches of the graphs of rational functions and occurs when the input value results in the function being undefined for that input.

In the investigation into Izzy's speed, the graph has two "branches," one in Quadrant I and the other in Quadrant III. We only consider the branch in Quadrant I since the input must be positive in the context of this problem.

Notice that, as the input gets close to 0, the output gets larger and larger (look at both the table and the graph). This means that as the time Izzy takes to get to school decreases, his average speed will increase. If he took less than 12 minutes (0.2 of an hour), he probably got a speeding ticket.

This idea illustrates an important feature of the behavior of a rational function near a vertical asymptote—the ***magnitude*** of the output increases without bound as the input nears the vertical asymptote.

Also, as the input gets large (5, 6, 7, etc.—that 7-hour, 15-mile trip must have been in a construction zone on the interstate), the output gets closer and closer to 0. This means that as time increases, the average speed must decrease.

The form of the function $s = \frac{15}{t}$ is a constant function divided by a linear function. Now we investigate a problem that results in a linear function divided by a linear function.

Concentration: The concentration of a given substance in a mixture is determined by the ratio of the amount of the substance to the total quantity.

$$\text{Concentration} = \frac{\text{Amount of substance}}{\text{Total quantity}}$$

For example, if 6 ounces of party punch contains 3 ounces of pure orange juice, the concentration of orange juice in the punch is $\frac{3}{6} = 0.5 = 50\%$. The punch is 50% orange juice. Consider the problem in which we begin with 6 ounces of punch that is 50% orange juice. Suppose we add x ounces of pure orange juice.

2. We wish to write a function that expresses the orange juice concentration after x ounces of pure orange juice have been added.

a. How much orange juice do you begin with? Write an expression containing x for the amount of orange juice present after x ounces have been added.

b. Write an expression containing x for the total amount of punch present after x ounces have been added.

c. Use your answers to parts **a** and **b** to write a rational function that defines the orange juice concentration as a function of x.

d. Ignoring the context, what is the domain of the function you wrote in part **c**? Why? How does the context change the domain?

e. Graph the function in part **c**. What is the equation of the vertical asymptote for the graph of the function?

f. Describe the behavior of the output when the input is very near the vertical asymptote.

3. Now let's consider the function from Investigation 2**c** globally as the input becomes either very large or very small.

a. Using the graph, zoom out only in the input direction (on the horizontal axis) by making Xmin = –100,000 and Xmax = 100,000. The graph should resemble a horizontal line. Write the equation of the line.

b. Display a table of values in which the increment on input is large (say 10,000). As the input gets very large, what happens to the output values? As the input gets very small, what happens to the output values?

c. Describe the behavior of the function as the input gets very large. Interpret this behavior in terms of the orange juice concentration.

If we begin with 6 ounces that is 50% orange juice and add x ounces of pure orange juice, the concentration of orange juice C is defined by the function

$$C = \frac{3 + x}{6 + x}.$$

Ignoring the context, the domain of this function is all real numbers except –6, since –6 will make the denominator 0. In the given context, the domain is all ***nonnegative real numbers***. The equation of the vertical asymptote is $x = -6$. Figure 2 displays the graph and table indicating that there is no output when the input is –6.

FIGURE 2

Plot1 Plot2 Plot3
\Y1=3+X
\Y2=6+X
\Y3=Y1/Y2
\Y4=
\Y5=
\Y6=
\Y7=

X	Y3
-8	2.5
-7	4
-6	ERROR
-5	-2
-4	-.5
-3	0
-2	.25

Y3=ERROR

WINDOW
Xmin=-9.4
Xmax=9.4
Xscl=1
Ymin=-5
Ymax=5
Yscl=1
Xres=1

Y3=Y1/Y2
X=-6 Y=

Notice that the numerator function was entered as Y1, the denominator function as Y2, and the rational function as Y1/Y2. This approach is helpful when we wish to analyze the numerator and denominator functions separately.

As you read the graph from left to right, notice that the "left branch" of the graph heads upward near $x = -6$ while the "right branch" of the graph heads downward near $x = -6$. To understand this behavior, we need to think of a dynamic, rather than static, situation in which we watch the output as the input gets closer and closer to the vertical asymptote.

Let's look at some notation that represents the situation for the function $C(x)$. First note the graph and table in Figure 3.

FIGURE 3

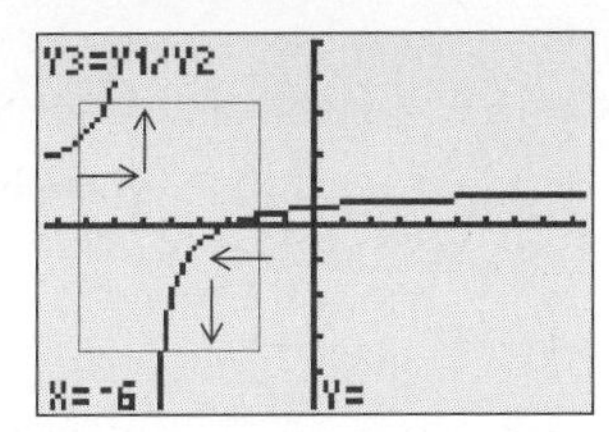

Change in input denoted by horizontal arrows.

Change in output denoted by vertical arrows.

X	Y3
-6.003	1001
-6.002	1501
-6.001	3001
-6	ERROR
-5.999	-2999
-5.998	-1499
-5.997	-999

Y3 = Y1/Y2

The arrows represent the word "approaches." Note that, for inputs just a little smaller than –6, the outputs get very large. For inputs just a little larger than –6, the outputs get very small.

Symbolically, we write

$$\text{If } (x \to -6^-), \text{ then } (C \to \infty).$$

$$\text{If } (x \to -6^+), \text{ then } (C \to -\infty).$$

This says, in words:

> If the input gets close to –6, but is smaller than –6, then the output gets larger and larger without bound.

and

> If the input gets close to –6, but is larger than –6, then the output gets smaller and smaller without bound.

The small "–" above and to the right of –6 indicates that the input is approaching –6 from smaller values, that is, from the left.

The small "+" above and to the right of –6 indicates that the input is approaching –6 from larger values, that is, from the right.

This discussion explains how to describe the behavior of rational functions near their vertical asymptotes.

We finish the discussion by looking at the ***global behavior of the function*** if we make the magnitude of the input very large. As we zoom out in the horizontal direction, we notice that the graph becomes a horizontal line (Figure 4).

FIGURE 4

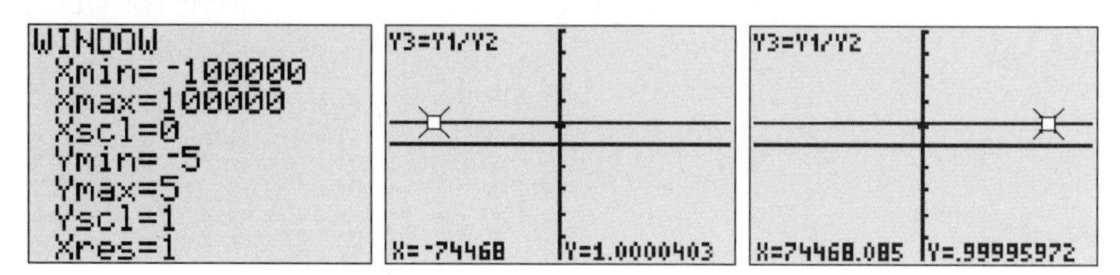

This horizontal line occurs at $y = 1$. Such a line is called a ***horizontal asymptote***. A more general name for this behavior is to call the function $y = 1$ an ***end behavior model*** for the rational function.

End behavior models for rational functions are functions whose graphs match the graph of a rational function globally; that is, when

$$x \to \infty \quad \text{or} \quad x \to -\infty.$$

If the end behavior model is a horizontal line, the line is called a horizontal asymptote.

Finally, if we return to the context of orange juice concentration in our punch, the end behavior suggests that the concentration approaches 1 as the number of ounces of orange juice added increases. This means that the concentration of orange juice is approaching 100% as x gets large. This makes sense since the more orange juice we add, the more concentrated the mixture becomes and the punch tastes like plain orange juice.

That is a lot of information about rational functions. Let's investigate a few more specific examples, putting these ideas into practice. Watch how the relationship between the degree of the numerator and the degree of the denominator affects the behavior of the rational function.

4. Let $y(x) = \dfrac{8x - 9}{2x + 3}$.

a. What two functions were used to form this quotient?

b. State the domain of the function. What function did you use to determine the domain? Why?

c. Identify the x-intercept of the graph of the function. What function did you use to determine the x-intercept? Why?

d. Where is the vertical asymptote located? Identify the behavior of the function on each side of the vertical asymptote using the notation given in the preceding discussion. Describe your answer in words also.

e. Use "zoom out" to determine an end behavior model for the rational function. Describe a way to determine this behavior by using the original algebraic form of the function.

f. Is the degree of the numerator larger, the same, or smaller than the degree of the denominator?

5. Let $y(x) = \dfrac{x^2 - x - 20}{x + 3}$.

a. What two functions were used to form this quotient?

b. State the domain of the function. What function did you use to determine the domain? Why?

c. Was the degree of the numerator larger, the same, or smaller than the degree of the denominator?

d. Determine if $y(x)$ can be reduced. To do so, display a table for the function in the numerator and the function in the denominator. Use the table to factor the function in the numerator. The rational function may be reduced if the numerator and denominator have a common zero.

e. Identify the x-intercepts of the graph of the function. What function did you use to determine the x-intercept(s)? Why?

f. Where is the vertical asymptote located? Identify the behavior of the function on each side of the vertical asymptote using the notation given in the preceding discussion. Describe your answer in words also.

g. Use zoom out both horizontally and vertically (in both x and y directions) together to determine an end behavior model for the rational function. Describe a way to determine this behavior by using the original algebraic form of the function.

Let's look at the two functions we just investigated in detail.

$$y(x) = \frac{8x - 9}{2x + 3}$$

This function was created from the quotient of the linear functions $f(x) = 8x - 9$ and $g(x) = 2x + 3$. Their degrees are the same. The numerator function $f(x)$ has a zero at $\frac{9}{8}$ and the denominator function $g(x)$ has a zero at $-\frac{3}{2}$.

The domain is the set of all real numbers except the zeros of $g(x)$.

So the domain is the set of all real numbers except $-\frac{3}{2}$.

The x-intercept is at $x = \frac{9}{8}$, the zero of $f(x)$.

There is a vertical asymptote at $x = -\frac{3}{2}$. A table and graph appear in Figure 5.

FIGURE 5

X	Y3
-1.8	39
-1.7	56.5
-1.6	109
-1.5	ERROR
-1.4	-101
-1.3	-48.5
-1.2	-31

Y3=ERROR

WINDOW
Xmin=-4.7
Xmax=4.7
Xscl=1
Ymin=-20
Ymax=20
Yscl=0
Xres=1

Y3=Y1/Y2
X=-1.5 Y=

The behavior near the asymptote follows.

If $\left(x \to -\frac{3}{2}^{-}\right)$, then $(y(x) \to \infty)$.

If $\left(x \to -\frac{3}{2}^{+}\right)$, then $(y(x) \to -\infty)$.

Finally, if we zoom out horizontally, the outputs approach 4. Thus an end behavior model is $y = 4$, the horizontal asymptote.

Note that if we consider the quotient of the highest degree terms in numerator and denominator, we get an end behavior model

$$\frac{8x}{2x} = 4.$$

This might suggest a way to determine an end behavior model from the algebraic definition of the function.

The other function we considered is more complex. A quadratic appears in the numerator. Let's take a look.

$$y(x) = \frac{x^2 - x - 20}{x + 3}$$

The numerator has zeros at $x = 5$ and $x = -4$, while the denominator has a zero at $x = -3$. Since the numerator and denominator have no common zeros, the function is irreducible. The domain is the set of all real numbers except for -3. The zeros of the function are 5 and -4. There is a vertical asymptote at $x = -3$. The graph and a table of values near the vertical asymptote appear in Figure 6.

FIGURE 6

X	Y3
-3.3	19.367
-3.2	32.8
-3.1	72.9
-3	ERROR
-2.9	-86.9
-2.8	-46.8
-2.7	-33.37

Y3=ERROR

WINDOW
Xmin=-6.7
Xmax=2.7
Xscl=1
Ymin=-20
Ymax=20
Yscl=0
Xres=1

Y3=Y1/Y2
X=-3 Y=

A new feature of this function is the end behavior. Zooming out results in a slant line passing very close to the origin (Figure 7).

FIGURE 7

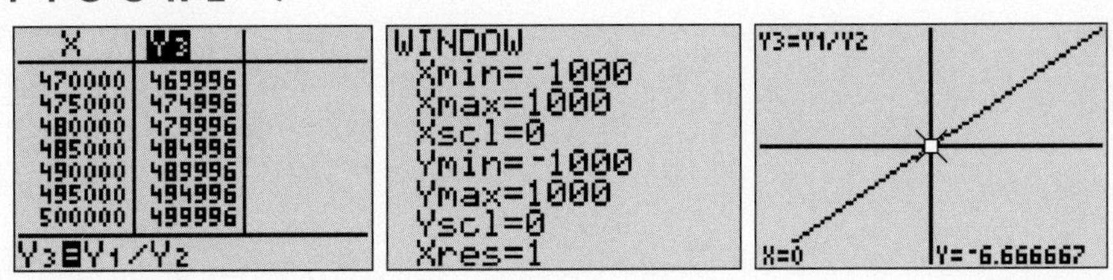

Looking at the table, we note that the inputs and outputs are almost identical for inputs that are large in magnitude. Thus we conjecture that an approximate end behavior model might be the identity function $y = x$. Overlaying the graph of $y = x$ on the graph of the original function further supports this answer. We could also obtain this result by finding the ratio of the highest degree terms of $y(x)$ as follows:

An approximate end behavior model $y = \frac{x^2}{x} = x$.

Of course, we can use linear regression on two points with large input values from the table in Figure 7. Rounding gives a better approximation of the end behavior model as $y = x - 4$. This equation is usually used since it can be derived by polynomial division (we'll look at that later) as well as regression, but for very large values of x, the difference between the two models is insignificant.

In this case, instead of having a horizontal asymptote, we have a ***slant asymptote***. Sometimes called the ***oblique asymptote***, the slant asymptote occurs whenever the degree of the numerator exceeds the degree of the denominator by 1. So $y(x) = \frac{x^5 - 3x - 2}{5x^4 + 3}$ has a slant asymptote.

We conclude this initial study of the behavior of rational functions with one that has a somewhat different behavior.

6. Let $y(x) = \frac{x^2 + 3x - 10}{x^2 + 2x - 8}$.

a. What two functions were used to form this quotient?

b. State the domain of the function. What function from part **a** did you use to determine the domain? Why?

c. Determine if $y(x)$ can be reduced. To do so, display a table for the function in the numerator and the function in the denominator. Use the table to factor the function in the numerator and the function in the denominator. The rational function may be reduced if the numerator and denominator have a common zero. Reduce by dividing out any factors common to numerator and denominator.

d. Identify the x-intercept(s) of the graph of the function. Be careful: x-intercepts must be in the domain of the function. What function did you use to determine the x-intercept(s)? Why?

e. Where is the vertical asymptote located? Identify the behavior of the function on each side of the vertical asymptote using the notation given in the preceding discussion. Describe your answer in words also.

f. The function $y(x)$ is undefined at two real number values. There is a vertical asymptote at one of these values. What happens to the graph at the other value?

g. Determine an end behavior model for the rational function. Describe how you found this answer and verify your answer graphically.

When we find the ***zeros of the denominator*** of a rational function, we are finding the real numbers that are *not* in the domain of the function. The behavior of the graph of the function near these values is interesting.

If the denominator has a zero that is not a zero of the numerator, the graph has a vertical asymptote at that zero.

However, if the denominator and the numerator share a zero, the rational function reduces and the graph has a hole at that zero.

Let's look at the function $y(x) = \dfrac{x^2 + 3x - 10}{x^2 + 2x - 8}$.

The zeros of $g(x) = x^2 + 2x - 8$ are 2 and –4 (Figure 8). The domain of the rational function is all real numbers except 2 and –4. The zeros of $f(x) = x^2 + 3x - 10$ are 2 and –5 (Figure 8).

FIGURE 8

X	Y2
-4	0
-3	-5
-2	-8
-1	-9
0	-8
1	-5
2	0

Y2=X²+2X-8

X	Y1
-5	0
-4	-6
-3	-10
-2	-12
-1	-12
0	-10
1	-6

Y1=X²+3X-10

X	Y1
-4	-6
-3	-10
-2	-12
-1	-12
0	-10
1	-6
2	0

Y1=X²+3X-10

The factored form is $y(x) = \dfrac{(x-2)\,(x+5)}{(x-2)\,(x+4)}$.

The reduced form of the function is

$$y(x) = \frac{x+5}{x+4}, \qquad x \neq 2.$$

The graph for the function appears in Figure 9.

FIGURE 9

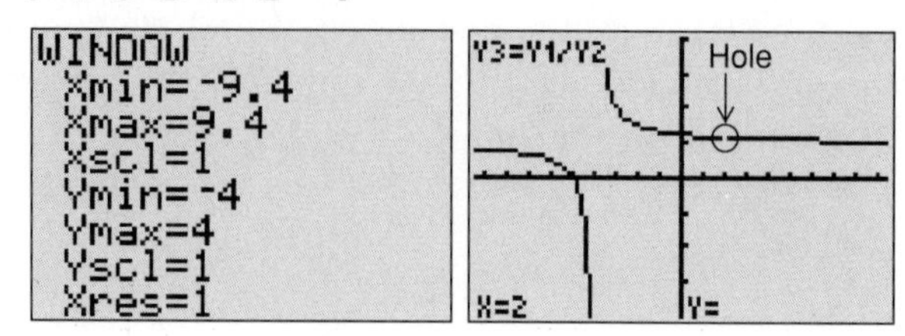

Notice that there is a vertical asymptote at $x = -4$. However, there is a "hole" at $x = 2$. When a graph has a hole, we say there is a ***removable discontinuity*** at the input where the hole occurred. The only x-intercept is at $x = -5$. A hole occurs when the rational function can be simplified (i.e., there is a common nonconstant factor in the numerator and in the denominator). The behavior near the asymptote is the opposite of what we noted before.

If $(x \to -4^-)$, then $(y(x) \to -\infty)$.

If $(x \to -4^+)$, then $(y(x) \to \infty)$.

An end behavior model is the horizontal asymptote $y = 1$ obtained from simplifying the quotient $\frac{x^2}{x^2}$ of terms with highest degree in numerator and denominator.

STUDENT DISCOURSE

Pete: These rational functions are sure different from any I've seen.

Sandy: How so?

Pete: The graphs have more than one piece. I would think that there would be two functions if I see two different graphs.

Sandy: I know what you mean, but they are parts of only one function. It's just that there are places where the graph is disconnected.

Pete: Yeah, the book called those discontinuities, I think.

Sandy: You're right.

Pete: Will a rational function always have a vertical asymptote?

Sandy: I don't know. We ought to think about this. Where do vertical asymptotes occur?

Pete: At places where an input value causes the denominator to be 0.

Sandy: Must there always be such a place?

Pete: Hmmm. What about a denominator like $x^2 + 1$?

Sandy: Good idea. The zeros of $x^2 + 1$ are nonreal numbers. The denominator will never be zero if the input is a real number.

Pete: So a function like $f(x) = \dfrac{2}{x^2 + 1}$ is a rational function, but it doesn't have any vertical asymptotes.

Sandy: Seems right. Let me see the graph.

Pete: Here it is.

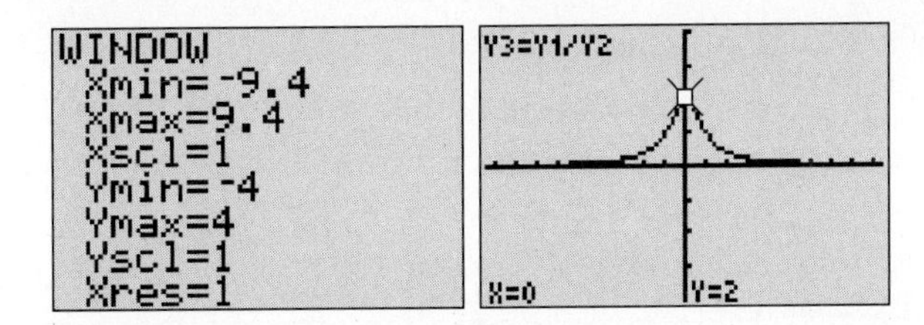

Sandy: That's interesting. It looks like it tried to have a vertical asymptote, but couldn't. I need to think some more about this.

1. In the Glossary, define the words in this section that appear in ***italic boldface*** type.

2. Given a rational function, what information can be determined by investigating the

a. numerator? **b.** denominator?

3. Graph each of the following rational functions and determine the vertical asymptote and end behavior model.

a. $f(x) = \dfrac{1}{x + 2}$ **b.** $f(x) = \dfrac{2x}{x - 1}$

4. Gina has 10 ounces of witch hazel, which is 10% alcohol. She adds x ounces of another face cleanser, which is 50% alcohol.

a. What amount of alcohol is in the witch hazel? What expression represents the amount of alcohol in the mixture? What expression represents the total amount of liquid?

b. Write a rational function that represents the concentration of alcohol in the final mixture.

c. Where is the vertical asymptote located? Is this a meaningful value in this problem situation?

d. Find an end behavior model for this function. Explain the meaning of the end behavior model in the context of the problem situation.

e. Use a table or graph to approximate how many ounces of liquid containing 50% alcohol must be added to change the concentration to 20% alcohol.

5. Sue plans a trip to visit a friend at her state university, which is 123 miles away.

a. Write a function that defines Sue's time for the trip as a function of her average speed.

b. Sue wishes to make the trip in about two hours. What must her average speed be? Describe how you found your answer.

c. Describe the meaning of the end behavior model in this problem situation.

6. Look at Pete's function $f(x) = \dfrac{2}{x^2 + 1}$ on your calculator.

a. Change the window to the following setting: Xmin = –4.7, Xmax = 4.7, Ymin = –5, Ymax = 10. Does Sandy's comment still seem appropriate? Why or why not?

b. Compare this graph to the graph of $g(x) = \dfrac{2}{x^2 - 1}$. How are they alike? How are they different?

7. Consider the rational function $y(x) = \dfrac{2x - 5}{x^2 + x + 1}$.

a. Explain why this function has no vertical asymptotes.

b. Make up your own rational function, using a different denominator, so that the function has no vertical asymptotes.

8. Create a rational function with a vertical asymptote at –2, a hole at 5, and an end behavior model of $2x$.

9. The graph of a rational function appears in Figure 10.

FIGURE 10

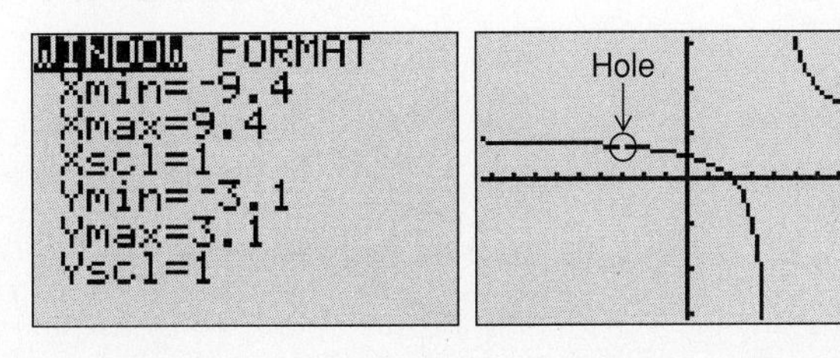

a. Identify the factors of the denominator.

b. Identify the factors of the numerator.

c. Write the algebraic representation of the function.

For each of the following rational functions in Explorations 10–14, find

a. the domain,

b. the zeros of the function,

c. the location of both vertical asymptotes and holes,

d. an end behavior model for the function.

10. $y(x) = \dfrac{x-6}{x+3}$

11. $y(x) = \dfrac{x^2-3x-10}{x^2+5x-6}$

12. $y(x) = \dfrac{x^2+3x-28}{x^3+5x^2-22x-56}$

13. $y(x) = \dfrac{x^3-3x^2-13x+15}{x^2-2x-15}$

14. $y(x) = \dfrac{ax+b}{cx+d}$ where a, b, c, and d are real number parameters. Assume $y(x)$ is not reducible.

15. Create several rational functions in which the degree of the numerator is smaller than the degree of the denominator. All these functions will have the same end behavior model. What is it?

16. Just as we can add, subtract, multiply, and divide numbers, we can perform these same operations on functions and produce new functions. A rational function is an example of a function formed by dividing two polynomial functions. Consider the functions $f(x) = 7x-9$, $g(x) = 2x^2-5x+3$, and $h(x) = 2(3)^x$. Create each of the following functions. Which results are polynomials? Rational? Exponential?

a. $f(x) + g(x)$

b. $f(x) - g(x)$

c. $f(x)g(x)$

d. $\frac{f(x)}{g(x)}$

e. $f(x) + h(x)$

f. $g(x)h(x)$

17. What operations are closed on the set of polynomial functions?

18. Use the functions defined in Exploration 16 to complete the following evaluations.

a. $f(2) + g(2)$

b. $f(-3)g(-3)$

c. $f(-5) - g(4)$

d. $g(2)h(1)$

19. Another operation on functions is called ***composition of functions***. In composition of functions, the output of one function acts as the input to another function. To explore this idea, define $f(x) = 2x - 5$ and $g(x) = x^2 - x + 2$.

a. Find $f(-1)$.

b. Evaluate $g(x)$ where $x = f(-1)$. The answer is $g(f(-1))$.

c. Evaluate $g(x)$ where $2x - 5$ is substituted for x. This is the function $g(f(x))$. Check by evaluating at $x = -1$.

d. Find $g(-1)$.

e. Evaluate $f(x)$ where $x = g(-1)$. The answer is $f(g(-1))$.

f. Evaluate $f(x)$ where $x^2 - x + 2$ is substituted for x. This is the function $f(g(x))$. Check by evaluating at $x = -1$.

g. Based on the functions above, are the functions $g(f(x))$ and $f(g(x))$ the same? Justify.

20. Sue plans a trip to visit a friend at her state university, which is 123 miles away. Then, as often happens, her car breaks down and she has to rent a car. Sue figures that paying by the hour would be cheaper than by the mile and negotiates a real deal. The rental costs her $24 for the day plus $5 per hour that she is driving.

a. Write a function that expresses her cost C in terms of hours spent driving t.

b. Recall from Exploration 5 the function that expressed her time t in terms of average speed s.

c. If Sue runs into road construction and her average speed is 20 mph, what will the cost of the car rental be?

d. Express the cost C in terms of a composite function whose input is average speed s.

21. Find the solutions to each of the following equations or systems of equations.

a. $5x - 9 = 4 - 7x$

b. $2x^2 - 5x = 11$

c. $4x - 9y = 12$
$3x + 2y = -7$

d. $3x^2 - 5 = 0$

REFLECTION

Describe the important features of a rational function. Compare and contrast rational functions with functions previously discussed in this text.

SECTION 7.2

Using Operations to Investigate Rational Functions

Purpose

- Combine several rational expressions into one rational function.
- Solve rational equations.

Now that we know what a rational function is and have a basic idea about how the function behaves, we can think about operations on rational functions such as addition, subtraction, multiplication, and division. Such problems are part of building up your skills with rational functions. Another related skill is solving equations that contain rational functions. We explore all of these skills in this section. As we do, think about how you operate on standard fractions. The basic concepts are the same for rational expressions.

This section is designed for those who wish to develop and understand the skills needed in manipulating rational expressions by hand. Many of the skills investigated in this section can be performed quickly on symbol manipulators such as the TI-92 or computer programs such as *Derive*, *Mathematica*, or *Maple*.

To investigate addition and subtraction of rational expressions, we will use two parallel problems: one involving subtraction of numeric fractions (an arithmetic problem) and one involving subtraction of algebraic fractions (an algebra problem). Consider the following subtractions.

a. $\dfrac{7}{22} - \dfrac{5}{33}$ **b.** $\dfrac{x+1}{x^2+x-12} - \dfrac{x-3}{x^2+7x+12}$

In each investigation, do part **a** on the arithmetic problem (given in part **a** above) and do part **b** on the algebraic problem (given in part **b** above).

1. We first must find a common denominator. To do so, factor each denominator into ***primes***.

a. $22 =$ $33 =$

b. $x^2 + x - 12 =$ $\qquad$ $x^2 + 7x + 12 =$

2. Write the ***best*** (***least common) denominator*** for each part. This is the smallest denominator that is divisible by each of the original denominators. Thus it contains all of their factors but no additional ones.

a. Best (least common) denominator for $\frac{7}{22} - \frac{5}{33}$:

b. Best (least common) denominator for $\frac{x+1}{x^2+x-12} - \frac{x-3}{x^2+7x+12}$:

3. For each fraction, multiply the numerator and denominator by the same quantity to create an equivalent fraction that has the common denominator found in Investigation 2. Multiply the numerators out. Leave the denominators in factored form. The problems are started for you.

a. $\frac{7}{22} = \frac{7}{2(11)} =$ $\qquad$ **b.** $\frac{5}{33} = \frac{5}{3(11)} =$

c. $\frac{x+1}{x^2+x-12} = \frac{x+1}{(x+4)(x-3)} =$

d. $\frac{x-3}{x^2+7x+12} = \frac{x-3}{(x+4)(x+3)} =$

4. Rewrite the original problems using the equivalent or expanded fractions from Investigation 3.

a. $\frac{7}{22} - \frac{5}{33} =$

b. $\frac{x+1}{x^2+x-12} - \frac{x-3}{x^2+7x+12} =$

5. Subtract the fractions by retaining the denominator and subtracting the numerators. Simplify numerators completely.

a. Arithmetic problem

b. Algebraic problem

6. Factor the numerator to determine if the fraction will reduce. Reduce if possible and record the final answer.

a. Arithmetic problem

b. Algebraic problem

7. Display a table for the rational function created in the previous investigation, part **b**, and the original problem $\frac{x+1}{x^2+x-12} - \frac{x-3}{x^2+7x+12}$. The outputs in the table should be the same if your answer is correct.

Adding and subtracting rational expressions involve performing steps equivalent to adding and subtracting fractions. Because traditional vocabulary seems to be very confusing, we have introduced some terms which may be a little more descriptive of the procedures involved.

Denominators are prime-factored to find the "best" denominator. Since the factors of 22 and 33 were 2 and 11, and 3 and 11, respectively, the best denominator is the product of the distinct prime factors 2, 3, and 11. Each rational expression is "expanded" to an ***equivalent fraction*** by multiplying its numerator and denominator by the same quantity so that both expressions have the same denominator, 66. By checking the factors of 22 and the factors of the best denominator, 66, we can tell that we must multiply the numerator and denominator by 3 since that factor is missing from 22. Actually that procedure is equivalent to multiplying by $\frac{3}{3}$ or 1, so even though $\frac{7}{22}$ looks different from $\frac{21}{66}$, they represent the same quantity.

The numerators are then added or subtracted while the denominators are unchanged. The denominator acts as the common unit in the computation.

Finally the numerator must be factored to determine if the rational expression reduces. Our task is completed when we have the reduced form.

Let's look at the derivation.

Subtraction: $$\frac{x+1}{x^2+x-12} - \frac{x-3}{x^2+7x+12}$$

Original expression: $$\frac{x+1}{x^2+x-12} - \frac{x-3}{x^2+7x+12}$$

Denominators factored: $$\frac{x+1}{(x+4)\ (x-3)} - \frac{x-3}{(x+4)\ (x+3)}$$

Best common denominator: $(x+4)\ (x-3)\ (x+3)$

Expanded expressions: $$\frac{(x+1)\,(x+3)}{(x+4)\,(x-3)\,(x+3)} - \frac{(x-3)\,(x-3)}{(x+4)\,(x+3)\,(x-3)}$$

Remove parentheses: $$\frac{x^2+4x+3}{(x+4)\ (x-3)\ (x+3)} - \frac{x^2-6x+9}{(x+4)\ (x+3)\ (x-3)}$$

Combine expressions: $$\frac{(x^2+4x+3) - (x^2-6x+9)}{(x+4)\ (x-3)\ (x+3)}$$

Remove parentheses in numerator: $$\frac{x^2+4x+3-x^2+6x-9}{(x+4)\ (x-3)\ (x+3)}$$

Combine like terms in numerator: $$\frac{10x-6}{(x+4)\ (x-3)\ (x+3)}$$

Factor numerator: $$\frac{2\,(5x-3)}{(x+4)\ (x-3)\ (x+3)}$$

The numerator and denominator share no zeros, so the resulting rational expression is irreducible. Another way to say that is that they have no non-constant common factors.

In these examples, each factor appeared only once in a denominator and although 11 and $x+4$ were in both denominators we used each only once in the best denominators. However, if a denominator were 8, we would have to use 2^3 as a factor in the best denominator.

The corresponding function $f(x) = \dfrac{2\,(5x-3)}{(x+4)\ (x-3)\ (x+3)}$ has vertical asymptotes at $x=-4$, $x=3$, and $x=-3$. The domain of the function is the set of all real numbers except -4, 3, and -3. This means that every other real number is an input into the function that will produce an output. An end behavior model for this function is the x-axis.

When transforming rational expressions, it helps to check answers using either a table or a graph. This can be done by entering the original expression as one function and the final answer as a second function. Table values and graphs will match if you have not made a mistake. If they do not match, it's time to check your work. Figure 1 displays the table and graphs for the problem just completed.

FIGURE 1

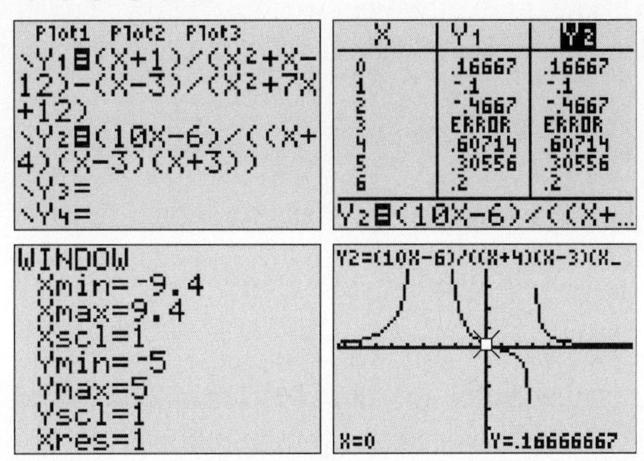

Notice that the quotient of the highest degree terms in the numerator and denominator in this problem was $\frac{10x}{x^3} = \frac{10}{x^2}$ and that an end behavior model was $y = 0$.

In Section 7.1 we analyzed the function $y(x) = \frac{x^2 - x - 20}{x + 3}$. Regression was used to find the end behavior model $y = x - 4$. Since rational functions represent a quotient of polynomials, we could perform long division to obtain a quotient and a remainder over the original divisor. Several computer software programs and the TI-92 calculator can perform this long division (Figure 2).

FIGURE 2

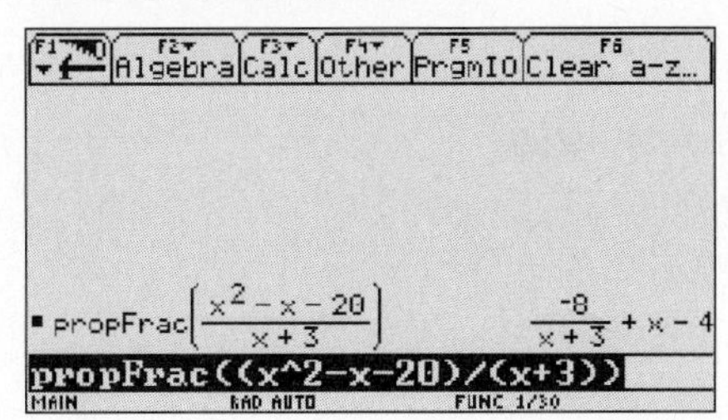

So the function $y(x) = \dfrac{x^2 - x - 20}{x + 3}$ is equivalent to

$$y(x) = x - 4 + \frac{-8}{x + 3}.$$

The $x - 4$ is the quotient and the -8 represents the remainder. Recall also that the equation $y = x - 4$ was a better approximation of end behavior found by regression. We could also find the equation of the slant asymptote by finding the quotient by long division.

In the next investigation, we look at how to reverse this process by continuing our study of the addition and subtraction of rational expressions.

8. a. Express $2 + \dfrac{3}{x - 4}$ as a rational function by adding the two expressions together. (Remember that 2 can be thought of as $\dfrac{2}{1}$.)

b. Identify the domain, location of vertical asymptotes, and an end behavior model for the rational function created in part **a**.

c. How could you determine the end behavior model using the original expression?

To express $2 + \dfrac{3}{x - 4}$ as one fraction, we use the denominator $x - 4$.

Original problem: $2 + \dfrac{3}{x - 4}$

Expand $\dfrac{2}{1}$: $\dfrac{2\,(x - 4)}{1\,(x - 4)} + \dfrac{3}{x - 4}$

Remove parentheses: $\dfrac{2x - 8}{x - 4} + \dfrac{3}{x - 4}$

Combine numerators: $\dfrac{2x - 8 + 3}{x - 4}$

Simplify: $\dfrac{2x - 5}{x - 4}$

This function $y(x) = \dfrac{2x-5}{x-4}$ has a domain consisting of all real numbers except 4, a vertical asymptote at $x = 4$, and an end behavior model of $y = \dfrac{2x}{x} = 2$. Notice that the end behavior model matches the original quotient 2 in the expression $2 + \dfrac{3}{x-4}$.

So far, we have investigated the addition and subtraction of rational expressions. The technique of writing rational expressions so they have a common denominator is useful when we wish to solve equations involving rational expressions. We return to the concentration problem situation introduced in Section 7.1. Recall the ***concentration*** of a given substance is represented by the ratio

$$\frac{\text{Amount of substance}}{\text{Total quantity}}.$$

Alcohol Content: A glass contains 10 ounces of a liquid that is 12% alcohol. Water is added to reduce the alcohol content. We'd like to determine how much water should be added to reduce the alcohol concentration to 9%.

9. Write a rational function that represents the concentration C of alcohol in the drink if x ounces of water are added.

10. Write an equation that represents the situation described in the alcohol content problem.

11. Solve the equation using a table or a graph. Describe what you did along with the solution to the problem.

The concentration of alcohol in the glass is represented by the ratio

$$\frac{\text{Amount of alcohol}}{\text{Amount of liquid in glass}}.$$

The amount of liquid in the glass is $10 + x$, the original amount plus the amount added.

The amount of alcohol in the glass is $0.12(10) + 0.0(x)$, the original amount of alcohol plus the amount of alcohol added. Note that water contains no alcohol—its alcohol content is 0%.

So the alcohol concentration C in the glass after x ounces of water are added is

$$C(x) = \frac{1.2}{10 + x}.$$

We wish to reduce the alcohol concentration to 9%, resulting in the equation

$$\frac{1.2}{10 + x} = 0.09.$$

Approximations of the solution appear in Figure 3.

FIGURE 3

X	Y1
3	.09231
3.1	.0916
3.2	.09091
3.3	.09023
3.4	.08955
3.5	.08889
3.6	.08824

X=3.3

```
Plot1 Plot2 Plot3
\Y1=1.2/(10+X)
\Y2=.09
\Y3=
\Y4=
\Y5=
\Y6=
\Y7=
```

```
WINDOW
 Xmin=0
 Xmax=20
 Xscl=1
 Ymin=0
 Ymax=.2
 Yscl=.1
 Xres=1
```

```
Intersection
X=3.3333333 Y=.09
```

We must add about 3.3 ounces of water to the glass to reduce the alcohol content of the liquid to 9%.

We can also solve this equation algebraically using the fact that the numerators of two equal fractions are equal if the denominators are equal.

12. Given the equation $\frac{1.2}{10 + x} = 0.09$, rewrite the right side of the equal sign so that both sides of the "=" have the same denominator.

13. Set the numerators equal to each other and solve the equation. Compare your answer to that obtained graphically and tabularly.

The equation $\frac{1.2}{10 + x} = 0.09$ is an example of a rational equation. We usually try to "legally" eliminate denominators when solving rational equations algebraically. In this example, we rewrite the equation so the denominators are the same on both sides of the "=". The numerators are then set equal and we solve the resulting equation.

Original equation: $\dfrac{1.2}{10+x} = 0.09$

Expand right side: $\dfrac{1.2}{10+x} = 0.09\left(\dfrac{10+x}{10+x}\right)$

Simplify: $\dfrac{1.2}{10+x} = \dfrac{0.9+0.09x}{10+x}$

Numerators are equal once we know the denominators are the same:

$$1.2 = 0.9 + 0.09x$$

Subtract 0.9 from both sides: $0.3 = 0.09x$

Divide both sides by 0.09: $3.\bar{3} = x$

This answer matches those obtained graphically and tabularly though $3\frac{1}{3}$ ounces is an exact result while 3.3 ounces is an approximation.

Although there are other algebraic techniques for solving this equation, the one shown here generalizes nicely to any rational equation.

To conclude this section, we look at solving a system of equations that involves rational functions.

14. A System of Rational Equations: Consider the rational functions $f(x) = \dfrac{2x-4}{x^2-4x+3} - \dfrac{2}{x+3}$ and $g(x) = \dfrac{1}{x-1}$. We'd like to determine when the outputs of the two functions are equal.

a. Use a table to approximate the value(s) of the input that causes the outputs of the two functions to be equal.

b. Write the equation defined by the problem statement.

c. Rewrite $f(x)$ as one rational expression; that is, perform the subtraction on the two rational expressions defining function f.

d. Expand each rational expression as necessary so that both functions f and g have a common denominator.

e. Set the numerators equal and solve for x.

f. Do your answers check with those obtained using a table or a graph? If not, explain why.

Let's look at what happened in the previous investigation. The table appears in Figure 4. We see that the outputs of the two functions are equal when $x = 9$. Scrolling though the table, we do not spot any other solutions, but that does not mean there aren't any more solutions. We look at the algebra to find all solutions.

FIGURE 4

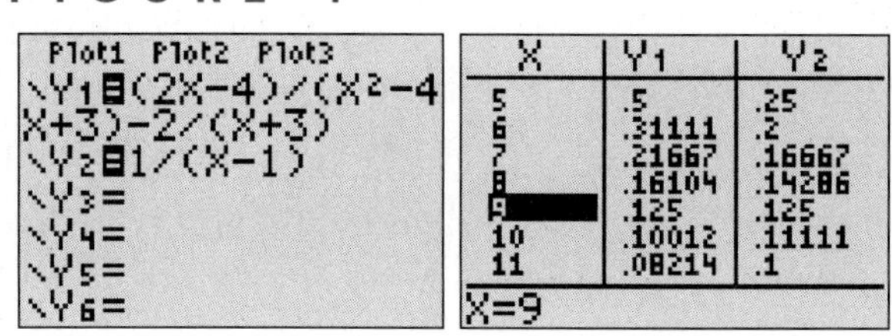

$$f(x) = \frac{2x-4}{x^2-4x+3} - \frac{2}{x+3} \quad \text{and} \quad g(x) = \frac{1}{x-1}$$

The outputs of these functions are equal when

$$\frac{2x-4}{x^2-4x+3} - \frac{2}{x+3} = \frac{1}{x-1}.$$

Let's rewrite all expressions with a common denominator.

Factor denominator: $$\frac{2x-4}{(x-3)(x-1)} - \frac{2}{x+3} = \frac{1}{x-1}$$

Expand:

$$\frac{(2x-4)(x+3)}{(x-3)(x-1)(x+3)} - \frac{2(x-3)(x-1)}{(x+3)(x-3)(x-1)} = \frac{1(x-3)(x+3)}{(x-1)(x-3)(x+3)}$$

Expand numerators:

$$\frac{2x^2+2x-12}{(x-3)(x-1)(x+3)} - \frac{2x^2-8x+6}{(x+3)(x-3)(x-1)} = \frac{x^2-9}{(x-1)(x-3)(x+3)}$$

Subtract fractions on left side:

$$\frac{2x^2+2x-12-2x^2+8x-6}{(x-3)(x-1)(x+3)} = \frac{x^2-9}{(x-1)(x-3)(x+3)}$$

Combine like terms in numerator on left side:

$$\frac{10x-18}{(x-3)(x-1)(x+3)} = \frac{x^2-9}{(x-1)(x-3)(x+3)}$$

Set numerators equal to each other: $10x-18 = x^2-9$

This equation is quadratic because of the x^2 term. We rewrite so that one side is zero and then either factor or use the quadratic formula to solve.

Rewrite so that one side is zero and find the zeros of the quadratic:

$$0 = x^2 - 10x + 9$$

The zeros of $y = x^2 - 10x + 9$ are 1 and 9.

We know that 9 was a solution, but what about 1? The table in Figure 5 indicates that 1 is not in the domain of either function. So 1 is not a solution to the equation.

FIGURE 5

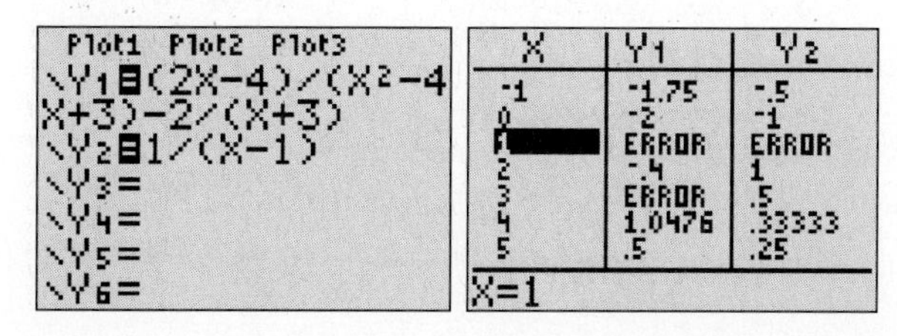

Solving rational equations algebraically requires us to check the solutions to make sure each solution is in the domains of the original expressions.

STUDENT DISCOURSE

Pete: I am still not sure why 1 is not a solution to the equation $\frac{2x-4}{x^2-4x+3} - \frac{2}{x+3} = \frac{1}{x-1}$. It is certainly a solution to the equation $0 = x^2 - 10x + 9$. Don't the two equations have the same solutions?

Sandy: No, they don't.

Pete: But I only did "legal" algebraic steps. What happened?

Sandy: At one point in the process you eliminated something from the equation.

Pete: Let's see. I dropped the denominators after they were the same.

Sandy: Exactly. By eliminating the denominators, you may no longer be aware of any restrictions on the values of the variable.

Pete: You mean like domain restrictions?

Sandy: Yes. The function $y = x^2 - 10x + 9$ has a domain of all real numbers, but the domains of the original functions were restricted because the denominators can't be 0.

Pete: I think I see. Even though we eliminate denominators to finish the solution process, we can't forget the restrictions on the variable.

Sandy: That's right.

1. In the Glossary, define the words in this section that appear in ***italic boldface*** type.

2. Refer to the **Alcohol Content** problem on page 529. Determine how much club soda, which has no alcohol, must be added to reduce the alcohol concentration to 7%. Justify your answer.

3. Write your own concentration problem. Create a rational function that represents the situation.

4. **Triathlon Time:** Sherry has trained long and hard for her county's local triathlon in which she must swim 1 mile, ride a bicycle 25 miles, and run 10 miles. While her actual speeds are dependent on the conditions and her health on the day of the race, she knows from experience that her average running speed is 5 miles per hour more than her average swimming speed. Her average biking speed is 15 miles per hour faster than her average swimming speed.

   a. Express the time it will take to swim 1 mile as a rational function of x, her average swimming speed.

b. Express the time it will take to bike 25 miles as a rational function of x, her average swimming speed.

c. Express the time it will take to run 10 miles as a rational function of x, her average swimming speed.

d. Express her total time as the sum of the swimming time, bicycling time, running time.

e. Determine her average speeds in each event if she completes the triathlon in 2 hours 40 minutes. Justify your result.

5. Rational expressions can be multiplied in the same way that rational numbers are multiplied. Remember to factor and remove pairs of matching factors that are in any numerator and denominator before multiplying. If you simplify by removing common factors before multiplying, the resulting answer will be in simplest form. Multiply the following and reduce completely.

a. $\left(\frac{42}{55}\right)\left(\frac{35}{27}\right)$

b. $\left(\frac{4x}{7x^3}\right)\left(\frac{35}{20x}\right)$

c. $\left(\frac{2x^2-3x-2}{3x-15}\right)\left(\frac{6x+30}{2x^2+3x+1}\right)$

d. $\left(\frac{x^2-9}{5x-15}\right)\left(\frac{5x+30}{2x^2+12x+18}\right)$

6. The quotient of two rational expressions can be found in the same way that division of rational numbers is performed. Factoring is used to reduce the resulting answer completely. Perform the operation and reduce completely.

a. $\frac{28}{39} \div \frac{21}{26}$

b. $\frac{5x-10}{x^2+2x-15} \div \frac{3x-6}{x^2-8x+15}$

7. Rational functions are sometimes difficult to capture in their entirety. In Figure 1 on page 527 we used a window that showed some characteristics of the graph very well, but omitted others. Change the window so that Xmin = -6, Xmax = 6, Ymin = -30, and Ymax = 5. Graph the function $f(x) = \dfrac{2(5x-3)}{(x+4)(x-3)(x+3)}$ again. Describe what you see and compare the two graphs of the function. Tell what the advantages of both representations are.

8. A ***complex fraction*** is a fraction whose numerator or denominator contains at least one fraction. A complex fraction is simply a division problem involving fractions. Write each as a division problem using parentheses appropriately and use order of operations to simplify to one noncomplex fraction. (Yes, you could find shortcuts to simplify complex fractions, but those techniques are probably useful to learn if you are going to be solving a lot of complex fractions in your life. Here you are just trying to apply basic division of fractions in a slightly more complex problem. Figuring out a procedure that works is the major objective.)

a. $\dfrac{\frac{1}{5}-\frac{1}{3}}{2}$

b. $\dfrac{\frac{1}{x+2}-\frac{1}{x}}{2}$

c. $x - \dfrac{x}{1-\frac{x}{1-x}}$

d. $\dfrac{3x}{2+\frac{1}{x}}$

9. Describe in words how you found an answer to Exploration 8 part **b**.

10. Write each of the following as one rational expression.

a. $x + \dfrac{3}{x+2}$

b. $x^2 + \dfrac{3}{x+2}$

c. $5 + \dfrac{x+1}{x^2+x+3}$

d. $x + \dfrac{2}{x-1} + \dfrac{3}{x^2-1}$

e. $x + \dfrac{2}{x}$

f. $x^2 + 2x - 5 + \dfrac{7}{x^2+2x+1} + \dfrac{5}{x+1}$

11. For each expression in Exploration 10, determine the domain, the location of vertical asymptotes, the location of any holes in the graph, and an end behavior model.

12. Find the solutions to each of the following equations or systems of equations.

a. $4 + x = \dfrac{12}{x}$

b. $\dfrac{1}{x-2} - \dfrac{1}{6} = \dfrac{2}{3x-6}$

c. $2 + \dfrac{3}{x-4} = \dfrac{2}{x+1}$

d. $\dfrac{3}{x-2} = \dfrac{7}{x+5}$

e. $\dfrac{4}{3x-1} = 2$

f. $\dfrac{4}{x-3} - \dfrac{7}{x+2} = \dfrac{1}{x-4}$

g. $4x - 5.7 = 2.8x + 4$

h. $3x^2 - x = 11$

i. $4x - y = -12$
$3x + 7y = 4$

j. $x - 7y + 5z = 1$
$3x + 2y - 2z = -7$
$4x + y + 4z = -2$

13. Consider the functions $f(x) = 3x^2 - x + 7$ and $g(x) = \dfrac{2}{3x-1}$. Find:

a. $f(g(-4))$

b. $g(f(-4))$

c. $f(g(x))$

d. $g(f(x))$

14. Building a Better Can: A cylindrical can of radius r and height h must hold 12 ounces of soda. Twelve ounces is equivalent to 354 milliliters, which equals a volume of 354 cubic centimeters. The manufacturer wishes to minimize the amount of aluminum used to make each can. Follow the steps below to determine the dimensions of the can that minimize the amount of aluminum needed.

a. Write an equation for the volume of the can. This can be done by multiplying the area of the base of the can by the can's height.

b. Use the equation from part **a** to express the height h of the can as a function of the radius r of the can.

To minimize the amount of aluminum used, you must minimize the surface area of the can. Determine a formula for the surface area S of the can as follows.

c. Write an expression for the area of the bottom and top of the can in terms of the radius r

d. Write an expression for the area of the side of the can in terms of the radius r and the height h. To do this, multiply the circumference of the can by its height. To help recall the formula for circumference, remember that *pi* is defined as the ratio of the circumference to the diameter of a circle. (This process makes more sense if you think about how you would manufacture a can. Take a rectangular sheet of paper and fasten two of the opposite ends, the widths, to make a tube. The area of the rectangle is length times width, so that is the area of the side of the can—the length has become the circumference and the width is the height. Make a sketch of the side area of a can to help you visualize what the area region looks like.)

e. Find the formula for surface area S by adding your answers to parts **c** and **d**. Write down this formula.

f. Use your answers to parts **b** and **e** to express the surface area S as a function of the radius r only (i.e., replace the h in the formula from part **e** with an expression involving r from part **b**).

g. Use a table or a graph to determine the value of the radius for which the surface area is the smallest. Compare the dimensions of a can with least surface area to the standard shape of soda cans.

h. What other considerations might influence the decision made by the manufacturer about the dimensions?

REFLECTION

Describe in words how to perform each of the four basic operations on rational expressions. Use a numerical example and an algebraic example to demonstrate what you have said, but write the explanation in complete sentences.

SECTION 7.3

Roots Extended

Purpose

- Investigate the meaning of roots, other than square roots.

In the last two sections, we investigated rational functions. We now turn our attention to ***radical functions***—functions that involve computing a root of a variable expression. In Section 6.3, we investigated the square root, which acts as the reverse process of squaring (raising to a power of 2). In this section, we investigate other types of roots, such as third roots, fourth roots, fifth roots, etc. These roots arise when we wish to reverse the process of cubing (raising to a power of 3), raising to a power of 4, raising to a power of 5, etc.

Doghouse Problem: Earlier we built a dog pen, but now we need a doghouse. Let's begin with a house that has a flat roof and is the same length in all three directions—in other words, a cube. We would like to investigate the relationship between the length of a side and the volume of the doghouse.

1. Complete Table 1 on page 540 by selecting some side lengths and computing the volume. Choose any appropriate unit as in the first example.

2. Draw a function machine using as input the length of a side and as output the volume of the doghouse.

TABLE 1 Volume of Doghouse

Length of Side s	Volume of Doghouse V
2 m	8 m^3

3. Write an equation expressing the volume V as a function of the length s of the side.

Now let's reverse the process.

4. Complete Table 2 by selecting a value for volume and finding the value for the length of a side. The first one is done as an example.

TABLE 2 Length as a Function of Volume

Volume of Doghouse V	Length of Side s
8 yd^3	2 yd

a. Were the outputs of this function all whole numbers? Why?

b. Suppose the input is 100 cubic feet. What is the length of the side? How did you find the answer?

c. Draw a function machine in which the input is volume and the output is the length of a side.

d. Write an equation expressing the length s of a side as a function of the volume V.

5. A real number a is the ***cube root*** of a real number b if $a^3 = b$.

a. 5 is the cube root of 125 since $5^3 = 125$. Which number, 5 or 125, is the cube and which number is the cube root? Defend your answer.

b. In the statement $a^3 = b$, which variable is the cube and which variable is the cube root? Defend your answer.

If s is the length of a side of the doghouse and V is the volume then $V(s) = s^3$. Previously we saw that the square root operation reversed the process of squaring. Likewise, the ***cube root*** or ***third root*** operation reverses the process of cubing (raising to the third power).

In $a^3 = b$, the letter a is acting as the cube root and the letter b is acting as the ***cube***. To obtain b we must raise a to the third power. To obtain a, we must reverse the cubing process, that is, take the cube root of b.

For example, 5 is the cube root of 125 since $5^3 = 125$. The number 125 is the cube and the number 5 is the cube root.

Symbolically, the cube root of b is written $\sqrt[3]{b}$.

Thus the statements $a^3 = b$ and $a = \sqrt[3]{b}$ are ***equivalent***.

Using numbers, $5^3 = 125$ and $5 = \sqrt[3]{125}$ provide identical information.

Recall that we used $\sqrt{b}$ to represent the principal square root of b. This symbolism actually represents $\sqrt[2]{b}$, but the "2" is seldom written because square roots are so common.

To complete Table 2 we had to calculate some cube roots. Many calculators have a cube root key that makes this task simple. Check your manual or ask someone to help you find and use the cube root key or the generic root key. Nevertheless, we will find a numerical approximation of a cube root based on calculating the cube of good guesses to better understand a procedure that is so complex that everyone resorts to technology, tables of cube roots, or the use of logarithms—the method of choice before technology. Logarithms are addressed in Chapter 8.

Consider the cubic doghouse with a volume of 100 cubic feet. What should the length of the side be? Figure 1 displays a series of three tables that numerically zoom in on the cube root of 100. Note that we are looking at a table for $y = x^3$ and locating an input for the output 100. In other words, given an output of 100, find the input that acts as the cube root of 100. We could approximate the side of the doghouse as 4.64 feet. The units give a hint about the cube, 100 cubic feet, and the cube root, 4.64 feet, which helps distinguish between these two terms when we measure volume based on length.

FIGURE 1

X	Y2
1	1
2	8
3	27
4	64
5	125
6	216
7	343

Y2=X^3

X	Y2
4.4	85.184
4.5	91.125
4.6	97.336
4.7	103.82
4.8	110.59
4.9	117.65
5	125

Y2=X^3

X	Y2
4.61	97.972
4.62	98.611
4.63	99.253
4.64	99.897
4.65	100.54
4.66	101.19
4.67	101.85

Y2=X^3

Now we write a general definition of nth roots. Assume that n is a whole number greater than 1.

The real number a is the nth root of real number b if $a^n = b$.

This is written symbolically as $a = \sqrt[n]{b}$. The n is called the ***index*** and b is called the ***radicand***.

Let's look at some of the characteristics of powers and roots in more detail.

6. Write the first six positive whole-number cubes. For example, the seventh positive whole-number cube is 343 since $7^3 = 343$.

7. Are there any negative-integer cubes? If your answer is yes, write four of them. If no, explain why.

8. Complete Table 3 on page 543. Write the numbers from the previous two investigations in the left column. Record the cube roots in the right column.

9. Is the cube (third) root of a negative number defined? Why or why not?

10. Write the first six positive whole-number fourth powers. For example, 2401 is a whole-number fourth power since $7^4 = 2401$.

TABLE 3 Cube Roots

Cubes	Cube Roots

11. Are there any negative-integer fourth powers? If yes, write four of them. If no, explain why.

12. Complete Table 4. Write the numbers from the previous two investigations in the left column. Record the fourth roots in the right column.

TABLE 4 Fourth Roots

Fourth Powers	Fourth Roots

13. Is the fourth root of a negative number defined? Why or why not?

14. Is the fifth root of a negative number defined? Why or why not?

15. Is the sixth root of a negative number defined? Why or why not?

16. Let n be a whole number greater than 1.

a. If n is odd, what is the domain of the function $y = \sqrt[n]{x}$? Why?

b. If n is even, what is the domain of the function $y = \sqrt[n]{x}$? Why?

When we raise a real number to an odd power or take a root with an odd index, the result has the same sign as the original number. Cubes and cube roots, fifth powers and fifth roots, seventh powers and seventh roots, etc. all share this property.

The odd power of a positive number is positive. For example, $5^3 = 125$.

The odd root of a positive number is positive. For example, $\sqrt[3]{125} = 5$.

The odd power of a negative number is negative. For example, $(-5)^3 = -125$.

The odd root of a negative number is negative. For example, $\sqrt[3]{-125} = -5$

On the other hand, when we raise a real number to an even power, the result is never negative.

There are two even roots for every positive number. The two roots are opposites. For example, the fourth roots of 625 are 5 and –5. However, the notation $\sqrt[n]{x}$ represents the **principal *n*th root**. If n is even, then $\sqrt[n]{x}$ is the positive nth root of x. For example, $\sqrt[4]{625} = 5$ and $-\sqrt[4]{625} = -5$, but $\sqrt[4]{-625}$ is not a real number.

Given the function $y = \sqrt[n]{x}$ where n is odd, the domain is all **real numbers**.

Given the function $y = \sqrt[n]{x}$ where n is even, the domain is all **nonnegative (zero and positive) real numbers**.

Let's conclude the section by looking at the graphs of functions involving roots.

17. Graph each of the following functions. Sketch the graph in the space provided. (You will need a calculator or computer that allows the entry of general nth roots.)

a. $y(x) = \sqrt[2]{x}$

b. $y(x) = \sqrt[3]{x}$

c. $y(x) = \sqrt[4]{x}$

d. $y(x) = \sqrt[5]{x}$

18. Based on the previous investigation, sketch the graph of the general nth root function when:

a. n is odd

b. n is even

For odd roots, the input may be any real number. The output may also be any real number, but the sign of the output must match the sign of the input. This means the graph lies in the first and third quadrants.

When working with even roots, the input must be nonnegative if we restrict ourselves to the real number outputs and the output is nonnegative because the use of the radical implies the principal root. Thus the entire graph is restricted to quadrant I.

Figure 2 displays generic graphs for odd and even roots.

FIGURE 2

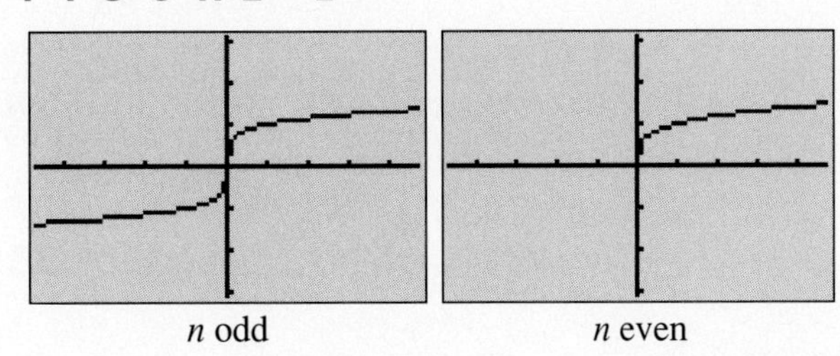

n odd $\qquad$ n even

In a later section, we will investigate some operations on radical expressions.

STUDENT DISCOURSE

Pete: While I think I understand this section, the most basic thing always confuses me.

Sandy: What's that?

Pete: Well, when I see something like the "cube root of 10," I don't know whether to find 10^3, which is 1000, or to find a number whose third power is 10.

Sandy: Don't feel bad. I had the same kind of problem. I don't have any magic solution, but I will tell you that I look for the word "root."

Pete: How does that help?

Sandy: If I see a phrase like "cube 10," the fact that the word "root" is missing tells me that the word "cube" means to "raise to the third power."

Pete: So you see the word "cube" as meaning an exponent of 3?

Sandy: That's right. But if I see the phrase "cube root of 10," the word "root" tells me that 10 is the **result** of raising a number to the third power.

Pete: So the "cube root of 10" means we want a number, so that the number times the number times the number is 10?

Sandy: Exactly. So, you see "cube 10" means 10^3, which is 1000, but...

Pete: ...but, "cube root of 10" means some number between 2 and 3 since 2^3, which is 8 is a little smaller than 10.

Sandy: That's right, in fact, my calculator says that the cube root of 10 is about 2.15.

Pete: Thanks for the help. I can see how the word "root" is the key. I need to think about this a bit more.

1. In the Glossary, define the words in this section that appear in ***italic boldface*** type.

2. **Volume of a Cylinder:** The volume of a circular cylinder, such as a soda can, is given by the formula $V = \pi r^2 h$, where r is the radius of the circular base and h is the height of the cylinder.

   a. Suppose the height is twice the radius. Rewrite the formula so that it contains only the variables V and r.

b. Draw a function machine to illustrate the formulas in part **a**.

c. Rewrite the formula so that the radius r is defined as a function of the volume V.

d. Draw a function machine to illustrate the formula in part **c**.

e. Approximate the diameter of the cylinder if the volume is 132 cubic centimeters.

3. Volume of a Sphere: The volume of a sphere, such as a basketball, is determined by the formula $V = \frac{4}{3}\pi r^3$, where r is the radius of the sphere.

a. Draw a function machine for this formula.

b. Rewrite the formula so that the radius r is expressed as a function of the volume V.

c. Draw a function machine for this formula. How is it different from part **a**?

d. Use the function you wrote in part **b** to approximate the radius if the volume of the sphere is 237 cubic inches.

4. Complete Table 5 on page 548. Select any values you like for the first column. Triple these values for the second column and cube the values for the third column.

5. Refer to Table 5.

a. Write functions for ***tripling*** a number and cubing a number.

TABLE 5 Tripling and Cubing

Number	Triple Number	Cube Number

b. What operation is the inverse of tripling? Write a function for the inverse of the tripling function.

c. What operation is the inverse of cubing? Write a function for the inverse of the cubing function.

d. Izzy frequently confuses the functions of tripling and cubing. Using complete sentences, explain to Izzy the difference. Use this problem to illustrate: Tom built a cubic doghouse for his beagle which was 2 feet on a side. He tripled the dimensions for the cubic doghouse for his Saint Bernard. How do the volumes of the two doghouses compare?

6. Which is larger: x or $\sqrt[n]{x}$? To answer this question, select some values for x and compute various roots comparing size to the original choice of x. Be sure to choose values for x that are greater than 1 and values between 0 and 1. Organize your data in a table.

We can learn a lot about functions by playing with their graphs. What happens when you add a number here or there? Explorations 7–9 give you some things to try on your calculator. Don't be confused by the h's and k's. Just replace them with numbers.

7. Given the function $y(x) = \sqrt[3]{x}$:

a. Graph this function in a "friendly" window.

b. Overlay the graphs of $y(x) = \sqrt[3]{x-h}$, where h is replaced with one positive number and one negative number. What effect does the choice of h have on the graph of $y(x) = \sqrt[3]{x}$?

c. Regraph $y(x) = \sqrt[3]{x}$ alone. Overlay the graphs of $y(x) = \sqrt[3]{x} + k$, where k is replaced with one positive number and one negative number. What effect does the choice of k have on the graph of $y(x) = \sqrt[3]{x}$?

8. Given the function $y(x) = \sqrt[4]{x}$:

a. Graph this function in a "friendly" window.

b. Overlay the graphs of $y(x) = \sqrt[4]{x-h}$, where h is replaced with one positive number and one negative number. What effect does the choice of h have on the graph of $y(x) = \sqrt[4]{x}$?

c. Regraph $y(x) = \sqrt[4]{x}$ alone. Overlay the graphs of $y(x) = \sqrt[4]{x} + k$, where k is replaced with one positive number and one negative number. What effect does the choice of k have on the graph of $y(x) = \sqrt[4]{x}$?

9. Based on the previous two explorations, describe the effect of h and k on the graph of $y(x) = \sqrt[n]{x-h} + k$ as compared to the graph of $y(x) = \sqrt[n]{x}$. Does the shape of the graph change? Does the position of the graph change?

10. A Better Doghouse: Remember that doghouse? Instead of a flat roof we are going to use a triangular roof. Also, the length will be twice the width, and the height from the ground to the base of the triangle will be the same as the width. The height of the triangle will be one-half the width. (Figure 3 displays a diagram showing the width and height.)

a. Write a formula that expresses the volume as a function of the width.

FIGURE 3

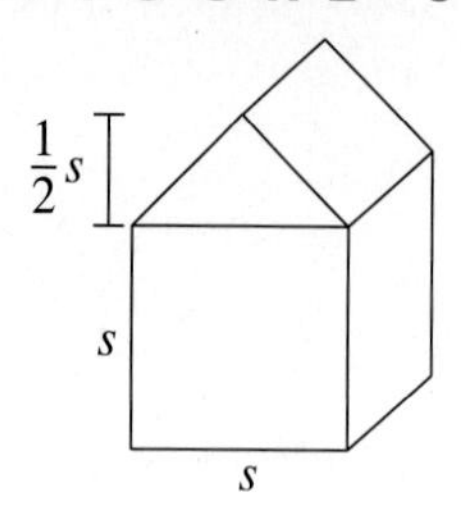

b. Rewrite the formula from part **a** so that the width is defined as a function of the volume.

c. Determine the width if the volume is 30 cubic feet. Defend your answer.

11. Assume n is odd. Simplify $\sqrt[n]{(x)^n}$. Use a graph or table with a specific root to check your answer.

12. Assume n is even. Simplify $\sqrt[n]{(x)^n}$. Use a graph or table with a specific root to check your answer.

13. Previously we saw how to simplify square roots, without approximating, by prime-factoring the radicand. Use a similar approach to simplify each of the following. For example,

$$\sqrt[5]{1152} = \sqrt[5]{(2^7 \cdot 3^2)} = 2 \cdot \sqrt[5]{(2 \cdot 3)^2}$$

a. $\sqrt[3]{343}$

b. $\sqrt[4]{80}$

c. $\sqrt[5]{486}$

d. $\sqrt[3]{-6000}$

e. $\sqrt[6]{128}$

f. $\sqrt[3]{3150}$

14. Find the solution to the following equations or systems of equations.

a. $3x^2 - 7 = 0$

b. $\frac{2}{3}x - \frac{5}{2} = 7 - \frac{5}{6}x$

c. $2x + 3y = -17$
$x = 9 - 7y$

d. $\frac{x}{x+1} = 2$

15. Rewrite the following as one rational expression.

a. $\frac{4x^2 - 9}{x^2 - x} \div \frac{2x^2 + x - 6}{x^2 + 4x}$

b. $x - 1 + \frac{1}{x+1}$

16. Consider the rational function $y = \frac{x^2 + 2x + 1}{x^2 + 3x + 2}$.

a. What is the domain of the function? Why?

b. Where does the graph of the function have vertical asymptotes?

c. Where does the graph of the function have holes?

17. A linear function has slope $-\frac{2}{5}$ and contains the point $(6, 1)$. Write the equation for the function.

18. Consider the quadratic function in Figure 4.

FIGURE 4

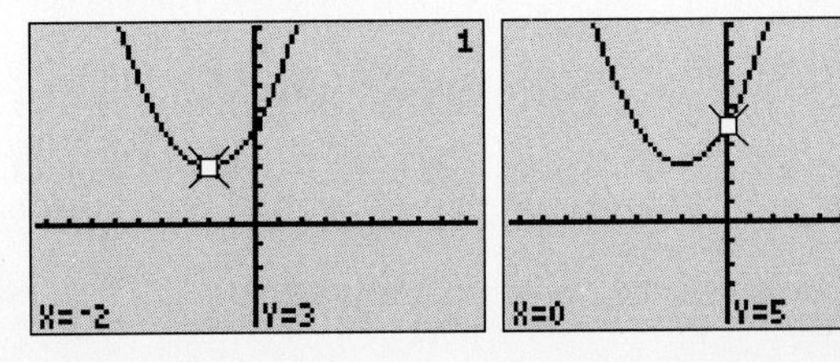

a. Write the equation of the function in "vertex" form.

b. Write the equation of the function in the form $y = ax^2 + bx + c$.

REFLECTION

Write a paragraph describing all the critical features of the general *n*th root function.

SECTION 7.4

Rational Exponents

Purpose

- Investigate the relationship between rational exponents and radicals.

Now that we have investigated roots in general, we turn our attention to exponents. In the past, you have learned about exponents that were counting numbers, about 0 exponents, and about negative-integer exponents. The expression a^n is called an ***exponential expression***. The letter a is called the ***base*** and the letter n is called the ***exponent***. The meaning of an exponent depends on the type of number used as an exponent. For example, 2^3 means that we multiply three factors of 2 to get 8. Also, 2^0 means 1, and 2^{-1} means the reciprocal of 2. We will extend our understanding of exponents to include the meaning of fractional exponents. Interestingly, there is a close connection between roots and fractional exponents.

We begin the section by investigating the meaning of integer exponents and some important properties of exponents. Then we will use the calculator to help us investigate the meaning of fractional exponents.

1. Explain what the exponent 20 means in the expression a^{20}.

2. Assume that the letter n represents any positive integer. Explain what the exponent n means in the expression a^n.

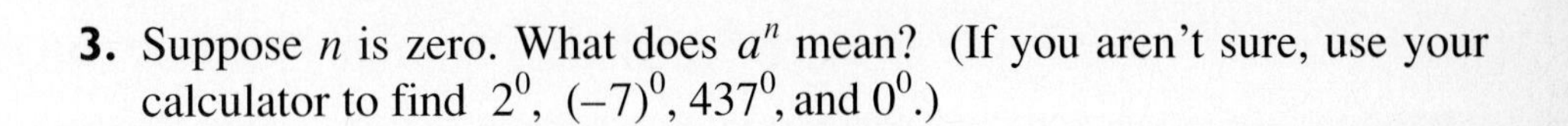

3. Suppose n is zero. What does a^n mean? (If you aren't sure, use your calculator to find 2^0, $(-7)^0$, 437^0, and 0^0.)

4. Explain what the exponent -20 means in the expression a^{-20}.

5. Assume that the letter n represents any whole number. Explain what the exponent $-n$ means in the expression a^{-n}.

DISCUSSION

Previously you have studied the meaning of integer exponents. For example, you know that 2^5 means $2 \cdot 2 \cdot 2 \cdot 2 \cdot 2$, the product of five 2's. Likewise, a^{20} means the product of 20 a's.

In general, a^n is shorthand for the product of n a's.

Also recall that $a^0 = 1$ provided that a itself is not zero. For example, $17^0 = 1$.

Similarly, you know that a^{-20} represents the ***reciprocal*** of the product of 20 a's. A negative-integer exponent provides two bits of information.

- The "oppositing" symbol says "take the reciprocal of the base."
- The numerical part says "number of factors of the base."

For example, 2^{-5} directs us to take the reciprocal of the base of 2 and to use five factors of the base to compute the result. This may be done in either order.

$$2^{-5} = \left(\frac{1}{2}\right)^5 = \frac{1}{2} \cdot \frac{1}{2} \cdot \frac{1}{2} \cdot \frac{1}{2} \cdot \frac{1}{2} = \frac{1}{32}$$

or

$$2^{-5} = (2 \cdot 2 \cdot 2 \cdot 2 \cdot 2)^{-1} = (32)^{-1} = \frac{1}{32}.$$

Next we investigate some basic properties of exponents.

6. Predict the exact simplification of each exponential expression. Use a symbol manipulator, if available, to simplify each expression. Modify your prediction as necessary.

a. $(x^9)(x^4)$

b. aa^6

c. x^5y^4

d. $x^4 + x^3$

7. Using the results of Investigation 6, what is another way to write $y^m y^n$? Why?

8. What happened when you multiplied exponential expressions that had different bases in Investigation 6**c**?

9. What happened when you added, rather than multiplied, exponential expressions in Investigation 6**d**?

10. Predict the exact simplification of each exponential expression. Use a symbol manipulator, if available, to simplify each expression. Modify your prediction as necessary.

a. $(x^9)^4$

b. $(a^2)^4$

11. Using the results of Investigation 10, what is another way to write $(y^m)^n$? Why?

12. Predict the exact simplification of each exponential expression. Use a symbol manipulator to simplify each expression. Modify your prediction as necessary.

a. $\frac{x^9}{x^4}$

b. $\frac{y^5}{y}$

c. $\frac{t^3}{t^7}$ **d.** $\frac{x^4}{y^5}$

13. Using the results of Investigation 12, what is another way to write $\frac{y^m}{y^n}$? Why?

14. What happened when we divided exponential expressions that had different bases?

The preceding Investigation introduced three important properties of exponential expressions.

Property 1: If we multiply exponential expressions with like bases, we retain the base and add the exponents.

Symbolically, $(a^m)(a^n) = a^{m+n}$.

Why is this true?

The expression a^m means the product of m factors of a, and a^n means the product of n factors of a. The product results in the multiplication of $m + n$ factors of a (Figure 1).

FIGURE 1

$$\underbrace{(a \cdot a \cdot a \cdot \ldots \cdot a)}_{m \text{ factors of } a}\ \underbrace{(a \cdot a \cdot a \cdot \ldots \cdot a)}_{n \text{ factors of } a} = \underbrace{a^{m+n}}_{m+n \text{ factors of } a}$$

For example, $(2^4)(2^3) = 2^{4+3} = 2^7 = 128$. Check it out on your calculator.

Verifying, $2^4 2^3 = (2 \cdot 2 \cdot 2 \cdot 2)(2 \cdot 2 \cdot 2) = 2 \cdot 2 \cdot 2 \cdot 2 \cdot 2 \cdot 2 \cdot 2 = 2^7$

Two characteristics must be present to use this property.

- The bases in the two expressions must be the same.
- The operation on the expressions must be multiplication.

We can't simplify x^5y^4 using Property 1 because the bases are different.

We can't simplify $x^4 + x^3$ using Property 1 because the expressions are being added, not multiplied.

Property 2: If we raise an exponential expression to a power, we retain the base and multiply the exponents.

Symbolically, $(a^m)^n = a^{mn}$.

Why is this true?

The expression $(a^m)^n$ means the product of n factors of a^m. Since we are multiplying exponential expressions with like bases, we can add the exponents. This means that the base is a and the exponent is the sum of n m's. This exponent is the product of m and n. (See Figure 2.)

FIGURE 2

$$(a^m)^n = \underbrace{a^m \cdot a^m \cdot a^m \cdot \ldots \cdot a^m}_{n \text{ factors of } a^m} = a^{\overbrace{m+m+m+\ldots+m}^{n\ m\text{'s in exponent}}} = a^{mn}$$

For example, $(7^3)^5 = 7^{(3)(5)} = 7^{15}$. Check it out on your calculator.

Does $(7^3)^5$ give the same result as 7^{15}? Verifying:

$$(7^3)^5 = 7^3 \cdot 7^3 \cdot 7^3 \cdot 7^3 \cdot 7^3 = 7^{3+3+3+3+3} = 7^{3 \cdot 5} = 7^{15}$$

Property 3: If we divide exponential expressions with like bases, we retain the base and subtract the exponents.

Symbolically, $\frac{a^m}{a^n} = a^{m-n}$ provided that a is not zero.

Let's see why this is true. The numerator contains the product of m factors of a and the denominator contains the product of n factors of a. If we reduce the fraction, we are left with $|m - n|$ factors of a. If m is larger than n, the remaining a's are in the numerator. If m is smaller than n, the remaining a's are in the denominator.

For example, $\frac{2^7}{2^3} = 2^{7-3} = 2^4 = 16$. Check it out on your calculator.

To verify: $\frac{2^7}{2^3} = \frac{2 \cdot 2 \cdot 2 \cdot 2 \cdot 2 \cdot 2 \cdot 2}{2 \cdot 2 \cdot 2} = \frac{2 \cdot 2 \cdot 2 \cdot 2}{1} = 2^4$

Also: $\frac{2^3}{2^7} = 2^{3-7} = 2^{-4} = \frac{1}{2^4} = \frac{1}{16}$

To verify: $\frac{2^3}{2^7} = \frac{2 \cdot 2 \cdot 2}{2 \cdot 2 \cdot 2 \cdot 2 \cdot 2 \cdot 2 \cdot 2} = \frac{1}{2 \cdot 2 \cdot 2 \cdot 2} = \frac{1}{2^4}$

Two characteristics must be present to use this property.

- The bases in the two expressions must be the same.
- The operation on the expressions must be division.

We can't simplify $\frac{x^4}{y^5}$ using Property 3, because the bases are different.

15. Predict the exact simplification of each exponential expression. Use a symbol manipulator, if available, to simplify each expression. Modify your prediction as necessary.

a. $(x^5y^3)^4$ **b.** $(a^2b^9)^3$

c. $(xy)^3$ **d.** $(x + y)^3$

16. Using the results of Investigation 15, what is another way to write $(x^my^n)^p$? Why?

17. How is Investigation **15d** different from Investigations **15a–c**? Why?

18. Predict the exact simplification of each exponential expression. Use a symbol manipulator to simplify each expression. Modify your prediction as necessary.

a. $\left(\frac{x^5}{y^3}\right)^2$ **b.** $\left(\frac{a^8}{b^5}\right)^3$

19. Using the results of Investigation 18, what is another way to write

$\left(\frac{x^m}{y^n}\right)^p$? Why?

20. Apply the result of the above investigations to simplify—that is, remove the negative exponents. Verify your answers using your calculator and appropriate parentheses.

a. $\frac{1}{2^{-2}}$

b. $\left(\frac{2}{3}\right)^{-1}$

c. $2^{-3} \cdot 2^{4}$

d. $\frac{2^{3}}{2^{-3}}$

The preceding investigation introduces two more properties of exponents.

Property 4: Exponentiation distributes over a product.

Symbolically, $(a^m b^n)^p = (a^m)^p (b^n)^p = a^{mp} b^{np}$.

This is true since the commutative and associative properties for multiplication can be used to reorganize the product so that all factors with base a are grouped together and all factors with a base of b are grouped together.

This is demonstrated in Figure 3.

FIGURE 3

$$(a^m b^n)^p = \underbrace{a^m b^n \cdot a^m b^n \cdot a^m b^n \cdot \ldots \cdot a^m b^n}_{p \text{ factors of } a^m b^n}$$

$$= \underbrace{(a^m \cdot a^m \cdot a^m \cdot \ldots \cdot a^m)}_{p \text{ factors of } a^m} \underbrace{(b^n \cdot b^n \cdot b^n \cdot \ldots \cdot b^n)}_{p \text{ factors of } b^n} = a^{mp} b^{np}$$

For example, $(2^4 3^5)^2 = (2^4 3^5)(2^4 3^5) = (2^4 2^4)(3^5 3^5) = 2^8 3^{10}$.

Another way to rewrite the expression is

$$(2^4 3^5)^2 = (2^4)^2 (3^5)^2 = (2^{(4)2})(3^{(5)2}) = (2^8)(3^{10}) = 2^8 3^{10}.$$

Check it out on your calculator.

Be careful: Exponents do not distribute over sums or differences. For example, $(x + y)^3$ is not the same as $x^3 + y^3$. If you let $x = 3$ and $y = 5$, you can confirm numerically that this is true.

$$(3+5)^3 = 8^3 = 512 \quad \text{but} \quad 3^3 + 5^3 = 27 + 125 = 152$$

In fact, $(x+y)^3 = (x+y)(x+y)(x+y) = x^3 + 3x^2y + 3xy^2 + y^3$

Property 5: Exponentiation distributes over a quotient.

Symbolically, $\left(\frac{a^m}{b^n}\right)^p = \frac{(a^m)^p}{(b^n)^p} = \frac{a^{mp}}{b^{np}}$ if b is not zero. Why?

This is true because we multiply fractions by multiplying numerators to obtain the new numerator and multiply denominators to obtain the new denominator as demonstrated in Figure 4.

FIGURE 4

$$\left(\frac{a^m}{b^n}\right)^p = \underbrace{\frac{a^m}{b^n} \cdot \frac{a^m}{b^n} \cdot \frac{a^m}{b^n} \cdot \ldots \cdot \frac{a^m}{b^n}}_{p \text{ factors of } \frac{a^m}{b^n}} = \frac{\overbrace{a^m \cdot a^m \cdot a^m \cdot \ldots \cdot a^m}^{p \text{ factors of } a^m}}{\underbrace{b^n \cdot b^n \cdot b^n \cdot \ldots \cdot b^n}_{p \text{ factors of } b^n}} = \frac{(a^m)^p}{(b^n)^p} = \frac{a^{mp}}{b^{np}}$$

For example,

$$\left(\frac{2^4}{3^5}\right)^2 = \frac{2^4 2^4}{3^5 3^5} = \frac{2^8}{3^{10}}.$$

Another way to rewrite the expression is

$$\left(\frac{2^4}{3^5}\right)^2 = \frac{(2^4)^2}{(3^5)^2} = \frac{2^{(4)2}}{3^{(5)2}} = \frac{2^8}{3^{10}}.$$

Check it out on your calculator. Finally, note that these rules apply with negative exponents as well if the original problem is rewritten using positive exponents. For example, $\frac{1}{2^{-2}} = 4$ can be rewritten as

$$\frac{1}{2^{-2}} = 2^2 = 4, \quad \left(\frac{2}{3}\right)^{-1} = \frac{3}{2} \quad \text{as} \quad \frac{1}{\frac{2}{3}} = \frac{3}{2},$$

$$2^{-3} \cdot 2^4 = 2 \quad \text{as} \quad \frac{1}{2^3} \cdot 2^4 = \frac{2^4}{2^3} = 2, \quad \text{and} \quad \frac{2^3}{2^{-3}} = 2^6 \quad \text{as} \quad 2^3 \cdot 2^3 = 2^6.$$

Next we investigate the meaning of rational number exponents. For example, we'd like to interpret the meaning of $a^{2/3}$. We begin by looking at the meaning of exponents like $\frac{1}{2}, \frac{1}{3}, \frac{1}{4}, \frac{1}{5}$, etc.

One word of warning about fractional exponents and calculators: enclose the **exponent in parentheses** when you enter the exponential expression.

21. The table will organize your investigation of the exponent $\frac{1}{2}$.

a. Use your calculator to approximate each expression in Table 1. Predict the exact simplification of each expression given. Use a computer or a calculator capable of exact results to simplify each expression. Modify your prediction as necessary. Square the approximation .

TABLE 1 Using One-half as an Exponent

Expression	Calculator Approximation	Exact Simplification	Square of Approximation
$1^{1/2}$			
$2^{1/2}$			
$3^{1/2}$			
$4^{1/2}$			
$5^{1/2}$			
$6^{1/2}$			

b. Based on the results in Table 1, what does $a^{1/2}$ mean?

22. a. Complete Table 2 on page 561 by approximating the expression in the left column. Record the answer in the Approximation column. To complete the third column, raise the value in the Approximation column to the third power.

b. Based on the results in Table 2, what does $a^{1/3}$ mean?

TABLE 2 Using One-third as an Exponent

Expression	Approximation	Cube of Approximation
$2^{1/3}$		
$5^{1/3}$		
$8^{1/3}$		
$11^{1/3}$		
$20^{1/3}$		
$27^{1/3}$		

23. Based on the previous investigations, what does $a^{1/4}$ mean? Defend your answer.

24. Based on the previous investigations, what does $a^{1/5}$ mean? Defend your answer.

25. Based on the previous investigations, what does $a^{1/n}$ mean, where n is a whole number greater than 1? Defend your answer.

From the results of the previous investigations, we can formally describe the meaning of an exponent of the form $\frac{1}{n}$, where n is a whole number greater than 1.

The results of Table 1 are shown in Figure 5.

FIGURE 5

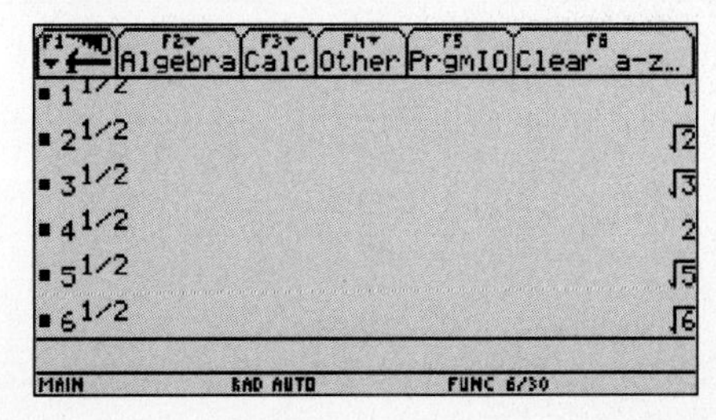

So Figure 5 suggests that $a^{1/2} = \sqrt{a}$. Similarly $a^{1/3} = \sqrt[3]{a}$. Generalizing, we can write $a^{1/n} = \sqrt[n]{a}$. So the exponent $\frac{1}{n}$ is another way of saying nth root. Therefore,

$$5^{1/2} = \sqrt{5}$$

$$5^{1/3} = \sqrt[3]{5}$$

$$5^{1/4} = \sqrt[4]{5}.$$

Figure 6 is a function machine for $a^{1/n}$ where a is the input.

FIGURE 6

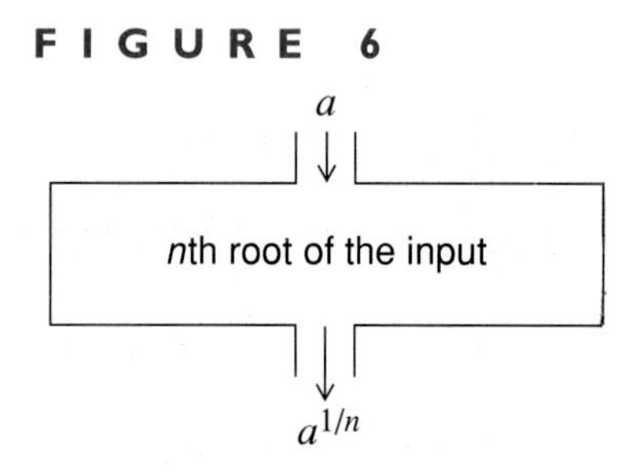

We now know the meaning of a positive rational number exponent if the numerator of the exponent is 1. Let's investigate what happens if the numerator of the exponent is not 1.

26. Complete Table 3 without the aid of a calculator or computer. Record exact answers only.

TABLE 3 **More Rational Exponent Data**

Expression	Simplification
$8^{1/3}$	
$27^{1/3}$	
$4^{1/2}$	
$81^{1/4}$	
$32^{1/5}$	

27. Predict the exact simplification of each expression given in Table 4. Use a computer or calculator capable of exact results to simplify each expression. Modify your prediction as necessary.

TABLE 4 Even More Rational Exponent Data

Expression	Simplification
$8^{2/3}$	
$27^{2/3}$	
$4^{3/2}$	
$81^{3/4}$	
$32^{4/5}$	

28. Compare the results of Tables 3 and 4. Describe how you could use the answers in Table 3 to find the answers in Table 4.

29. Use $32^{4/5}$ as an example.

a. What is the role of the 5 in the exponent?

b. What is the role of the 4 in the exponent?

30. Consider $a^{m/n}$.

a. What is the role of the n in the exponent?

b. What is the role of the m in the exponent?

c. Describe the meaning of the fractional exponent $\frac{m}{n}$.

The results of Table 4 appear in Figure 7.

FIGURE 7

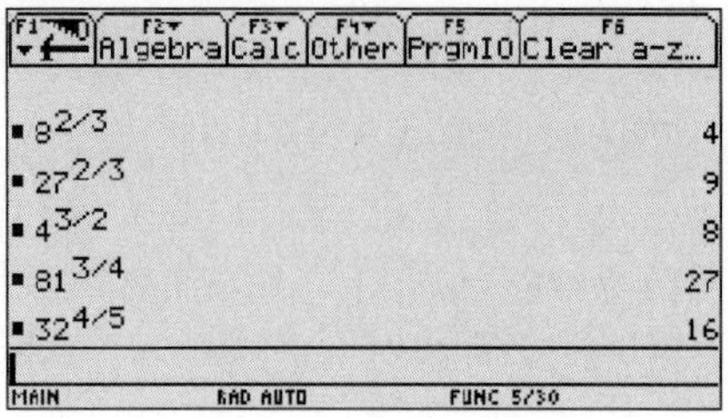

F1 F2 Algebra F3 Calc F4 Other F5 PrgmIO F6 Clear a-z...

■ $8^{2/3}$	4
■ $27^{2/3}$	9
■ $4^{3/2}$	8
■ $81^{3/4}$	27
■ $32^{4/5}$	16

MAIN RAD AUTO FUNC 5/30

When we see $a^{1/n}$, we interpret this as the nth root of a. We can interpret the fraction $\frac{m}{n}$ as the product of $\frac{1}{n}$ and m.

The value $8^{2/3}$ can be interpreted as $(8^{1/3})^2$. But $8^{1/3} = 2$, the cube root of 8. So, $8^{2/3} = (8^{1/3})^2 = (2)^2 = 4$.

In general, $a^{m/n} = (a^{1/n})^m = (\sqrt[n]{a})^m$

or

$$a^{m/n} = (a^m)^{1/n} = \sqrt[n]{a^m}.$$

This means that we interpret the exponent $\frac{m}{n}$ in two parts. The denominator n represents the root of the base and the m represents the power on the base. We calculate $a^{m/n}$ by first finding the nth root of a and then raising the answer to the mth power. We could also calculate this by raising a to the mth power and then finding the nth root of the answer. The expressions $(\sqrt[n]{a})^m$ and $\sqrt[n]{a^m}$ are equivalent if a is nonnegative or if the root n is odd.

Let's construct a function machine for $a^{m/n}$ where a is the input (Figure 8).

FIGURE 8

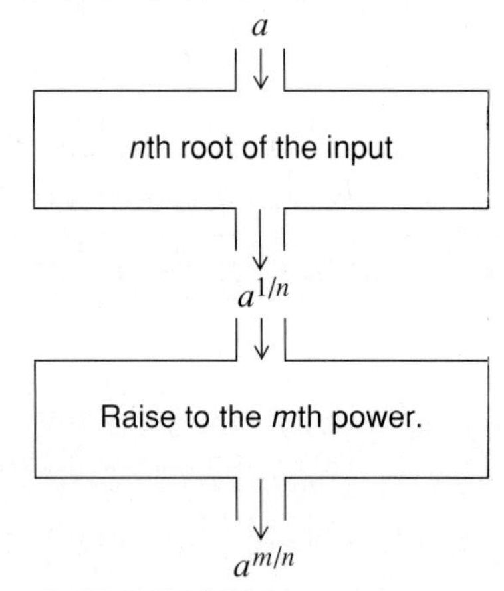

For example, $32^{4/5}$ means that we first find the fifth root of 32. The answer is 2. Raising this answer to the fourth power gives a final answer of 16. We could also raise 32 to the fourth power first and then find the fifth root of the answer, but the numbers get so large! We usually make the arithmetic easy.

Symbolically,

$$32^{4/5} = (\sqrt[5]{32})^4 = (2)^4 = 16$$

or

$$32^{4/5} = \sqrt[5]{32^4} = \sqrt[5]{1048576} = 16.$$

We conclude the section by interpreting the meaning of a negative rational exponent.

31. Predict the exact simplification of each expression given in Table 5. Use a computer or calculator capable of exact results to simplify each expression. Modify your prediction as necessary.

TABLE 5 Negative Rational Exponents

Expression	Simplification
$8^{-2/3}$	
$4^{-3/2}$	
$81^{-3/4}$	
$32^{-4/5}$	

32. Compare the results of Table 5 with the results from Table 4. How does the answer change when the sign of the exponent is changed?

33. Consider $a^{-m/n}$.

a. What is the role of n in the exponent?

b. What is the role of m in the exponent?

c. What is the role of the "–" in the exponent?

The results of Table 5 appear in Figure 9.

FIGURE 9

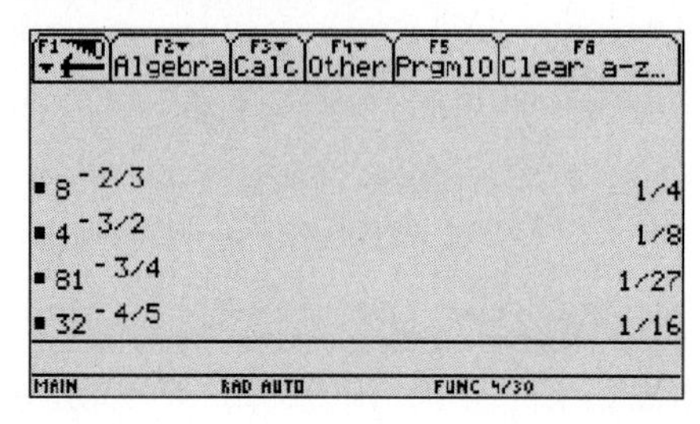

When we mentioned negative exponents earlier, we saw that the "–" in the exponent meant that we had to find the reciprocal of the base.

Thus $a^{-m/n}$ means that we must perform three operations in succession:

- Find the nth root of a.
- Raise the answer to the mth power.
- Take the reciprocal of the answer.

These operations may be performed in any order provided that the base is positive. Experiment to see which order you are most comfortable with.

For example, to calculate the value of $32^{-4/5}$,

- Find the fifth root of 32, which is 2.
- Raise the answer to the fourth power to get 16.
- Take the reciprocal of the answer to get $\frac{1}{16}$.

Symbolically,

$$32^{-4/5} = (\sqrt[5]{32})^{-4} = (2)^{-4} = (16)^{-1} = \frac{1}{16}$$

or

$$32^{-4/5} = \frac{1}{(\sqrt[5]{32})^4} = \frac{1}{(2)^4} = \frac{1}{16}$$

The function machine (Figure 10) displays the process for $a^{-m/n}$ assuming that a, the base, is the input.

FIGURE 10

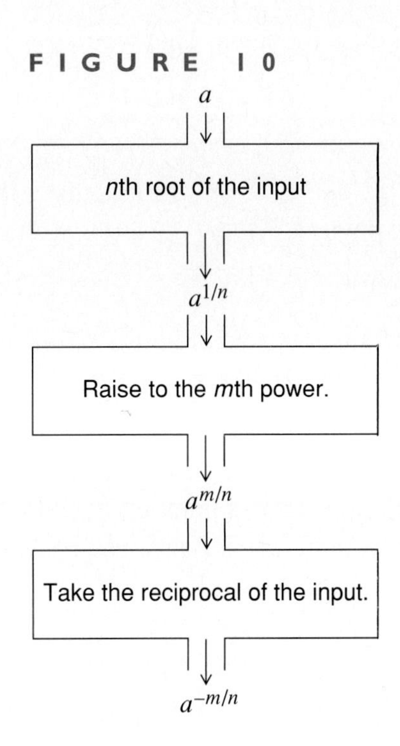

STUDENT DISCOURSE

Pete: This sure was new to me. I never thought about fractional exponents before. This all made sense to me, but what can we use fractional exponents for?

Sandy: I don't know all the uses, but having already completed this course, I know that exponential functions in Chapter 8 use the real numbers as exponents, not just whole numbers. This section helps you understand all the things exponents can mean. I also know they are helpful in simplifying radical expressions sometimes.

Pete: You mean because fractional exponents are a way of indicating a radical expression? How do they help in simplifying?

Sandy: Well, consider the expression $\sqrt{2} \cdot \sqrt[3]{2}$. How would I write that using only one radical?

Pete: I don't know. I've never mixed square and cube roots together.

Sandy: That's exactly the problem. Rewrite the expression using exponents rather than radicals.

Pete: Okay, that would be $2^{1/2} \cdot 2^{1/3}$. Correct?

Sandy: Exactly. Now I can use Property 1 of exponents.

Pete: Oh yeah, I see. We are multiplying like bases so we keep the base and add the exponents. That would be $2^{\frac{1}{2}+\frac{1}{3}}$.

Sandy: Great. Then $\frac{1}{2} + \frac{1}{3} = \frac{3}{6} + \frac{2}{6} = \frac{5}{6}$.

Pete: So $\sqrt{2} \cdot \sqrt[3]{2} = 2^{5/6} = \sqrt[6]{2^5}$ and that's the sixth root of 2 raised to the fifth power.

Sandy: Very good.

Pete: So now I have two different ways to express roots and one form, such as the exponent form, might be helpful in simplifying expressions.

Sandy: That's the way I see it. Plus, if your calculator won't accept *n*th roots, you can always enter them using fractional exponents.

1. In the Glossary, define the words in this section that appear in ***italic boldface*** type.

2. List the different ways in which exponents can be used, depending on the type of number used as the exponent.

3. Complete Table 6.

There is only one case in which $\sqrt[n]{a^m}$ and $(\sqrt[n]{a})^m$ are not equal. Clearly describe this case.

TABLE 6 **Roots and Powers**

m	n	a	Simplification of $\sqrt[n]{a^m}$	Simplification of $(\sqrt[n]{a})^m$
5	3	8		
5	3	−8		
5	2	4		
5	2	−4		
6	2	4		
6	2	−4		

4. Both Table 7 below and Table 8 on next page result in a conclusion about 0^0. Complete the tables and then discuss the difficulty defining 0^0.

TABLE 7 **Zero to a Power**

Expression	Answer
0^4	
0^3	
0^2	
0^1	
0^0	

TABLE 8 Number to Zero Power

Expression	Answer
4^0	
3^0	
2^0	
1^0	
0^0	

5. Find exact values.

a. $16^{3/4}$

b. $9^{5/2}$

c. $(-8)^{4/3}$

d. $81^{-3/4}$

e. $(-64)^{-2/3}$

6. Use the fact that $a^m \cdot a^n = a^{m+n}$ to rewrite the following.

a. $2^{1/3} \cdot 2^{3/4}$

b. $7^5 \cdot 7^{1/5}$

c. $x^{4/5} \cdot x^{-3/4}$

d. 3^{x+2}

7. Use the fact that $\frac{a^m}{a^n} = a^{m-n}$, a not zero, to rewrite the following.

a. $\frac{7^{1/4}}{7^{1/5}}$

b. $\frac{11^{-2/3}}{11^{3/5}}$

c. $\frac{x^2}{x^{1/2}}$

d. 3^{x-2}

8. Use the fact that $(a^m)^n = a^{m \cdot n}$ to rewrite the following.

a. $(13^{2/3})^{3/5}$

b. $(3^{-3/4})^{6/5}$

c. $(x^{1/4})^{1/4}$

9. Use exponential form to multiply the following radical expressions. Write the answer in radical notation.

a. $\sqrt[3]{5^2} \cdot \sqrt[3]{5^4}$

b. $\sqrt[4]{7} \cdot \sqrt[3]{7^2}$

c. $\sqrt[m]{a} \cdot \sqrt[n]{a}$

10. Use exponent properties to simplify the following. Do not leave negative exponents in your answers.

a. $\dfrac{x^{-2}y^{-3}}{x^1y^{-4}}$

b. $(4x^2y^3)\,(2x^5y^7)$

c. $(-2x^2y^3)^4\,(-3xy^2)^{-1}$

d. $\left(\dfrac{3x^2y}{2x^{-3}y^2}\right)^0$

e. $\left(\dfrac{a^4b^2}{a^2b^5}\right)^3$

f. $a^{-1} + b^{-1}$

g. $(a + b)^{-1}$

h. $\left(\dfrac{x^{-2}}{y^{-4}}\right)^{-3}$

11. Consider the rational function $y = \dfrac{x^2 - x}{x^2 + 2x - 3}$

a. What is the domain of the function? Why?

b. Write the equations of any vertical asymptotes.

c. Where, if any, will there be holes in the graph? Why?

12. Find solutions to the following equations.

a. $\frac{1}{2}x + \frac{1}{3}x - 7 = \frac{1}{4}x$

b. $3x^2 + x - 4 = 0$

c. $\frac{5}{x+2} = \frac{3}{x-3}$

d. $\frac{x+3}{x-2} - \frac{5}{x+2} = \frac{15}{x^2-4}$

13. Rewrite as one rational expression in simplest form.

a. $\left(\frac{3x}{x^2 - 7x + 12}\right)\left(\frac{x-4}{x}\right)$

b. $\frac{2}{x-5} - \frac{2}{x+5}$

14. What is the equation of the linear function that contains the points $(2, 5)$ and $(7, -4)$?

15. What is the equation of the exponential function that contains the points $(2, 47)$ and $(4, 32)$?

16. What is the equation of a quadratic function with zeros 2 and $-\frac{3}{4}$ and with a "stretch" factor $a = 5$?

17. A jar contains 30 ounces of liquid that is 20% grape juice. We then add x ounces of a liquid that is 60% grape juice.

a. Write a function that expresses the concentration C of grape juice in the jar after the x ounces are added.

b. Find x so that the concentration of grape juice in the jar is at least 25%. Briefly describe what you did.

18. Pete was looking for a use of rational exponents. One use of rational exponents is an area of mathematics called fractals. The basic idea relates to dimensions of geometric figures. A line has one dimension called length. A square has two dimensions, length and width, and its area is denoted with an exponent of 2. Similarly a cube has three dimensions and its volume is denoted by a side raised to the third power. Fractals have a dimension that is a fraction. Look up the Koch snowflake to see such a figure. A publication by the National Council of Teachers of Mathematics, *The Mathematics Teacher*, had an article that gives more detail. (Draw a Koch snowflake and ask for extra credit!!)

Write several paragraphs discussing the meaning of exponents. Specifically address whole-number exponents, integer exponents, and rational-number exponents.

SECTION 7.5

Manipulating Radicals

Purpose

- Use radicals to measure distances.
- Solve equations containing radicals.
- Add radical expressions.

In the preceding sections, we introduced radical expressions and fractional exponents as a way to write radical expressions. We close the chapter and the investigation of radicals by looking at some simplification techniques for radicals and at how to solve equations involving radicals. As part of the coverage, we will introduce the distance formula for measuring how far apart two points in the plane are. This is appropriate since the distance formula involves a square root.

We begin the section by creating a function for a contextual situation that requires the use of the distance formula. Analysis of this function will allow us to investigate radical expressions in more depth.

Infinite Channels Cable Systems Problem: Being valued employees of Infinite Channels, our team has been asked to determine routes for laying cable between satellite dishes and switching stations. The current assignment relates to determining the placement of a switching station C to service two new satellite dishes at points A and B in Figure 1.

FIGURE 1

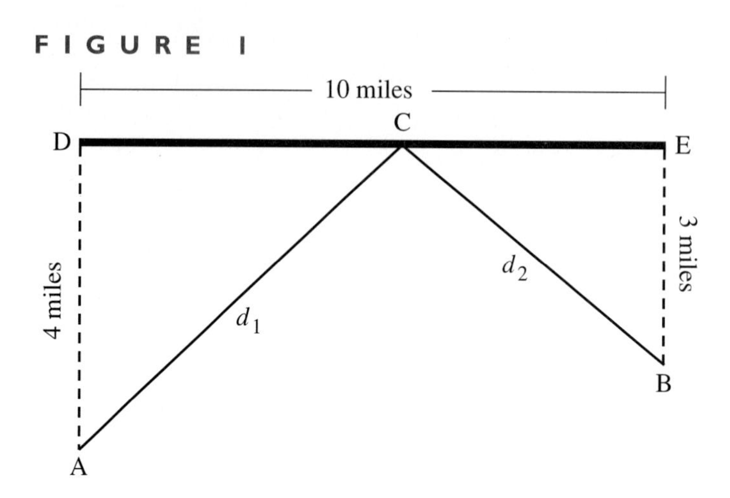

The dark horizontal line containing point C is a road. The two satellite dishes must be connected at a switching station located at some point C. The problem revolves around where to locate point C. This station can be located anywhere along the road from points D to E. The best location for C would be the one that minimizes the amount of cable used to connect C to the two satellite dishes at A and B. First, we want to make sure the problem is clear.

1. We want to minimize the amount of cable used. Write an algebraic expression using letters from Figure 1 that represents the quantity we want to minimize. Justify your answer.

2. The location of what point (A, B, C, D, or E) is not known in Figure 1? Why?

The total length of cable is represented by the sum of the distances between A and C and C and B. These distances are labeled d_1 and d_2. So the quantity that must be minimized is $d_1 + d_2$. The lengths of the segments between A and C and between C and B depends on the placement of point C. Thus, the position of C is not fixed. It is the location of point C that we wish to determine.

Before attempting to solve the problem, we need to explore the idea of distance in the plane. That's the purpose of the next investigation.

3. How far apart are the points? Plot on graph paper to justify your answer.

a. (2, 0) and (5, 0)

b. (3, 1) and (–8, 1)

c. (–5, 4) and (–7, 4)

d. (x_1, b) and (x_2, b)

4. How far apart are the points? Justify your answer by plotting the points on graph paper.

a. (0, 3) and (0, 8)

b. (–2, –5) and (–2, 3)

c. (7, –1) and (7, –4)

d. (a, y_1) and (a, y_2)

When two points share the same vertical component, the distance between them is the absolute value of the difference of the input components. This is called the absolute value of change in input.

The distance between (x_1, b) and (x_2, b) is

$$|x_2 - x_1| = |\Delta x|.$$

In this case, the two points are on the same horizontal line.

For example, the distance between (3, 1) and (–8, 1) is

$$|-8-3| = |-11| = 11.$$

When two points share the same horizontal component, the distance between them is the absolute value of the difference of the output components. This is the absolute value of change in output.

The distance between (a, y_1) and (a, y_2) is

$$|y_2 - y_1| = |\Delta y|.$$

In this case, the two points are on the same vertical line.

For example, the distance between (–2, –5) and (–2, 3) is

$$|3-(-5)| = |8| = 8.$$

What happens if the two points do not share the same input or output components? Let's investigate.

5. Given the two right triangles in Figure 2, write an equation that demonstrates the relationship among the three sides.

FIGURE 2

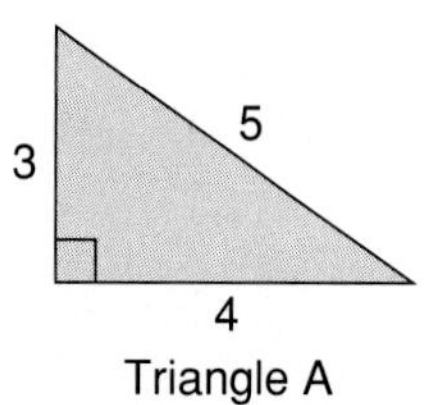

Triangle A

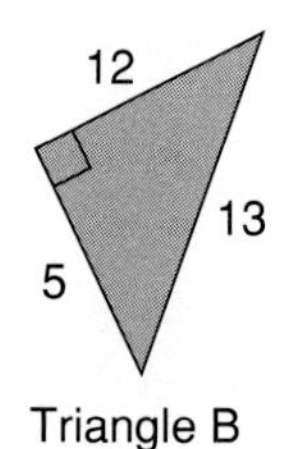

Triangle B

Triangle A:

Triangle B:

6. Using the same relationship, write an equation that demonstrates the relationship among the three sides of the right triangles in Figure 3.

FIGURE 3

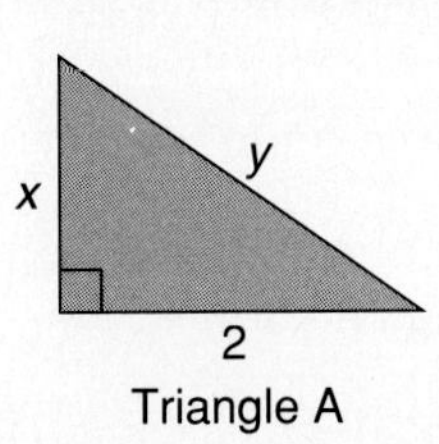

Triangle A

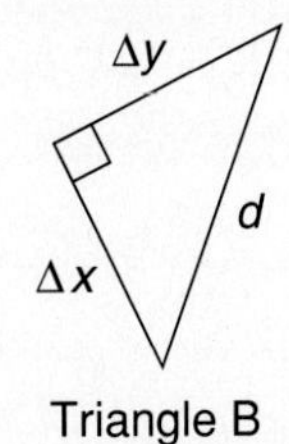

Triangle B

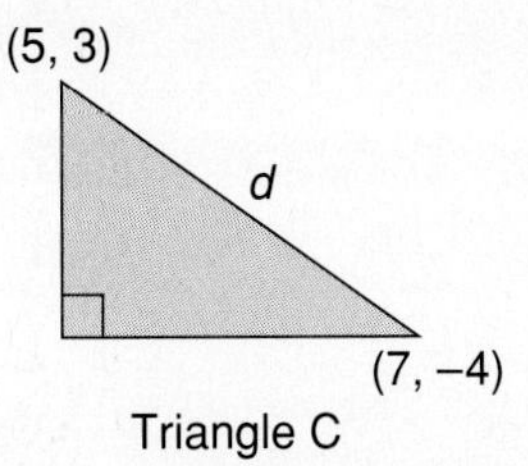

Triangle C

Triangle A:

Triangle B:

Triangle C: (Assume this right triangle is in a coordinate plane with its sides parallel to the axes.)

In Figure 2, we find the sides related as follows.

Triangle A: $3^2 + 4^2 = 5^2$

Triangle B: $5^2 + 12^2 = 13^2$

In Figure 3, the relationships follow.

Triangle A: $x^2 + 2^2 = y^2$

Triangle B: $(\Delta x)^2 + (\Delta y)^2 = d^2$

Triangle C: $(7 - 5)^2 + (-4 - 3)^2 = d^2$

In all of these, we have used a famous theorem from geometry called the ***Pythagorean theorem,*** which states that the sum of the squares of the legs of a right triangle is equal to the square of the ***hypotenuse***.

When measuring lengths on a right triangle, this theorem is commonly used. If we take the square root of both sides of any of the equations above, we have the length of the hypotenuse defined as a function of the length of the legs.

Specifically, the distance between any two points in the plane is given by

$$d = \sqrt{(\Delta x)^2 + (\Delta y)^2}$$

where Δx is the change in the input components and Δy is the change in the output components.

Note that we are not using absolute value. Both Δx and Δy are squared in the formula and thus their sign is irrelevant to the computation. Likewise, we assume d to be a nonnegative quantity.

We are ready to return to our cable system problem and begin to solve it. Let's look at the geometry again. We need to introduce a variable to represent the position of C. Let's use x to represent the distance between D and C. (See Figure 4.)

FIGURE 4

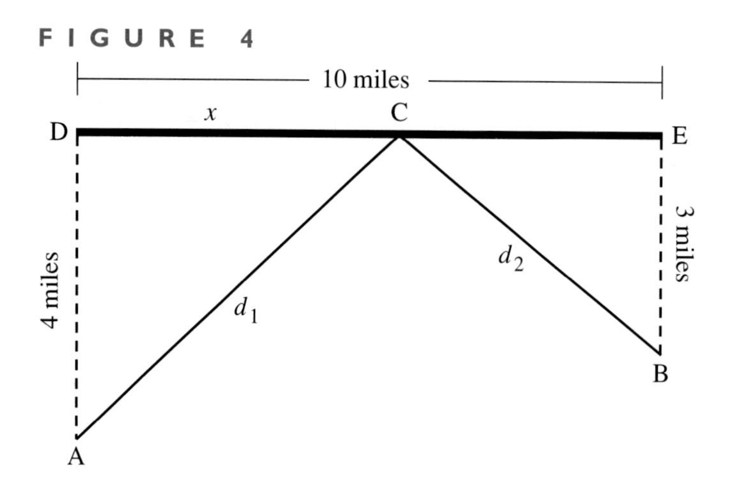

7. What is the domain of variable x? In other words what is the smallest x can be? Where would C be located in this case? What is the largest x can be? Where would C be located in this case?

8. If x is the distance between D and C, write an expression containing x that represents the distance between C and E.

9. Use the distance formula to define

a. d_1 as a function of x

b. d_2 as a function of x

DISCUSSION

Recall that point C can be located anywhere between points D and E. If C is located at D, then $x = 0$ (Figure 5).

FIGURE 5

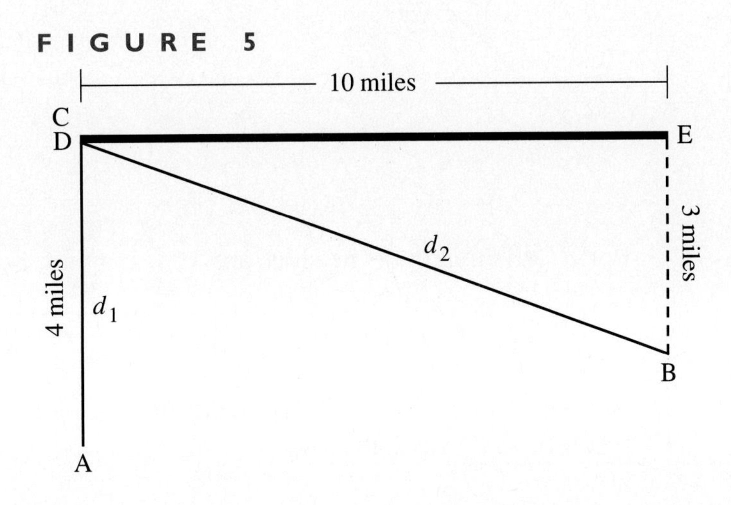

If C is located at E, then $x = 10$ (Figure 6).

FIGURE 6

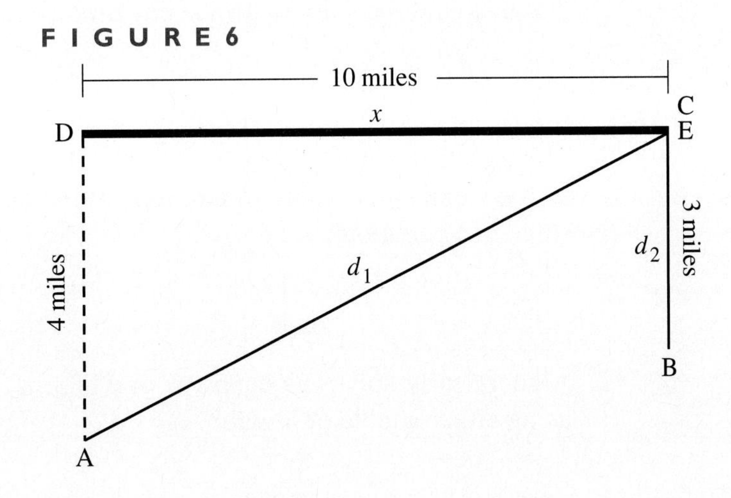

Thus x can take on all real number values between 0 and 10 inclusive.

We can use the distance formula to define both d_1 and d_2 in the original picture (Figure 7 on page 580).

For d_1, the lengths of the sides are x and 4. So

$$d_1 = \sqrt{x^2 + 4^2} = \sqrt{x^2 + 16}.$$

FIGURE 7

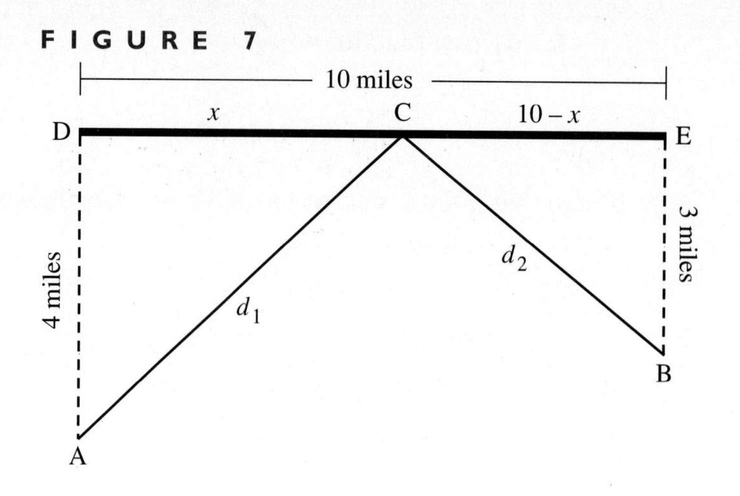

For d_2, the length of the sides are $10 - x$ and 3. So

$$d_2 = \sqrt{(10-x)^2 + 3^2} = \sqrt{100 - 20x + x^2 + 9} = \sqrt{x^2 - 20x + 109}.$$

Before we try to solve the original problem, let's look at some equations that result from these distances.

10. Write an equation for the case where d_1 would be 5. Use a table or a graph to find the solution to the equation.

11. The equation you just wrote is a radical equation. To solve it algebraically, we must undo the square root. What can be done to both sides to solve this equation?

12. Algebraically solve the equation in which d_1 is 5. Check your answer using either a table or a graph.

13. **a.** Write an equation in which d_2 is 5.

b. Solve this equation algebraically. Check your answer using either a table or a graph.

DISCUSSION

If $d_1 = 5$, then $\sqrt{x^2 + 16} = 5$. To solve using either a table or graph, we determine the input so that the output of $y(x) = \sqrt{x^2 + 16}$ is 5 (Figure 8).

FIGURE 8

To solve a ***radical equation*** algebraically, we raise both sides of the equation to a power that reverses the radical process. In particular, if the equation involves a square root, we square both sides to reverse the square root process. Usually, we try to isolate the radical on one side of the "=" before raising both sides to a power. Here are the steps to solve the equation $\sqrt{x^2 + 16} = 5$ algebraically.

Square both sides:	$(\sqrt{x^2 + 16})^2 = (5)^2$
Simplify:	$x^2 + 16 = 25$
Isolate the square:	$x^2 = 9$
Take square roots:	$\sqrt{x^2} = \sqrt{9}$
Simplify:	$\lvert x \rvert = 3$
Remove absolute value:	$x = 3$ or $-x = 3$
Solve for:	$x = 3$ or $x = -3$

We reject the negative solution since x represents a distance. In fact, it is always necessary to check solutions to a radical equation since the process of raising both sides to a power can create ***extraneous*** solutions.

If C is located 3 miles from D then the distance between the satellite dish at A and the switching station is 5 miles.

If $d_2 = 5$, then $\sqrt{(10 - x)^2 + 9} = 5$.

Square both sides:	$(\sqrt{(10 - x)^2 + 9})^2 = (5)^2$
Simplify:	$(10 - x)^2 + 9 = 25$
Isolate the square:	$(10 - x)^2 = 16$

Take square roots: $\sqrt{(10-x)^2} = \sqrt{16}$

Simplify, remembering $10-x$ is not negative:

$$10-x = 4$$

Solve for x: $x = 6$

If C is located 6 miles from D, then the distance between the satellite dish at B and the switching station is 5 miles.

Finally we're ready to finish the problem that started this section.

14. Write a function containing x that represents the total distance from A to C and C to B.

15. Our assignment is to minimize the output of this function.

a. Describe how this can be done using a table.

b. Describe how this can be done using a graph.

16. Where should C be located in order to minimize the total cable used? Justify your answer.

The total length of cable is the sum of d_1 and d_2. Thus the total distance function is

$$D = \sqrt{x^2+16} + \sqrt{(10-x)^2+9}\,.$$

Using a table, we would look for the smallest output for inputs between 0 and 10 inclusive.

Using a graph, we would graph the function for inputs between 0 and 10 inclusive. We then find the x value at the lowest point on the graph. Figure 9 displays the solution graphically.

FIGURE 9

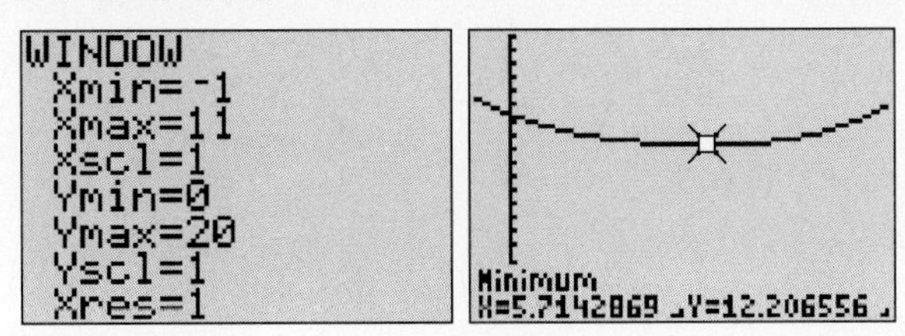

The solution suggests that the switching station C should be located approximately 5.71 miles from point D in order to minimize the total cable used, which will be approximately 12.21 miles (Figure 10).

FIGURE 10

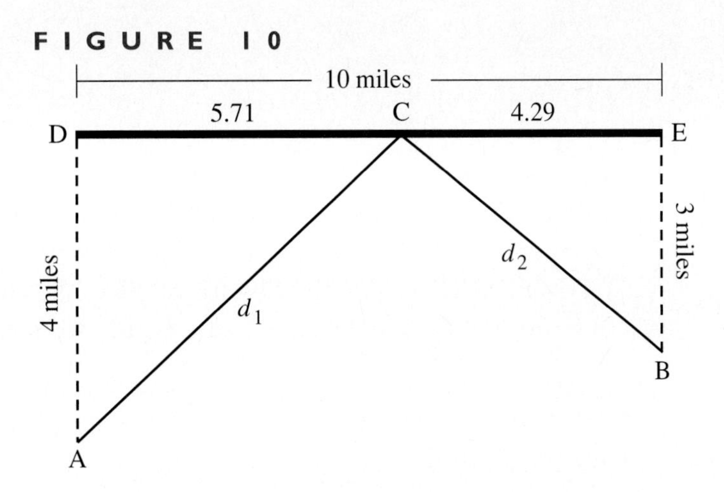

We've seen radical equations we can solve by hand along with others that can be analyzed graphically or with a table. Before we end our discussion of radicals, we should investigate the addition of radicals. There are some rules that govern the addition and subtraction of radicals. Let's find them.

17. Use a computer or calculator capable of exact simplifications to complete Table 1.

TABLE 1 Operations on Radicals I

Expression	Simplification
$5\sqrt{2} + 3\sqrt{2}$	
$11\sqrt{5} - 17\sqrt{5}$	
$2\sqrt{5} + 3\sqrt{2} + 7\sqrt{5}$	
$6 - 2\sqrt{6} + 3\sqrt{6}$	

18. Do the following without technology. Use the results of Table 1 to help you understand the operations.

a. $7\sqrt{13} + 2\sqrt{13} =$

b. $4x\sqrt{3y} - 7x\sqrt{3y} =$

c. $2\sqrt{10} - 3\sqrt{5} + 7\sqrt{10} =$

d. $4x\sqrt{2y} + 5\sqrt{2y} + 7x\sqrt{2} + 5x\sqrt{2y} =$

The results of Table 1 suggest that addition and subtraction of radical expressions follow the same rules as addition and subtraction of polynomials. To add or subtract polynomials, we use the distributive property of multiplication over addition or subtraction to combine like terms. Recall that like terms are terms that are identical, except possibly for their numerical coefficients.

In a similar vein, we define ***like radicals*** as radical expressions that are identical except possibly for their numerical coefficients. Let's look at the problems you did.

a. $7\sqrt{13} + 2\sqrt{13} = (7+2)\sqrt{13} = 9\sqrt{13}.$

b. $4x\sqrt{3y} - 7x\sqrt{3y} = (4-7)x\sqrt{3y} = -3x\sqrt{3y}.$

c. $2\sqrt{10} - 3\sqrt{5} + 7\sqrt{10} = (2+7)\sqrt{10} - 3\sqrt{5} = 9\sqrt{10} - 3\sqrt{5}.$

Note that the like radicals are $2\sqrt{10}$ and $7\sqrt{10}$.

d. $4x\sqrt{2y} + 5\sqrt{2y} + 7x\sqrt{2} + 5x\sqrt{2y} = (4+5)x\sqrt{2y} + 5\sqrt{2y} + 7x\sqrt{2}$

$$= 9x\sqrt{2y} + 5\sqrt{2y} + 7x\sqrt{2}.$$

Note that the like radicals are $4x\sqrt{2y}$ and $5x\sqrt{2y}$.

Sometimes radical expressions must be simplified before attempting to add or subtract them. Let's look at some examples.

19. Use a computer or calculator capable of exact simplifications to complete Table 2.

TABLE 2 Operations on Radicals 2

Expression	Simplification
$\sqrt{8} + \sqrt{32}$	
$3\sqrt{75} - 2\sqrt{27}$	
$5\sqrt{108} + 7\sqrt{18} - 4\sqrt{48}$	

To understand the answers in Table 2, work the same problems without technology.

20. Given $\sqrt{8} + \sqrt{32}$,

a. Simplify each radical completely.

b. Perform the addition on the simplified radical expressions.

21. Given $3\sqrt{75} - 2\sqrt{27}$,

a. Simplify each radical completely.

b. Perform the addition on the simplified radical expressions.

22. Given $5\sqrt{108} + 7\sqrt{18} - 4\sqrt{48}$,

a. Simplify each radical completely.

b. Perform the addition on the simplified radical expressions.

We see there are two steps when adding or subtracting rational expressions.

1. Simplify each radical expression completely.

2. Combine like radicals.

Here's the last problem from Table 2.

$$5\sqrt{108} + 7\sqrt{18} - 4\sqrt{48}$$

First, simplify each radical completely.

$$5\sqrt{108} = 5\sqrt{2^2 \cdot 3^2} = 5 \cdot 2 \cdot 3\sqrt{3} = 30\sqrt{3}$$

$$7\sqrt{18} = 7\sqrt{2 \cdot 3^2} = 7 \cdot 3\sqrt{2} = 21\sqrt{2}$$

$$4\sqrt{48} = 4\sqrt{2^4 \cdot 3} = 4 \cdot 2^2\sqrt{3} = 16\sqrt{3}$$

Now combine like radicals.

$$5\sqrt{108} + 7\sqrt{18} - 4\sqrt{48} = 30\sqrt{3} + 21\sqrt{2} - 16\sqrt{3} = 14\sqrt{3} + 21\sqrt{2}$$

That completes our brief study of radical expressions.

STUDENT DISCOURSE

Pete: I have a real basic issue that has always bothered me. When I see something like $2\sqrt{3}$, what does that really mean?

Sandy: What do you think?

Pete: Wait a minute! I asked you, but I'll give it a try because I think the work in this section helped me with this.

Sandy: Go ahead.

Pete: Well, I think it means that I am multiplying two numbers, 2 and $\sqrt{3}$.

Sandy: See, you did know what it means.

Pete: Wait, now let me think about this numerically. The square root of 3 is somewhere between 1.5 and 2 and I'm multiplying that by 2. So is $2\sqrt{3}$ just a number somewhere around 3.5?

Sandy: Why don't you check on your calculator?

Pete: My calculator says about 3.46. Okay, I think I understand. So a number next to a radical just means we are multiplying the number and the value of the radical together.

Sandy: That's right. It is similar to $2x$ or $2i$. We are just writing a product of 2 and another number.

1. In the Glossary, define the words in this section that appear in ***italic boldface*** type.

2. Write an explanation to a friend describing how to

a. multiply two radical expressions.

b. add two radical expressions.

3. Refer to Figure 1 on page 574.

a. Find the total length of the cable if point C is placed on point D.

b. Find the total length of the cable if point C is placed on point E.

4. Are $\sqrt{a^2 + b^2}$ and $a + b$ equivalent? To help answer this question, create some values for a and b to test for equivalence.

5. Given an equation, does squaring both sides result in an equivalent equation? To answer, complete the following problems.

a. How many solutions does the equation $x = 5$ have?

b. Square both sides of the equation given in part *a*. How many solutions does the resulting equation have?

6. Consider the identity $2 + 3 = 5$.

a. Square the left side $2 + 3$ and the right side 5. Do you still have an identity?

b. Square individual terms 2, 3, and 5. Do you still have an identity?

c. Based on the results, when you need to square to solve an equation, do you square sides or do you square terms?

7. Is $(a + b)^2 = a^2 + b^2$? Why or why not? Which previous exploration asked you to answer essentially the same question?

8. Is $\sqrt{a + b} = \sqrt{a} + \sqrt{b}$? Why or why not? Which previous exploration asked you to answer essentially the same question?

9. Consider the ***radical function*** $y(x) = \sqrt{2x - 3} - \sqrt{x + 7} + 2$.

a. Use a table to locate the zeros of $y(x)$ numerically.

b. Use a graph to locate the zeros of $y(x)$ graphically.

10. Consider $y(x) = \sqrt{2x - 3} - \sqrt{x + 7} + 2$.

a. Set the function equal to zero.

b. Rewrite the equation so that one radical term is alone on one side of the equation.

c. Square both sides of the equation. Be sure to square *sides*, not individual terms.

d. There should be a radical remaining in the equation. Rewrite the equation so that the radical is the only term on one side of the equation.

e. Square both sides again. This should eliminate all radicals from the equation.

f. Solve the resulting equation for x.

g. Compare your results with those of the previous exploration. Eliminate any extraneous solutions.

11. Consider the radical function $y(x) = \sqrt[3]{2x - 3}$.

a. Write an equation that represents the problem of finding the input so that the output is –5.

b. Solve the equation algebraically. Check your result numerically or graphically.

12. Consider the radical function $y(x) = \sqrt[4]{x - 5} - 1$.

a. Write an equation that represents the problem of finding the input so that the output is 2.

b. Solve the equation algebraically. Check your result numerically or graphically.

13. Simplify the following radical expressions completely.

a. $7\sqrt{2} + 5\sqrt{2}$

b. $(7\sqrt{2})(5\sqrt{2})$

c. $3\sqrt{5} - 9\sqrt{5}$

d. $5\sqrt{2} + 2\sqrt{5}$

e. $(5\sqrt{2})(2\sqrt{5})$

f. $5\sqrt{12} + 7\sqrt{75}$

g. $5\sqrt{7}\,(3 + 6\sqrt{2})$

h. $4\sqrt{20} + \sqrt{45} + 3\sqrt{60}$

14. Use exponent properties to simplify the following (no negative exponents).

a. $16^{-1/4}$

b. $\dfrac{(a^{-2})^{-5}}{(a^{3})^{4}}$

c. $(-2x^{3}y^{2})(xy^{-2})^{3}$

d. $\left(\dfrac{a^{4}b^{-3}c^{2}}{ab^{-4}c^{-1}}\right)^{2}$

15. Find solutions to the following equations.

a. $3\sqrt{2x-7} = 8$

b. $2\sqrt{x+3} = -1$

c. $x^{2} = 3x - 5$

d. $\dfrac{x}{x+1} = 2$

e. $\dfrac{x}{x+4} - \dfrac{2}{x+3} = \dfrac{3x+7}{x^{2}+7x+12}$

f. $\sqrt{2x-1} - 5 = -2$

g. $\sqrt{x-1} = x + 1$

h. $\sqrt{11x-9} = \sqrt{5x+2}$

REFLECTION

Write a paragraph describing the various operations you have learned to perform on radical expressions. Provide examples that demonstrate each operation described.

SECTION 7.6

Nonpolynomial Algebraic Functions: Review

Purpose

- Reflect on the key concepts of Chapter 7.

In this section you will reflect upon the mathematics in Chapter 7—what you've done and how you've done it.

1. State the most important ideas in this chapter. Why did you select each?

2. Identify all the mathematical concepts, processes, and skills you used to investigate the problems in Chapter 7?

There are a number of important ideas that you might have listed, including addition and subtraction of functions, multiplication of functions, division of functions, rational functions, asymptotes, holes in a graph, discontinuous functions, end behavior models, composition of functions, operations on rational functions, *n*th roots, rational exponents, exponent properties, distance formula, and operations with radical expressions.

1. Consider the problem in which we have 35 ounces of liquid, 20% of which is grape juice. Suppose we add x ounces of liquid that is 75% grape juice.

a. Write a rational function that represents the concentration of grape juice in the final mixture.

b. Where is the vertical asymptote located? Is this a meaningful value in this problem situation?

c. Find an end behavior model for this function. Explain the meaning of the end behavior model in the context of the problem situation.

d. Use a table or graph to approximate how many ounces of liquid containing 75% grape juice must be added to raise the concentration to 30% grape juice.

e. Write the equation that represents the problem in part **d**.

f. Use algebraic techniques to solve the equation in part **e**. Compare your solution(s) with those obtained in part **d**.

2. Kelly plans a trip to visit a friend at her state university which is 187 miles away.

a. Write a function that defines Kelly's time for the trip as a function of her average speed.

b. Kelly wishes to make the trip in about three hours. What must her average speed be? Explain how you found your answer.

c. Describe the meaning of the end behavior model in this problem situation.

3. Create a rational function with a vertical asymptote at 7, a hole at –1, and a horizontal asymptote at $y = 3$.

4. The graph of a rational function appears in Figure 1.

FIGURE 1

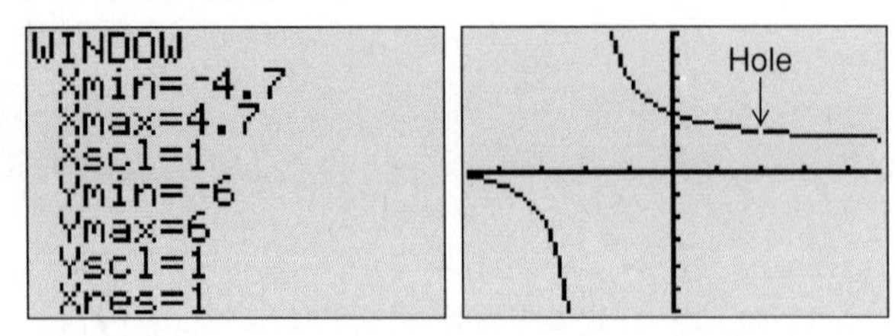

a. Identify the factors of the denominator.

b. Identify the factors of the numerator.

For each of the following rational functions (Review Exercises 5–7), find

a. the domain,

b. the zeros of the function,

c. the location of vertical asymptotes and holes,

d. an end behavior model for the function.

5. $y(x) = \dfrac{3x-5}{2x+7}$

6. $y(x) = \dfrac{x^2-5x-6}{2x^2+9x+7}$

7. $y(x) = \dfrac{2x^2-7x+6}{2x^3+x^2-6x}$

8. Consider the functions $f(x) = 2x+3$ and $g(x) = 3x^2 - x + 2$. Find:

a. $f(g(-4))$

b. $g(f(-4))$

c. $f(g(x))$

d. $g(f(x))$

9. Write each of the following as one rational expression.

a. $7 - \dfrac{3x}{2x-5}$

b. $\left(\dfrac{4x-6}{3x+9}\right)\left(\dfrac{15}{8x+16}\right)$

c. $\frac{3}{x+2}+\frac{4}{x-6}$

d. $\frac{4x}{x^2+x-42}+\frac{x-1}{x^2+9x+14}$

e. $\frac{5x+10}{3x-5}\div\frac{6x+12}{5x+5}$

f. $4-\frac{3}{x+2}-\frac{2x-3}{x^2+x-2}$

10. Given the function $y(x) = x^2 - x - 2$, find the inputs if the output is 1. Explain what you did.

11. Given the function $y(x) = 6x - 1$, find the input if the output is $\frac{2}{3}$. Explain what you did.

12. A quadratic function has a vertex at $(2, -5)$ and a y-intercept at $(0, -3)$. Find the equation of the function. Explain what you did.

13. Write the equation for a rational function that has a vertical asymptote at $x = 7$ and a hole at $x = -3$.

14. Find the zeros and factor completely. If the polynomial is prime, say so.

a. $4x^2 + 16x - 15$

b. $8x^2 - 4$

c. $4x^2 + 20x - 24$

d. $5x^2 + 24x - 5$

15. Volume of a Cylinder: The volume of a circular cylinder, such as a soda can, is given by the formula $V = \pi r^2 h$, where r is the radius of the circular base and h is the height of the cylinder.

a. Suppose the radius is twice the height. Rewrite the formula so that it contains only the variables V and h.

b. Rewrite the equation so that the radius h is defined as a function of the volume V.

c. Approximate the height of the cylinder if the volume is 115 cubic centimeters.

16. Volume of a Sphere: The volume of a sphere, such as a basketball, is determined by the formula $V = \frac{4}{3}\pi r^3$, where r is the radius of the sphere. Approximate the radius if the volume of the sphere is 153 cubic inches. Explain what you did.

17. Simplify each of the following as much as possible without approximating.

a. $\sqrt[3]{13{,}720}$

b. $\sqrt[4]{324}$

c. $\sqrt[5]{-192}$

d. $\sqrt[3]{61{,}250}$

e. $6\sqrt{72} - 5\sqrt{98}$

f. $(9\sqrt{2})(5\sqrt{6})$

18. Find the solutions to each of the following equations, inequalities, or systems of equations.

a. $\frac{5}{7}x - 8 < \frac{2}{3} - \frac{5}{6}x$

b. $5x^2 - 3x > 2$

c. $\frac{y-1}{y-4} = \frac{y+1}{y+3}$

d. $5t - 3q + 14 = 0$
$2t = 9 - q$

e. $\frac{x}{x+3} - \frac{1}{2x-1} = \frac{3x-5}{2x^2+5x-3}$

f. $\sqrt{3x-5} = 2$

g. $3\sqrt{x-1} = 5x-2$

h. $\dfrac{x}{x+1} + \dfrac{2}{x-3} = \dfrac{8}{x^2-2x-3}$

i. $\sqrt{7-2x} = \sqrt{x+5}$

j. $x+2y = 7$
$-4x-8y = 0$

19. Use exponential form to multiply the following radical expressions. Write the answer in radical notation.

a. $(3\sqrt{5})(2\sqrt[3]{5})$

b. $\dfrac{\sqrt[4]{7}}{\sqrt{7}}$

20. Use exponent properties to simplify the following. Do not leave negative exponents in your answers.

a. $\dfrac{x^3}{x^{-3}}$

b. $\dfrac{x^3y^4}{x^5y}$

c. $x^{2/3}x^{4/5}$

d. $(2a^{-1})^{-2}(4a^2)$

e. $\dfrac{y^{1/2}}{x^{-3}y^{3/2}}$

f. $16^{-5/4}$

g. $\left(\dfrac{2a^3b^{-2}}{2^{-3}a^{-1}b}\right)^3$

h. $\left(\dfrac{a^{1/3}}{ab^{3/4}}\right)^2$

i. $(x^{-5/9}y^{-6/7})^0$

j. $(x^{-1})^{-1}$

21. Write in the form $y(x) = a(x-h)^2 + k$. Identify the vertex.

a. $y(x) = 3x^2 - 24x + 36$

b. $y(x) = 5x^2 + 24x - 5$

22. If an earthly object is projected straight up into the air, then the height h (in feet) of the object above the Earth is given by a quadratic function of the time t (in seconds) elapsed since the object was shot or thrown into the air. The general model is $h(t) = -16t^2 + v_0 t + h_0$, where v_0 is the initial velocity (in feet per second) and h_0 is the initial height (in feet) above the Earth. Suppose the initial velocity is 60 feet per second and the initial height is 0 feet.

a. Write the specific model for the given information.

b. What are the zeros of this function? What do the zeros mean in the context of the time and height of the object?

c. What is the vertex of the graph of this function? What does the vertex mean in the context of the time and height of the object?

d. Write the function in factored form.

e. Find the exact times when the height will be 20 feet.

23. Write the equation of the function satisfying the given conditions.

a. A linear function through $(-4, 0)$ and $(1, 2)$.

b. A quadratic function with vertex $(4, 1)$ and y-intercept $(0, -2)$.

c. An exponential function with base 2.3 and y-intercept $(0, 8)$.

d. A quadratic function with zeros 4 and –8 with y-intercept $(0, 1)$.

24. Perform the following complex number operations.

a. $(5 - i) - (3 - 2i)$

b. $(5 - 3i)(6 + 2i)$

25. Write the equation for the function given the following graphs. Briefly describe what you did.

a.

FIGURE 2

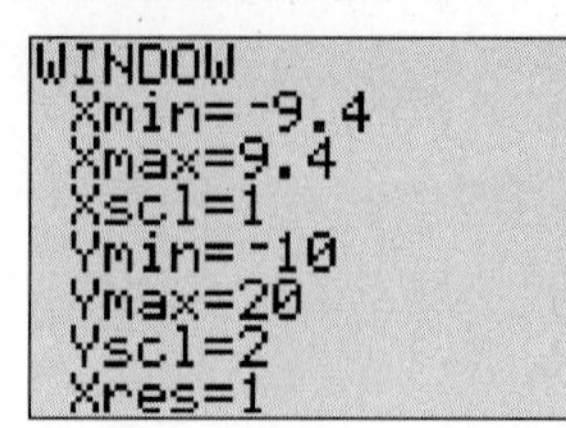

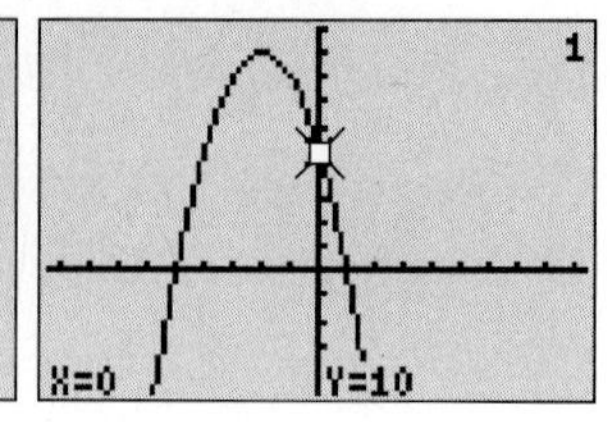

b.

FIGURE 3

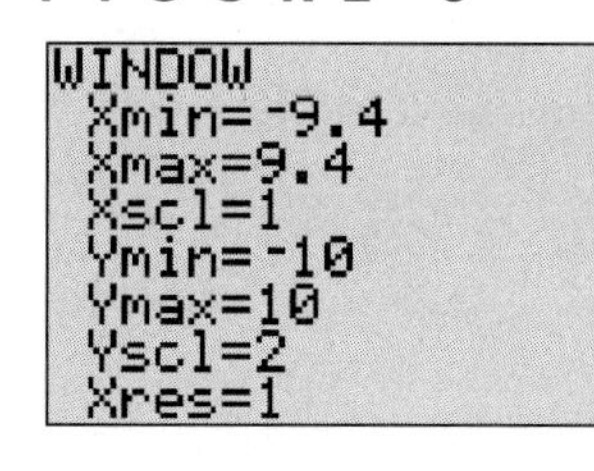

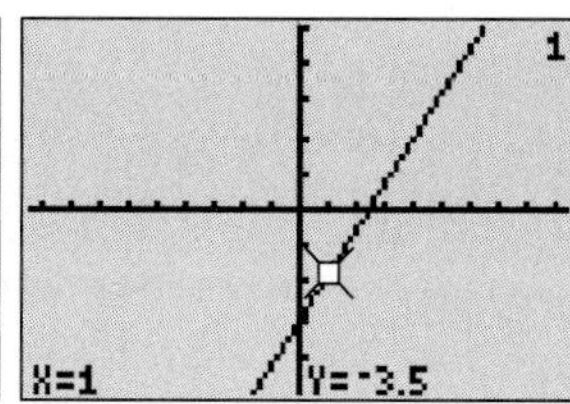

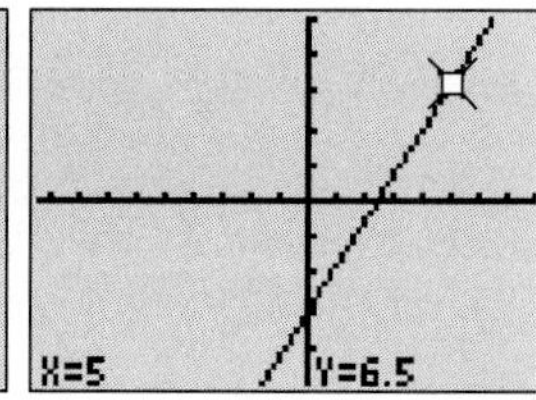

CONCEPT MAP

Construct a Concept Map centered on one of the following:

a. rational functions

b. operations on rational functions

c. *n*th root of a real number

d. rational exponents

e. radical expressions

REFLECTION

Select one of the important ideas you listed in Investigation 1. Write a paragraph to Izzy explaining your understanding of this idea. How has your thinking changed as a result of studying this idea?

8 Transcendental Models

SECTION 8.1

Inverse Relationships

Purpose

- Introduce one-to-one functions.
- Identify inverse operations.
- Investigate inverses numerically, algebraically, and graphically.

Linear and quadratic functions are members of a class of functions called polynomial functions. Rational and radical functions, the focus of Chapter 7, are not polynomial functions, but instead are called algebraic functions. Exponential functions are neither polynomial nor algebraic. Exponential functions belong to a class of functions called ***transcendental***. Other examples of transcendental functions include logarithmic functions and trigonometric functions.

The purpose of this chapter is to investigate exponential functions in more depth and to introduce logarithmic functions. In fact, just as addition and subtraction can be considered inverse operations of each other, exponential functions and logarithmic functions are inverses of each other. Therefore, to begin this chapter, we look at the idea of inverse relationships. We begin by looking at the mathematics involved in converting money between U.S. dollars and British pounds.

Money Conversion: We're planning on taking some courses at a British university. As a result, we need to mail a tuition payment. Recently, the exchange rate was one British pound to $1.63 U.S.

1. a. Complete Table 1 converting British pounds to U.S. dollars.

TABLE 1 **Pounds to Dollars**

British Pounds	U.S. Dollars
100	
200	
300	
400	
500	

b. State the process being performed on the number of British pounds to obtain the number of U.S. dollars.

c. Let d represent the number of U.S. dollars and p represent the number of British pounds. Write an equation for the function that has the number of British pounds as input and the number of U.S. dollars as output.

2. a. Complete Table 2 by converting U.S. dollars to British pounds.

TABLE 2 **Pounds to Dollars**

U.S. Dollars	British Pounds
100	
200	
300	
400	
500	

b. State the process being performed on the number of U.S. dollars to obtain the number of British pounds.

c. Let d represent the number of U.S. dollars and p represent the number of British pounds. Write an equation for the function that has the number of U.S. dollars as input and the number of British pounds as output.

3. Graph both conversion functions (Investigations 1**c** and 2**c**). Where do they intersect?

Given a relationship that involves the conversion of units, such as between two monetary systems, we can use a proportion to write the conversion functions.

The ratio of dollars to pounds is 1.63 to 1. Thus, if d represents the number of U.S. dollars and p represents the number of British pounds, we can write the proportion

$$\frac{d}{p} = \frac{1.63}{1}.$$

Therefore, the function $d = 1.63p$ converts from pounds to dollars.

Given an amount in dollars, we could substitute for d and solve for p. This is fine if we only need to do the process once or twice. However, if we need to do this often, it would be helpful to have a general function for the conversion. This function must reverse the work done by the original function.

Since the process in the original function is "multiply by 1.63," the reverse process would be "divide by 1.63."

Pounds to dollars function: $d = 1.63p$

Divide both sides by 1.63: $\frac{d}{1.63} = p$

Thus the function that converts dollars to pounds is

$$p = \frac{d}{1.63} \quad \text{or} \quad p = 0.61349693d.$$

Reversibility of operations is an important capability. When we looked at powers and roots, it was important to reverse processes being done to variables. We have seen it is usually fairly easy to evaluate a function; that is, to find the output given the input. It can be much harder to find the input given the output. If we could reverse the process defined by the function, we would reduce the problem of solving (finding input) to an evaluation.

We begin by investigating the reversibility of certain mathematical operations.

4. For each operation in the left column of Table 3, record the inverse operation in the right column.

TABLE 3 Operations and Their Inverses

Operation	Inverse
Add 7	
Subtract 8	
Multiply by 4	
Divide by 15	
Square	
Raise to the fifth power	
Fourth root	
Third root	
Opposite	
Reciprocal	

5. Recall the linear salary function $y = 2000x + 33{,}000$ for working at CDs-R-Us. In this function, x is the number of years worked and y is that year's salary.

a. Draw a function machine for y. Clearly label both the input and the output. Inside the machine, list each operation in the order in which it is performed.

b. List the operations that would reverse the process defined by the given function *y*. Be sure to list them in the correct order.

c. Draw a function machine for the inverse function of *y*.

d. Write the inverse function algebraically.

e. Record five ordered pairs that satisfy function *y*.

TABLE 4 Numeric Samples for Relationship y

Year	Salary

f. Record the outputs from Table 4 as inputs in Table 5. Use Investigation 5**d** to evaluate the inverse function at each input. Record the answers.

TABLE 5 Numeric Samples for Relationship Inverse of y

Salary	Year

g. What is the relationship between the ordered pairs represented in Table 4 and the ordered pairs represented in Table 5?

A relation that reverses the process performed by any function $f(x)$ is called the ***inverse*** of $f(x)$. The inverse is a function if each input has exactly one output. It is important to realize that the inverse function is a function in its own right. It has input and output, can be manipulated to find its zeros, vertical intercept, and other important features. However, when the output of $f(x)$ is evaluated in its inverse, the new output is the original input. Furthermore if we reverse the process and begin with an input into the inverse function and use its output into $f(x)$, we again return the original input. Pretty amazing!

Here are three function machines for the operations given in Table 3.

Notice that the ordered pair (2, 9) satisfies the function in Figure 1 and that the ordered pair (9, 2) satisfies the inverse function. The inverse is a function if each in input has exactly one output

FIGURE 1

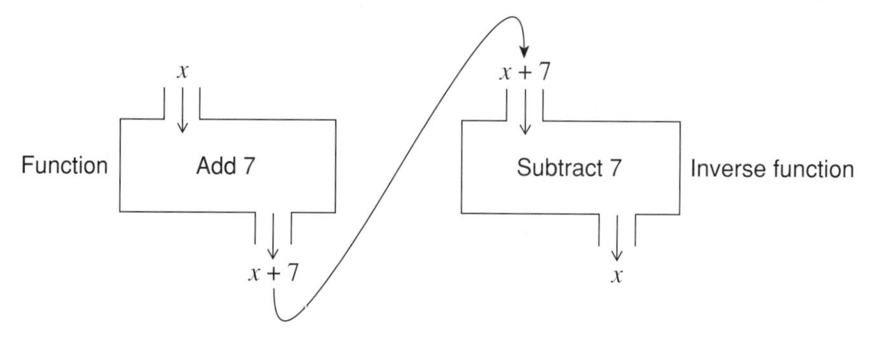

The ordered pair (30, 2) satisfies the function in Figure 2 and the ordered pair (2, 30) satisfies the inverse function.

FIGURE 2

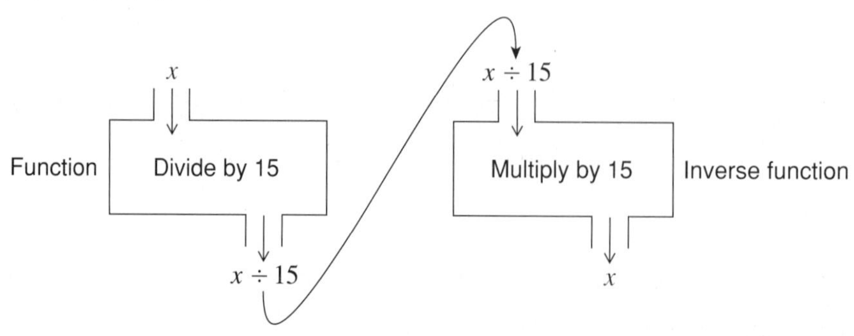

The ordered pair (64, 4) satisfies the function in Figure 3 and the ordered pair (4, 64) satisfies the inverse function.

FIGURE 3

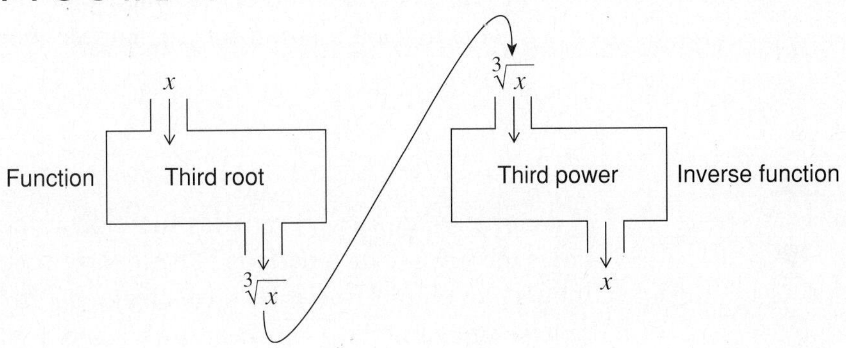

Now let's look at the salary function $y(x) = 2000x + 33{,}000$ and its inverse function using function machines (Figure 4).

FIGURE 4

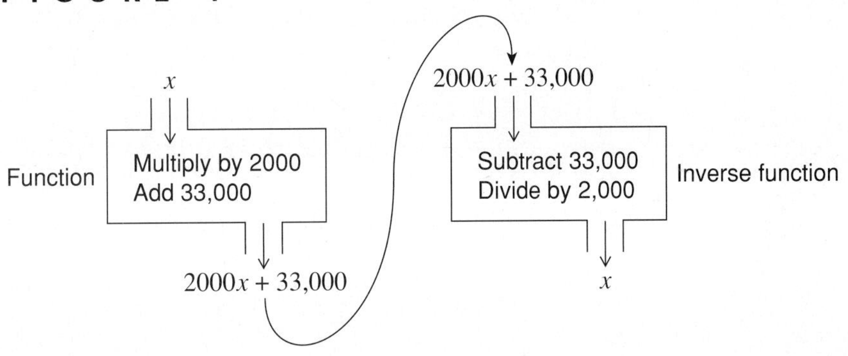

The ordered pair (2, 37,000) satisfies the function in Figure 4 and the ordered pair (37,000, 2) satisfies the inverse function.

In each of the previous examples, notice that the inverse function interchanges the input and output of the given function. For example, the ordered pairs for the CDs-R-Us salaries function have form (year, salary). The ordered pairs for the inverse function have form (salary, year).

The idea of interchanging input and output gives us a method for creating an inverse function algebraically. Let's investigate.

6. A function f contains all ordered pairs (x, y) for which $y = 2000x + 33{,}000$. The inverse of this function reverses the role of the variables; that is, x (year) becomes y (salary) and vice versa.

a. Write an equation representing the inverse of f by interchanging the variables.

b. Since y is the output variable, solve the equation for y. The result is the inverse relation for f. Compare the result with that of Investigation 9.

If the ordered pair (x, y) satisfies function S, then the ordered pair (y, x) satisfies the inverse of function f. The inverse function reverses the roles of the input and output variables. To create an inverse function algebraically, reverse the variables (replace x with y and y with x) and solve the resulting equation for the output variable y.

For example, if function f represents all ordered pairs (x, y) for which $y = 2000x + 33{,}000$, let's use algebra to find the inverse function.

Original function: $y = 2000x + 33{,}000$

Interchange variables (the equation will be the inverse relationship):

$$x = 2000y + 33{,}000$$

Solve for y, the output as follows.

Subtract 33,000 from both sides: $x - 33{,}000 = 2000y$

Divide both sides by 2000: $\dfrac{x - 33{,}000}{2000} = y$

The inverse function is expressed as $y = \dfrac{x - 33{,}000}{2000}$. This defines years on the job in terms of annual salary.

One notation used for **inverses** in mathematics is a **raised –1**. For example, if A is a matrix, we use A^{-1} to represent the inverse matrix. Likewise, the inverse of function f is written as f^{-1}. From the previous example we can write

$$f(x) = 2000x + 33{,}000 \text{ and } f^{-1}(x) = \frac{x - 33{,}000}{2000}.$$

Note that x^{-1} represents the multiplicative inverse, called reciprocal, of x and that f^{-1} does not mean the reciprocal of f—rather, it represents the inverse of f.

We have looked at the idea of inverse relations using both tables and equations. We don't want to slight the graphs. To see the symmetry in these graphs, which is the visual way to recognize inverse functions, you need to square the grid. This means the range on values on the input axis should be about 1.5 times the range of the output values. For example if Xmin is –15 and Xmax is 15, then Ymin should be about –10 and Ymax about 10.

7. Graph the pounds to dollars function $d = 1.63p$. Select an output value and draw a horizontal line through it.

a. How many times does the horizontal line intersect the graph of the function?

b. Regardless of the chosen output value, how many times will a horizontal line through a given output intersect the graph?

8. Graph the inverse relationship $p = \frac{d}{1.63}$ on the same grid as the original relationship. Sketch both graphs below. Where do the two graphs intersect? Is it possible to draw a line of symmetry between the two graphs? If so, draw the line of symmetry.

9. Consider the most basic quadratic function $Q(x) = x^2$.

a. Graph the function. Select a positive output value and draw a horizontal line through it. How many times does the horizontal line intersect the graph of the function?

b. Algebraically solve for the inverse relationship.

c. Is the inverse relationship a function? More specifically, is there only one output for each input? How does your answer relate to your answer to part **a**?

The graph of the function $d = 1.63p$ is a slant line through the origin with a slope of 1.63. Any horizontal line through a given output intersects the function's graph at exactly one point. This means there is exactly one input for each output.

Compare this to the graph of $Q(x) = x^2$. The graph is a parabola with vertex at the origin. Any horizontal line through a positive output intersects the function's graph at two points. This means there are two inputs for a given output.

Figure 5 summarizes this idea.

FIGURE 5

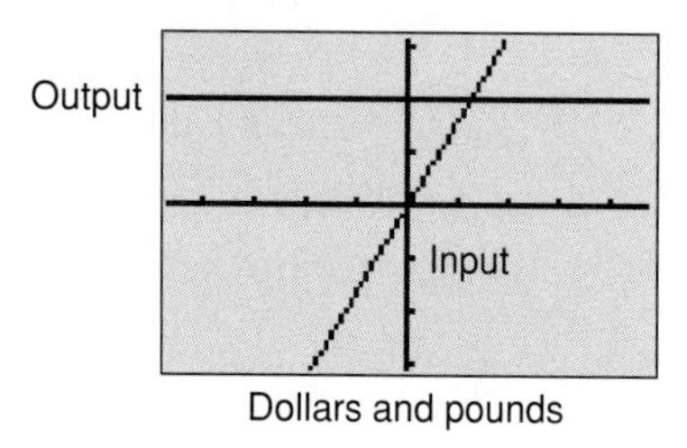

Dollars and pounds

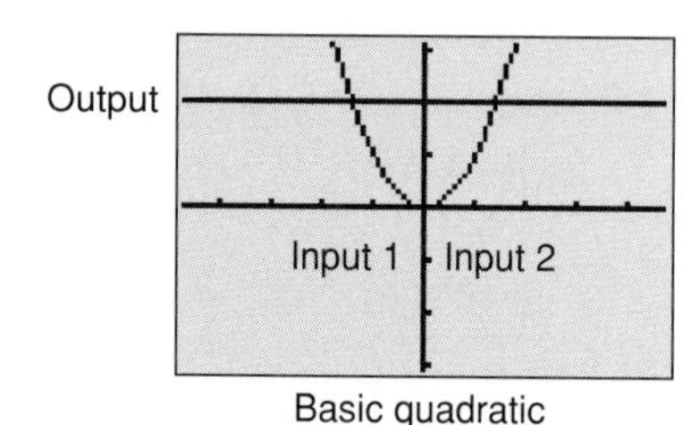

Basic quadratic

In order for a relationship to be a function, there must be exactly one output for each input. In a rectangular coordinate system, this means that a vertical line through a specific input intersects the function's graph at exactly one place.

Figure 6 displays the graphs of the functions and their inverses for both the dollars and pounds conversion and the basic quadratic function.

FIGURE 6

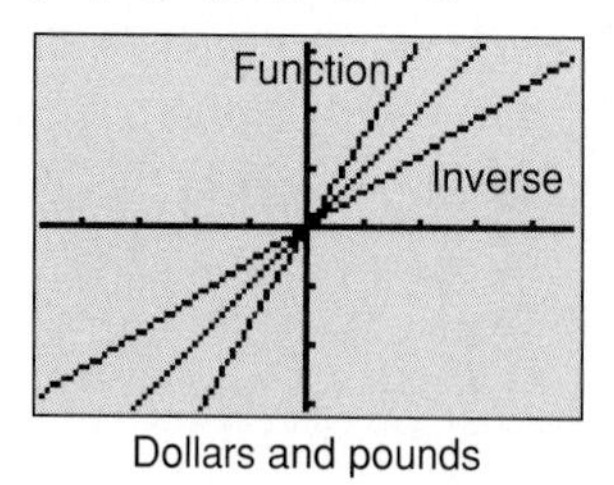

Dollars and pounds

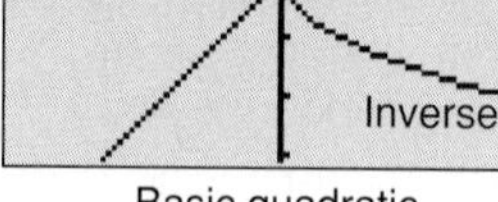

Basic quadratic

Notice the line of symmetry $y = x$ between the graphs of the original functions and the graphs of their inverses. Also, note that the graph of the inverse of the quadratic function is not a function. For each positive input, there are two outputs—one positive and one negative. This can be demonstrated by drawing a vertical line through any positive input and seeing that the vertical line intersects the graph at two different outputs.

A function in which there is only one input for each output is called a ***one-to-one function***. Each domain value has one range value and each range value has one domain value.

One-to-one functions have inverse functions. The dollars-to-pounds function is an example of a one-to-one function.

A function that is not one-to-one has an inverse, but the inverse is a relation rather than a function. The basic quadratic function is an example of a function that is not one-to-one.

You may have wondered how we can graph a function and its inverse on the same axes system since in reversing the x and y we are switching the units as well. Does the input axes represent pounds or dollars when both graphs are graphed together? Good question!! Mathematicians got a little sloppy about this, especially since they often looked at functions without a context. So as long as the units on each axis were approximately the same, the issue of units was ignored. This is equivalent to saying that when we investigate a problem with a context, we step back and look at the mathematical model to determine the features that apply to the context and ignore what doesn't. In the next section we look at a context in which the values are too different to use the same axes system.

We complete this investigation of inverses by looking at relationships from a parametric point of view. This is particularly helpful when the algebraic form of the inverse is difficult to derive.

Super Dog Burger Bits Salary Function: Recall the salary function at Super Dog Burger Bits: $y(x) = 23364.48598(1.07)^x$, where y is the annual salary during the xth year of work.

10. Graph this function in a good rectangular coordinate window.

a. Is this function one-to-one? Why?

b. Is the inverse of this function a function? Why?

11. Are you aware of any method for finding the inverse of this exponential function algebraically?

12. Enter the function in parametric mode. That is, let $X_{1T} = T$ and $Y_{1T} = 23364.48598(1.07)^T$.

a. If (X_{1T}, Y_{1T}) is an ordered pair satisfying the original salary function, write an ordered pair satisfying the inverse relation.

b. We want the ordered pair (X_{2T}, Y_{2T}) satisfying the inverse relation. How would you define X_{2T} and Y_{2T} in terms of X_{1T} and Y_{1T}? Use a table or a graph of the function defined parametrically by X_{2T} and Y_{2T} to determine if you have the correct inverse.

c. Consider the original problem situation in which the input is the year and the output is the salary. What quantity does the input to the inverse function represent? What quantity does the output of the inverse function represent?

Given the function $y(x) = 23364.48598(1.07)^x$, it is not trivial to find an algebraic definition for this inverse function. While we will be able to do so soon, we are not able to solve the equation for an exponent yet. This means we cannot solve for the output variable in the inverse yet.

The original function's graph increases over its domain (the output always gets larger as the input gets larger), so the original exponential function is one-to-one and thus has an inverse that is a function.

Given the rectangular coordinate function $y(x) = 23364.48598(1.07)^x$, the parametric definition would be

$$X_{1T} = T \text{ and } Y_{1T} = 23364.48598(1.07)^T.$$

With parametrics, we can define the inverse using ordered pairs.

If (X_{1T}, Y_{1T}) satisfies the original function, then (Y_{1T}, X_{1T}) satisfies the inverse function. We can define the inverse parametrically as

$$X_{2T} = Y_{1T} \quad \text{and} \quad Y_{2T} = X_{1T}.$$

The nice thing about this definition is that these definitions work for any function whose parametric form is entered as X_{1T} and Y_{1T}. Thus we can easily create a table and graph of the inverse of any function, regardless of whether we know the inverse's algebraic representation.

Figure 7 displays a parametric graph of the salary function and its inverse.

Note that the two functions are graphed in different windows. Trying to graph in the same window distorts both graphs adversely. The parametric form of the line of symmetry has also been entered, but, because of either very large inputs or very large outputs; it does not appear in the viewing window.

Finally, recall that the original function defined annual salary in terms of the number of years on the job. The inverse function defines the number of years on the job in terms of annual salary. The input represents annual salary. The output represents the year when we will receive that annual salary.

FIGURE 7

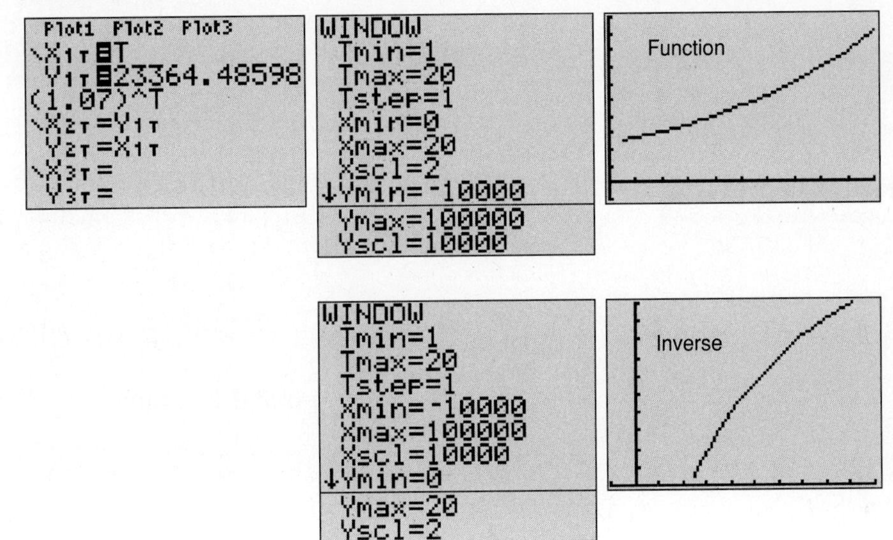

STUDENT DISCOURSE

Pete: One thing I don't understand is why is it so hard to find the inverse of an exponential function using algebra.

Sandy: It isn't hard if you know something about logarithms, but that is coming up later in the chapter.

Pete: I think I can do it based on what I know now.

Sandy: Okay, why don't you give it a try with $y = 23364.48598(1.07)^x$?

Pete: To create the inverse, I first interchange x and y to get $x = 23364.48598(1.07)^y$.

Sandy: So far, so good.

Pete: Now I need to solve for y by getting y alone on one side of the equal sign. Well, first I'd divide by 23364.48598. That "undoes" the multiplication.

Sandy: I'm with you so far.

Pete: So I have $\frac{x}{23364.48598} = (1.07)^y$. Then I . . . . Wait a minute. How do I get y out of the exponent position?

Sandy: That's exactly the problem.

Pete: I can't add, subtract, multiply, or divide to do it. I can't even use a root since I wouldn't know what root to use.

Sandy: I think you've hit a brick wall. Now do you see the problem?

Pete: I think so. If a variable is in the exponent position, I don't know how to change its position. Is that what logarithms are all about?

Sandy: That's part of it, but you're getting ahead of yourself. You now see the problem. Watch for the solution to the problem in upcoming sections.

1. In the Glossary, define the words in this section that appear in ***italic boldface*** type.

2. Is a function one-to-one if its graph

a. is strictly increasing? Why or why not?

b. is strictly decreasing? Why or why not?

c. is a horizontal line? Why or why not?

d. changes from increasing to decreasing at $x = 4$? Why or why not?

3. Given two functions $f(x)$ and $g(x)$, the composite function $f(g(x))$ is found by using the process for $g(x)$ as input to $f(x)$. The result of $f(g(x))$ is interesting when $f(x)$ and $g(x)$ are inverse relations. Find the composite function of the function

a. that converts pounds to U.S. dollars and its inverse.

b. for the CDs-R-Us salary function and its inverse.

4. For each setting given in Exploration 3, graph $f(x)$, $g(x)$, and $f(g(x))$ in the same window. Describe the relationships among the graphs of $f(x)$, $g(x)$, and $f(g(x))$.

5. In Chapter 1, we defined both *mean* and *median* as functions.

a. Are either of these one-to-one functions? Why or why not?

b. Describe problems that might arise if we tried to define the inverses of these functions.

6. Alcohol Content: Ten ounces of a mixed drink is 12% alcohol. Club soda is added to reduce the alcohol content. If C is the concentration of alcohol and x is the number of ounces of club soda added, then the function is $C(x) = \frac{1.2}{10 + x}$.

a. Find the inverse function algebraically.

b. What does the input of the inverse function represent? What does the output of the inverse function represent?

7. Hardware Sale: A hardware store marks down every item in the store 15%. Consider the relationship between the original price and the sale price. If s is the sale price and p is the original price, then the function is $s(p) = 0.85p$.

a. Find the inverse function algebraically.

b. What does the input of the inverse function represent? What does the output of the inverse function represent?

8. Telephone Bill: Table 6 displays the charges by a local telephone company based on the number of calls made per month.

TABLE 6 Local Telephone Charges

Number of Calls	7	12	15	23	31	33	40
Charges ($)	12.65	13.65	14.25	15.85	17.45	17.85	19.25

a. Create an appropriate function that models this data if number of calls is the input and charges ($) is the output.

b. Using Table 6, create a table that represents the inverse function.

c. Find the inverse function algebraically.

d. What does the input of the inverse function represent? What does the output of the inverse function represent?

9. Recall the following functions from Section 4.2.

$LP(x) = -0.01488x^4 + 0.737x^3 - 12.9x^2 + 60.9x + 257$ expresses the relationship between the number of LPs sold (in millions) in the United States and the time x since 1975 ($x = 0$ in 1975).

$CD(x) = 2.626x^2 - 22.55x + 12.25$ expresses the relationship between the number of CDs sold (in millions) in the United States and the time x since 1975 ($x = 0$ in 1975). This expression cannot be used for years before 1984.

$Cass(x) = -0.38x^3 + 8.67x^2 - 15.6x + 16.7$ expresses the relationship between the number of cassettes sold (in millions) in the United States and the time x since 1975 ($x = 0$ in 1975).

a. Are any of these functions one-to-one? Justify.

b. Create the parametric form of each of these functions and their inverses. Graph each inverse relation in an appropriate window and sketch the graphs.

c. For each function's inverse, identify the meaning of the input and the meaning of the output with respect to the problem situation.

10. Which of the following basic functions have inverses that are functions? Justify your answers.

a. $L(x) = x$

b. $Q(x) = x^2$

c. $C(x) = x^3$

d. $E(x) = 10^x$

e. $R(x) = \frac{1}{x}$

f. $A(x) = |x|$

g. $S(x) = \sqrt{x}$

h. $R(x) = \sqrt[3]{x}$

11. Recently, the U.S. dollar was worth 5.41 French francs.

a. Write a function that defines the number of francs as a function of the number of dollars.

b. Create the inverse function for the function you wrote in part **a**.

c. For the inverse function, what does the input represent? What does the output represent?

12. Are the two functions whose graphs appear in Figure 8 inverses of each other? Why or why not?

FIGURE 8

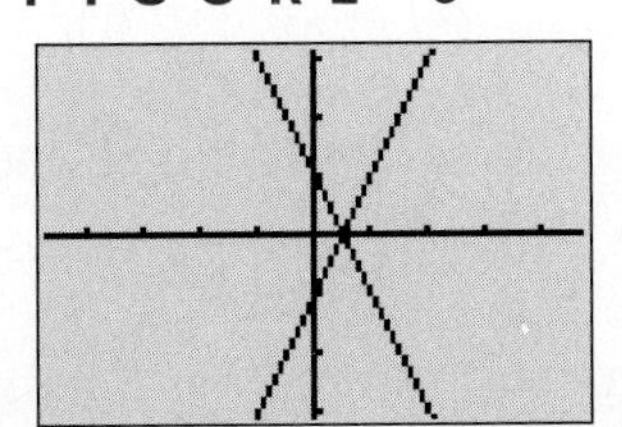

13. Each of the following expresses a relationship between x and y. Let the input be x and the output be y. Find the inverse function/relation algebraically. Check your result numerically and graphically.

a. $y = 2x - 7$

b. $y = \frac{2}{3}x + 5$

c. $3x + 2y = 7$

d. $y = x^2 - 5$

e. $y = \frac{2x - 3}{x + 4}$

f. $y = \sqrt{3x - 5}$

14. A partial table for a function appears below. Create a table for the inverse of this function.

TABLE 7 A Random Function

Input	Output
3	79
–9	18
53	–11
–23	–15
7	17

15. Which of the inverse relations in Exploration 14 are functions? Why?

16. Housing Prices: In 1985, the average price of a house in a certain county was $76,000. During the period 1985–1997, prices rose at an average rate of 3% per year.

a. Create a table with input the number of years since 1985 and output the average price of a house. Calculate the outputs for inputs 0, 1, 2, 3, and 4.

b. What kind of function should be used to model the data?

c. Create an algebraic equation for the function in which the input is the number of years since 1985 and the output is the average price ($) of a house.

d. In what year did the average price of a house pass $100,000? Explain how you found the answer.

17. Computer Depreciation: The value of a certain brand of computer depreciates at a rate of 10% per year. The original cost of the computer was \$1200. Let the input be the number of years since the computer was purchased and the output be the value (\$) of the computer.

a. Create a table for this function using inputs 0, 1, 2, 3, and 4.

b. What kind of function should be used to model the data?

c. Create an algebraic equation for the function.

d. How long will it take for the computer's value to fall below \$500? Explain how you found the answer.

e. Find out how the IRS suggests computing depreciation on personal computer equipment. What is a reasonable number of years to completely depreciate a personal computer?

18. Find the solution to the following equations.

a. $x^2 + 5x - 3 = -9$

b. $\frac{5}{3}x - \frac{2}{7} = 3 - x$

c. $\frac{x}{x+1} + \frac{2}{x-3} = \frac{8}{x^2 - 2x - 3}$

d. $\sqrt{7x+5} = 2$

REFLECTION

Write a paragraph describing how to find the inverse of a function numerically, algebraically, and graphically.

SECTION 8.2

Exponential Models Revisited

Purpose

- Revisit the concept of exponential models.
- Study functions of the form $P(t) = a(b)^{f(t)}$.
- Investigate the nature of the natural exponential function e^x.

In Section 8.1, we looked at inverse functions. In the process, we mentioned that the inverse function of an exponential function is a logarithmic function. We will investigate logarithmic functions in detail in Sections 8.3 and 8.4. Before we do, however, we extend our understanding of exponential functions by looking at some additional examples. The following investigation provides the opportunity to refresh your understanding of exponential functions by recalling a couple we investigated earlier in the text.

1. Recall the **Super Dog Burger Bits** salary function

$$y = 23364.48598(1.07)^x$$

where x is the number of years worked and y is the annual salary ($).

a. What does the 1.07 represent?

b. What does the 23364.48598 represent?

c. Does the output increase or decrease as the input increases? How can you tell by looking at the equation that defines the function?

2. **Drug in Bloodstream:** Five milliliters of a drug is injected into the bloodstream. After one hour 60% of the original amount remains in the bloodstream. For each succeeding hour, 60% of the amount present in the previous hour remains in the bloodstream. The function defining the relationship is $A(t) = 5(0.6)^t$, where t is the number of hours since the

drug was injected and A is the amount (in milliliters) of drug remaining in the bloodstream.

a. What does the 0.6 represent?

b. What does the 5 represent?

c. Does the output increase or decrease as the input increases? How can you tell by looking at the equation that defines the function?

3. How can you determine from an input-output table that the defining function is exponential?

4. What kind of sequence is defined by an exponential function? What is an important feature about the relationship between successive terms of such a sequence?

Basic exponential functions have form $y(x) = a(b)^x$, where b represents the base and a represents the value of the output when the input is zero. The base b represents the constant ratio between successive outputs for each unit change in the input. Geometric sequences are modeled by exponential functions.

For the **Super Dog Burger Bits** salary function $y = 23364.48598(1.07)^x$, the annual salaries grow by 7% per year. The 1.07 represents the previous salary (1) plus the raise (0.07). Thus the ratio of change is $1 + 0.07 = 1.07$. The number 23364.48598 represents the hypothetical salary in year 0. Of course, salary is normally given with only two decimal places, but rounding errors accumulate quickly in exponential functions, so the decimals are all retained. (It has been rumored that bank employees figuring interest have managed to embezzle big bucks by putting the difference between the actual value and the rounded value into their own accounts.)

Since $b = 1.07$ is greater than 1, the output grows larger as the input increases. When $b > 1$, the function is called an ***exponential growth function***. The table and graph appear in Figure 1 on page 620.

For the **Drug in Bloodstream** problem, the function is $A(t) = 5(0.6)^t$. The 60% represents how much of the previous amount (in milliliters) of drug remains in the bloodstream one hour later. This is the previous amount of drug in the bloodstream (1) minus the amount lost in the following hour

FIGURE 1

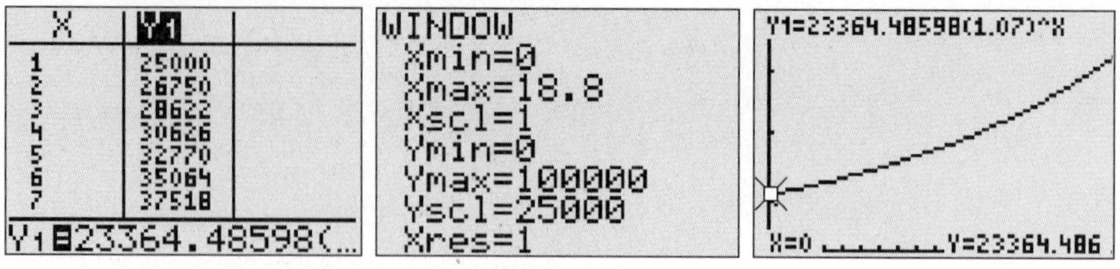

X	Y1
1	25000
2	26750
3	28622
4	30626
5	32770
6	35064
7	37518

Y1=23364.48598(...

WINDOW
Xmin=0
Xmax=18.8
Xscl=1
Ymin=0
Ymax=100000
Yscl=25000
Xres=1

Y1=23364.48598(1.07)^X
X=0 Y=23364.486

(40%). Thus the ratio of change is $1 - 0.4 = 0.6$. The 5 represents the original amount of drug taken.

Since $b = 0.6$ is between 0 and 1, the output decreases as the input increases. When $0 < b < 1$, the function is called an ***exponential decay function***. The table and graph appear in Figure 2.

FIGURE 2

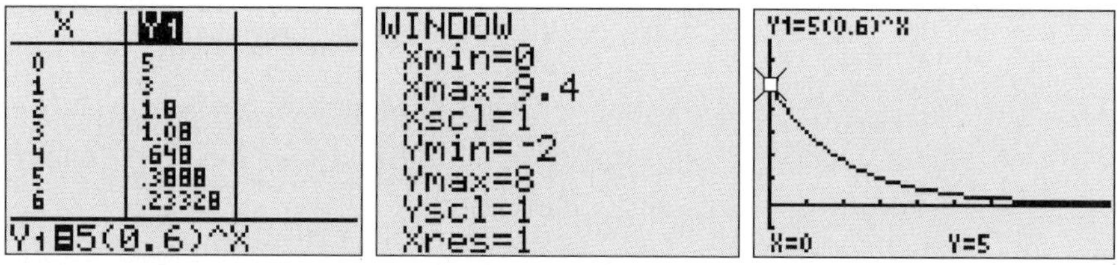

X	Y1
0	5
1	3
2	1.8
3	1.08
4	.648
5	.3888
6	.23328

Y1=5(0.6)^X

WINDOW
Xmin=0
Xmax=9.4
Xscl=1
Ymin=-2
Ymax=8
Yscl=1
Xres=1

Y1=5(0.6)^X
X=0 Y=5

The most general exponential growth model is $y = a(1+r)^{f(x)}$. The parameter r is the rate of growth and the base $b = 1 + r$. The exponent $f(x)$ is some function of the input variable. The point $(0, a)$ is the vertical intercept when $f(x) = x$. If $f(x)$ is more complex or other terms such as constants are added to the function, the vertical intercept may change. Evaluate the exponential growth function at $x = 0$ to be sure.

For example, for the salary function, $y = 23364.48598(1.07)^x$:

- $a = 23364.48598$,
- $r = 0.07$ and $b = 1 + r = 1 + 0.07 = 1.07$, and
- $f(x) = x$.

The most general exponential decay model is $y = a(1-r)^{f(x)}$. The parameter r is the rate of decay and the base $b = 1 - r$. The exponent $f(x)$ is some function of the input variable. The point $(0, a)$ is the vertical intercept when $f(x) = x$. If $f(x)$ is more complex or other terms such as constants are added to the function, the vertical intercept may change. Evaluate the exponential growth function at $x = 0$ to be sure.

For example, for the **Drug in Bloodstream** function, $A(t) = 5(0.6)^t$:

- $a = 5$,
- $r = 0.4$ and $b = 1 - r = 1 - 0.4 = 0.6$, and
- $f(t) = t$.

Next we look at some new situations in which the exponent is a function of the input variable. We will regularly use t as the input variable, instead of x, since the applications we look at involve functions of time.

5. Housing Value: The average value of a house in a local community in 1990 was approximately $93,000. Historically, housing values in the community have increased at a rate of 5% every three years. Let's create an exponential function to model this scenario.

a. There is a good year to use as the arbitrary zero. What is it? Define the input variable t to be the number of years since this arbitrary zero year.

b. Create a table with appropriate values from 1990 to 1999 to get a sense of the data.

c. What is the value of parameter a? Why?

d. Is this a growth or decay function? Why?

e. What is the value of parameter r? What is the value of parameter b? Why?

f. The rate of growth-decay r occurs over a three-year period. Since t is the number of years and the rate is given over a three-year period, the exponent must have a value of 1 when $t = 3$. Write an expression in t that satisfies these restrictions.

g. Write the exponential function that models this situation.

h. Use your modeling function to predict the average value of a house in 1995. Explain what you did.

i. Use your function to estimate the year when the average value of a house will pass \$110,000. Explain what you did.

Since the initial information is given for the year 1990, let t represent the number of years since 1990. This implies that the output (housing value) is 93,000 when $t = 0$. This is a growth function since housing values are increasing and the value of r is 0.05. Thus we will use the model

$$A(t) = a(1 + r)^{f(t)}$$

where $a = 93{,}000$, $r = 0.05$, $b = 1 + r = 1 + 0.05 = 1.05$, and A represents the average housing value t years after 1990.

The exponent $f(t)$ must be 1 when $t = 3$. Dividing t by 3 will accomplish this, so $f(t) = \frac{t}{3}$.

We have our modeling function $A(t) = 93{,}000(1.05)^{t/3}$. Be careful when you enter this on a calculator. The exponent must be in its own set of parentheses.

To determine the average housing value in 1995, let $t = 5$ and find the output.

$$A(5) = 93{,}000(1.05)^{5/3} \approx 100{,}878$$

The approximate value of an average house in 1995 is predicted to be about \$100,878.

Use either a table or graph to determine the year the average value will pass \$110,000 (Figure 3).

FIGURE 3

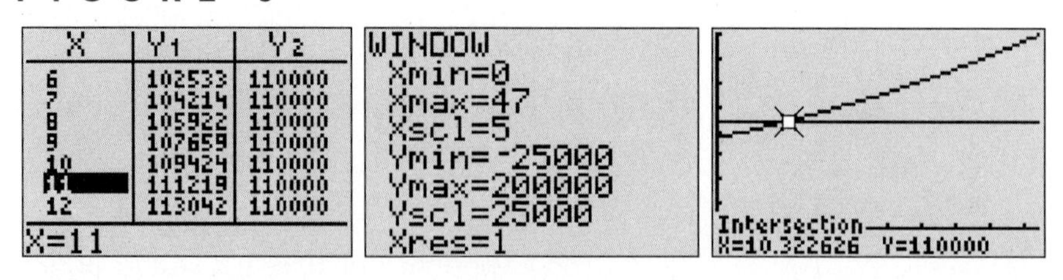

After 11 years, the average housing value will pass \$110,000. This would be the year 2001.

One of the more important applications of exponential decay relates to the concept of ***half-life***.

Half-life is a measure of time. It is the length of time it will take for one-half of a substance to decay. In a previous section, you measured the half-life of a drug in the body.

One application of half-life relates to the use of nuclear reactors. Plutonium, which causes cancer if inhaled or ingested, has a half-life of 24,000 years. Plutonium is created in fast-breeder nuclear reactors. Many people greatly fear the transport of plutonium because an accident could lead to exposure to a lethal substance that remains in the environment for such a very long time.

The algebraic model for half-life, based on the general exponential decay model, is

$$A(t) = a(0.5)^{t/h},$$

where a is the amount of the substance when time $t = 0$ and h is the half-life of the substance. The rate of decay r is 50% or 0.5. So $b = 1 - r = 1 - 0.5 = 0.5$.

$A(t)$ is the amount of the substance remaining after t time units.

6. Assume that a ship containing 10 tons of plutonium sinks, spilling its load into the ocean.

a. Write an algebraic model expressing the amount of plutonium in the ocean after t years as a result of this accident.

b. How many tons of the plutonium will remain in the ocean after 100 years? Justify your answer.

c. Graph the function in an appropriate window. Use the graph to determine the number of years it will take until less than 9 tons of the plutonium remain. Write a sentence describing your answer.

Since the initial amount of plutonium is 10 tons and the half-life is 24,000 years, the model for this situation is

$$A(t) = 10(0.5)^{t/24{,}000}.$$

To find the amount of plutonium remaining after 100 years, let $t = 100$.

$$A(100) = 10(0.5)^{100/24{,}000} \approx 9.9712$$

Of the original 10 tons, 9.9712 tons remains after 100 years.

Use a table or graph to determine how many years it will take until 9 tons remain (Figure 4).

FIGURE 4

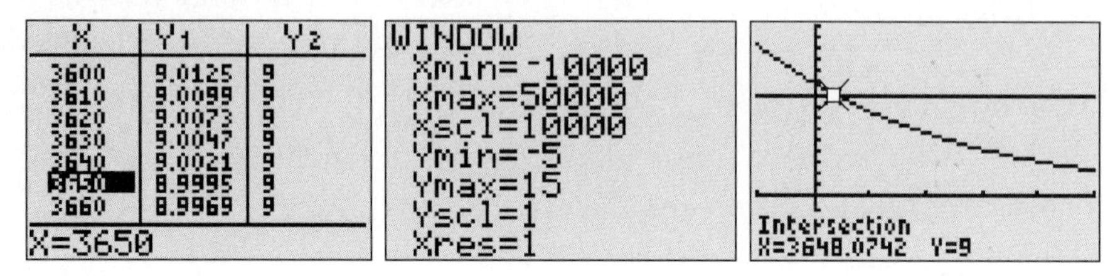

It will take about 3650 years for more than 1 ton of the original 10 tons of plutonium to decay.

We conclude the section by investigating an important new number.

7. Enter the number ***e*** on your calculator. (If you are unsure how to do this, ask a friend or your instructor, or consult your manual.) Write the approximate value of the number e.

8. Between what two consecutive integers does the number e lie?

We will investigate the rate of growth of three functions 2^x, 3^x, and e^x.

To do so, we will use a ***difference quotient*** function $dq(x)$ to measure the rate of change of each function over intervals of length 0.001. We can define a function that is essentially the formula for rate of change used in Chapter 2 except that the change in x has been set at 0.001, while the change in y is calculated by the function.

If we assume the original function is defined on the calculator as Y_1, then the difference quotient function $dq(x)$ is

$$dq(x) = \frac{Y_1(x+0.001) - Y_1(x)}{0.001}.$$

Enter this function on your calculator in a location other than Y_1.

9. Look at a table of outputs for $Y_1 = 2^x$ and for the difference quotient. Which increases faster: the output or the rate of change dq?

10. Look at a table of outputs for $Y_1 = 3^x$. Which increases faster: the output or the rate of change dq?

11. Look at a table of outputs for $Y_1 = e^x$. Which increases faster: the output or the rate of change dq?

12. For each function 2^x, 3^x, and e^x, compare the function's output to its rate of growth.

a. For which does the rate of growth exceed the output?

b. For which is the rate of growth less than the output?

c. For which does the rate of growth approximately equal the output?

The number e is an irrational number that naturally arises in many exponential functions. The approximate value of this number is

$$e \approx 2.7182818.$$

As a testament to its importance, the number e has its own key on most calculators. One important aspect of the exponential function with base e is its rate of growth.

Figure 5 contains the tables for the functions 2^x, 3^x, e^x and their difference quotients.

FIGURE 5

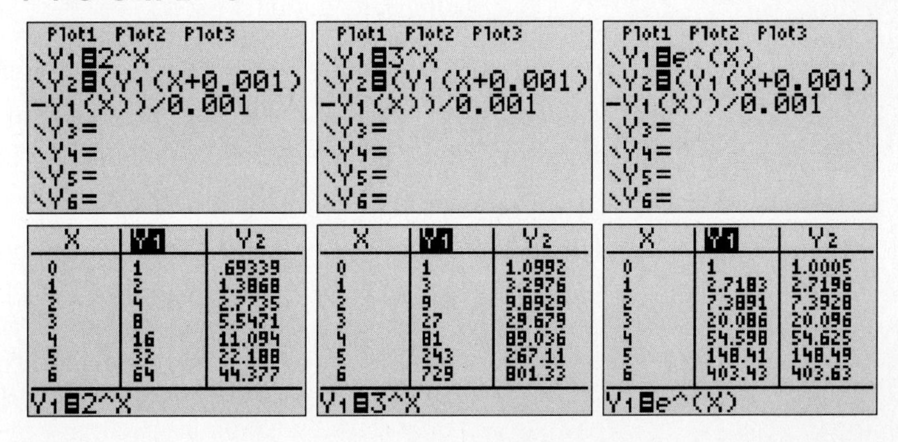

Notice that the rate of change dq is

- less than the output for the function 2^x,
- greater than the output for the function 3^x.
- about the same as the output for the function e^x.

A very important feature of the ***natural exponential function***

$$f(x) = e^x$$

is that $f(x) = e^x$ is the only function whose rate of change equals its output.

We will look at some applications involving this function in the explorations and the next few sections.

Pete: So they're trying to tell me that e is a number?

Sandy: That's right.

Pete: I never heard of such a number.

Sandy: Have you ever heard of the number π?

Pete: Sure. That's just a math way to write the number 3.14.

Sandy: Not exactly. The number π is an irrational number. It has no exact decimal representation. 3.14 is just an approximation. Do you know where π comes from?

Pete: I suppose someone just made it up. I do know it has something to do with circles, though.

Sandy: Correct. The number π is a ratio. Take any circle and divide the circumference by the diameter and you get π.

Pete: You mean that if I measure around a circle and then measure across a circle and divide, I get π?

Sandy: Kind of. Measurement by its very nature is inaccurate, so you will only get an approximation of π.

Pete: So I guess you are telling me that π is just a name for the ratio of circumference to diameter of a circle. Then what's e?

Sandy: That's a bit tougher. The natural number e is another irrational number. Actually scientists collecting data found this number occurring regularly when they investigated all kinds of natural phenomena like bacteria growth, electricity, and the cables on the Golden Gate Bridge in San Francisco. It is used a lot with exponential and logarithmic functions. As you saw in the last investigation, e to a power grows at a rate equal to the value of e to the power. That's important since rate of growth and decay are often dependent on how many objects there are.

Pete: What do you mean?

Sandy: Well, for example, the growth of a population may be dependent on how many people there are.

Pete: I can see that, but I'm still unsure about e.

Sandy: I think working on the explorations and learning about logarithms may help you feel more comfortable with e. Why don't you do that? I'll see you later.

1. In the Glossary, define the words in this section that appear in ***italic boldface*** type.

2. Graph the function $y(x) = a(2)^x$ for several values of parameter a where a is negative.

a. What is the domain and range of $y(x)$ when a is negative?

b. How does the graph of $y(x)$ when a is negative compare to the graph of $y(x)$ when a is positive?

3. Chocolate and Dogs: Another example of half-life occurs in many households. You may not know that chocolate can be lethal to dogs. Dogs are highly sensitive to caffeine and theobromine, both of which are found in chocolate. If dogs could quickly metabolize and eliminate chocolate from their systems, the chance of chocolate poisoning would be minimal. The half-life of caffeine and theobromines in humans is between 2 and 3 hours. The half-life of these substances in dogs is 18 hours. The fact that these substances remain in dogs so long causes a dog's system to continually repoison itself. As a result, a 55-pound dog could die from eating just 5 ounces of baker's chocolate. My labrador retriever just got hold of 4 ounces of chocolate.

a. Write an algebraic model expressing the amount of chocolate in the dog's system after t hours.

b. How much chocolate remains in the dog's system after 10 hours? Justify your answer.

c. Graph the function in an appropriate window. Use the graph or a table to determine the number of hours it will take until less than 1 ounce of chocolate remains in the dog's system. Write a sentence describing your answer.

4. A Growing City: The population of a city grows at a rate of 10% per year. The population in 1990 was 400,000. Let x represent the number of years since 1990 and y represent the population (measured in 100,000s) after x years.

a. Write a function that models this information.

b. What is the predicted population in the year 2000? Explain what you did.

c. In what year will the population reach 1 million? Explain what you did.

d. Compare the function you wrote in part **a** to the function $y(x) = 4(e)^{0.1x}$. As the input grows, which function grows faster?

5. A Shrinking City: The population of a city decreases at a rate of 10% per year. The population in 1990 was 400,000. Let x represent the number of years since 1990 and y represent the population (measured in 100,000s) after x years.

a. Write a function that models this information.

b. What is the predicted population in the year 2000? Explain what you did.

c. In what year will the population drop below 100,000? Explain what you did.

d. Compare the function you wrote in part **a** to the function $y(x) = 4(e)^{-0.1x}$. As the input grows, which function decreases faster?

6. You are given the function $y(x) = 5(2)^x$. Do the following:

a. Graph this function in a "friendly" window.

b. Overlay the graphs of $y(x) = 5(2)^{x-h}$, where h is replaced with one positive number and one negative number. What effect does the choice of h have on the graph of $y(x) = 5(2)^x$?

c. Graph $y(x) = 5(2)^x$. Overlay the graphs of $y(x) = 5(2)^x + k$, where k is replaced with one positive number and one negative number. What effect does the choice of k have on the graph of $y(x) = 5(2)^x$?

7. After having completed Exploration 6, do the following:

a. Describe the effect of h and k on the graph of $y(x) = a(b)^{x-h} + k$ as compared to the graph of $y(x) = a(b)^x$.

b. Does the shape of the graph change? Does the position of the graph change?

c. Assuming a is positive, what is the domain and range of

$$y(x) = a(b)^{x-h} + k?$$

8. Given that a, b, and c are parameters. Compare the exponential function $y(x) = a(b)^x + c$ to the quadratic function $q(x) = ax^2 + bx + c$.

a. How is the role of the input variable different for these two classes of functions.

b. How do the graphs differ?

9. Compound Interest: The amount A in an account t years after a dollars is deposited at an annual rate i (expressed as a decimal) compounded n times per year is given by the exponential function

$$A(t) = a\left(1 + \frac{i}{n}\right)^{nt}.$$

a. What letter is the input variable? What letter is the output variable? What letters are the parameters?

b. What is the rate of growth r in this exponential growth function?

c. Why is i divided by n in the formula?

d. What is the base b of this exponential function?

e. Why is the variable t multiplied by the parameter n?

f. How much is in an account after two years if $500 is deposited at an annual rate of 5% compounded monthly? Describe what you did to answer this question.

10. In the compound interest formula, $\left(1 + \frac{i}{x}\right)^{x}$ is the rate of growth for one year on an initial deposit. Create a table comparing $\left(1 + \frac{i}{x}\right)^{x}$ as x gets large to e^{i} for several values of annual interest rates i. How does $\left(1 + \frac{i}{x}\right)^{x}$ as x gets large compare to e^{i}?

11. Recently, the U.S. dollar was worth 145.56 Spanish pesetas.

a. Write a function that defines the number of pesetas as a function of the number of dollars.

b. Create the inverse function for the function you wrote in part **a**.

c. For the inverse function, what does the input represent? What does the output represent?

12. Monthly Payments on a Loan: Suppose you borrow A dollars at an annual rate of i percent. You will make monthly payments. The monthly payment P is an exponential function of the total number of payments n where

$$P(n) = \frac{\left(\frac{i}{12}\right)A}{1 - \left(1 + \frac{i}{12}\right)^{-n}}.$$

a. What is the monthly payment if you borrow \$10,000 for a car at an annual rate of 7.25% for three years? Describe how you found your answer.

b. Suppose you borrow \$10,000 at an annual rate of 7.25%. You must have a monthly payment less than \$250. What is the shortest time for the loan? Explain how you found your answer.

c. Your bank is charging 9.25% annual interest for a three-year new-car loan. What is the most you can borrow if your monthly payments must be less than \$300? Describe how you found your answer.

d. You plan to borrow \$5000 for three years. How does your monthly payment change as the interest rate changes from 7% to 7.25% to 7.5%?

13. Future Value Annuity: There is an exponential function that defines the future value of an investment, assuming regular deposits. Let R be the monthly deposit and i be the annual interest rate. The amount in the annuity after n months is given by

$$A(n) = R\left(\frac{\left(1 + \frac{i}{12}\right)^n - 1}{\frac{i}{12}}\right).$$

a. Suppose you deposit \$100 per month in an annuity account paying a guaranteed annual rate of 6%. How much will be in the annuity after 5 years? After 10 years? After 20 years?

b. You deposit \$100 monthly for 10 years in an annuity. How much will be available at the end of this period if the annual interest rate is 5%? 6%? 7%? 8%?

14. Normal Distribution: The normal distribution curve is defined by the function $y(x) = \frac{1}{\sqrt{2\pi}} e^{\frac{-x^2}{2}}$.

a. Graph this function using Xmin = –5, Xmax = 5, Ymin = – 0.1, and Ymax = 0.2. The graph is a famous shape. What is it commonly called?

b. For what inputs is the output less than 5%? Careful—there are two parts to the answer.

c. For what inputs is the output less than 1%? Describe what you did.

15. Given the points (1, 11) and (7, 4), create an exponential function that contains the points.

16. Figure 6 displays a table of an exponential function. Write the equation for this function. Explain what you did.

FIGURE 6

X	Y1
0	52
1	56.68
2	61.781
3	67.342
4	73.402
5	80.008
6	87.209

X=0

17. Figure 7 displays a graph of an exponential function. Write the equation for this function. Explain what you did.

FIGURE 7

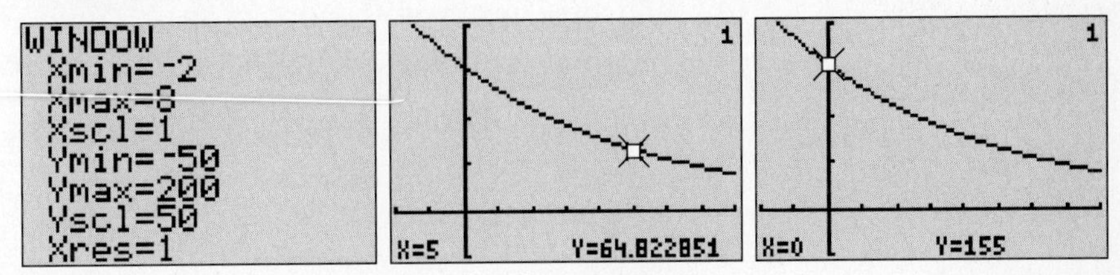

18. Find the inverse function for each of the following. Assume that x is the input and y is the output.

a. $y = 5x + 12$

b. $7x + 2y = -9$

Write a paragraph explaining the concept of half-life to a good friend who doesn't understand it.

SECTION 8.3

Logarithmic Function Introduction

Purpose

- Introduce the logarithm as the inverse of an exponential function.
- Investigate the meaning of a logarithm.
- Study properties involving logarithms.

We first studied the basic concepts behind exponential functions in Chapter 2 when we investigated geometric sequences. The class of exponential functions was further investigated in Chapters 3 and 4 concluding with exponential regression. In this chapter, we return and complete our study of exponential functions by introducing the inverse of an exponential function called a logarithmic function.

We will use logarithms as an aid in solving exponential equations in which the variable occurs in the exponents. Logarithms will help us analyze data and determine if the data can be modeled with an exponential function. Finally, we look at using logarithmic regression to model data.

In this section, we introduce the basic logarithmic function and focus on the meaning of a logarithm. We also investigate some basic properties of logarithmic expressions. Finally, we continue our introduction of the number e, which arises in important applications of exponential and logarithmic functions.

1. **Super Dog Burger Bits Salary Function:** Annual salary as a function of years worked at Super Dog is defined by the exponential function $y = 23364.48598(1.07)^x$. In Section 8.1, we investigated the inverse of this function numerically and graphically. Now we'll look at the inverse algebraically.

   a. Create the inverse relationship by replacing x with y and y with x.

   b. Isolate the exponential factor by dividing both sides by the salary in year zero.

2. We need to solve for the variable y. Do you know a way to do this? Why or why not?

Let's look at where the previous investigation leads us.

Original function:	$y = 23364.48598(1.07)^x$
Inverse function:	$x = 23364.48598(1.07)^y$
Isolate exponential factor:	$\dfrac{x}{23364.48598} = (1.07)^y$

We need to solve for the exponent y. This requires a new function called a ***logarithm***.

Another way to write the equation $\dfrac{x}{23364.48598} = (1.07)^y$ is

$$y = \log_{1.07}\left(\frac{x}{23364.48598}\right).$$

The equation $y = \log_{1.07}\left(\frac{x}{23364.48598}\right)$ is read "y equals the logarithm with base 1.07 of the argument x divided by 23364.48598."

In essence, logarithms are used to represent the inverse of exponential functions.

Before continuing with this problem, let's investigate the idea of a logarithm in more detail.

3. Consider the exponential function $f(x) = 10^x$.

a. Complete Table 1.

TABLE 1 Exponent Function $f(x) = 10^x$

Input	Output
−5	
−4	
−3	
−2	
−1	
0	
1	
2	
3	
4	
5	

b. Sketch a graph of the function $f(x) = 10^x$. Identify the domain and the range.

4. Consider the logarithmic function $g(x) = \log_{10}(x)$

a. Record the outputs from Table 1 as the inputs in Table 2. For each input in Table 2, find the output using the **LOG** key on your calculator. The **LOG** key on your calculator represents a logarithm with base 10.

TABLE 2 **Logarithmic Function $g(x) = \log_{10}(x)$**

Input	Output

b. Sketch a graph of the function $g(x) = \log_{10}(x)$. Identify the domain and the range.

5. Compare Tables 2 and 3. What do you notice?

6. Compare and contrast the domain and range of $f(x) = 10^x$ with the domain and range of $g(x) = \log(x)$.

For every exponential function with base b of the form $f(x) = a(b)^x$, there is a corresponding inverse function, called the ***logarithmic function with base b***.

Table 1 is a numerical representation of the exponential function $f(x) = 10^x$ with base 10 and $a = 1$. The input is an exponent and the output is the result of raising the base to that exponent.

Table 2 is a numerical representation of the logarithmic function $g(x) = \log_{10}(x)$ with base 10. The input is the result of raising the base to an exponent and the output is the exponent.

Table 3 displays the domain and range of the functions $f(x) = 10^x$ and $g(x) = \log_{10}(x)$.

TABLE 3 Domain and Range

Function	Domain	Range
$f(x) = 10^x$	all real numbers	all positive real numbers
$g(x) = \log_{10}(x)$	all positive real numbers	all real numbers

Next we discuss the meaning of logarithmic notation. The equation $y = \log_b(x)$ is another way of writing $b^y = x$.

The equation $y = \log_b(x)$ is called the ***logarithmic form*** of the relationship between x and y.

The equation $b^y = x$ is called the ***exponential form*** of the relationship between x and y.

Given the logarithmic form $y = \log_b(x)$:

- The base b must be positive and not 1. This is the same restriction we placed on the base of the exponential function.
- The value of x must be positive since x represents the input. The input must be positive because of the domain of the function. The variable x is often called the ***argument*** of the logarithm.
- The value of y represents the exponent on b needed to obtain x.
- $\text{Log}_b(x)$ does not represent multiplication—$\log_b$ represents the name of a function and x is the input to the function.

The next investigation provides an opportunity to investigate the meaning of logarithmic notation.

7. The exponential form $8 = 2^3$ is written in logarithmic form as $3 = \log_2(8)$. Consider the exponential form $\frac{1}{32} = 2^{-5}$.

a. Identify the base, the exponent, and the result of the exponentiation.

b. Express in logarithmic form.

8. The logarithmic form $4 = \log_2(16)$ is written in exponential form as $16 = 2^4$. Consider the logarithmic form $\log_8(4) = \frac{2}{3}$.

a. Identify the base, the exponent, and the result of the exponentiation.

b. Express in exponential form.

9. The expression $\log_3(81)$ represents the exponent on 3 to obtain 81. The exponent is 4. So $\log_3(81) = 4$. Use similar reasoning to calculate the value of the following logarithms.

a. $\log_2(16)$

b. $\log_4\left(\frac{1}{4}\right)$

c. $\log_9(3)$

To check your understanding, let's look at the previous investigations.

Investigation 7: $\frac{1}{32} = 2^{-5}$

The base is 2, the exponent is –5, and the result of the exponentiation is $\frac{1}{32}$.

Logarithmic form: $-5 = \log_2\left(\frac{1}{32}\right)$

Investigation 8: $\log_8(4) = \frac{2}{3}$

The base is 8, the exponent is $\frac{2}{3}$, and the result of the exponentiation is 4.

Exponential form: $4 = 8^{2/3}$

Investigation **9a**: $\log_2(16)$

Think: What exponent do we place on 2 to get a result of 16?

Name logarithm: $\log_2(16) = x$

Exponential form: $2^x = 16$

Rewrite so both sides have the same base:

$$2^x = 2^4$$

Exponents are equal when bases are equal:

$$x = 4$$

So: $\log_2(16) = 4$

Investigation **9b**: $\log_4\left(\frac{1}{4}\right)$

Think: What exponent do we place on 4 to get $\frac{1}{4}$?

Name logarithm: $\log_4\left(\frac{1}{4}\right) = x$

Exponential form: $4^x = \frac{1}{4}$

Rewrite so both sides have the same base:

$$4^x = 4^{-1}$$

Exponents are equal when bases are equal:

$$x = -1$$

So: $\log_4\left(\frac{1}{4}\right) = -1$

Investigation **9c**: $\log_9(3)$

Think: What exponent do we place on 9 to get 3?

Name logarithm: $\log_9(3) = x$

Exponential form: $9^x = 3$

Rewrite so both sides have the same base:

$$(3^2)^x = 3$$

Exponent property: $3^{2x} = 3$

Bases are the same, so exponents are the same. The exponent of 3 is 1.

$$2x = 1$$

Divide both sides by 2: $x = \frac{1}{2}$

Answer: $\log_9(3) = \frac{1}{2}$

Check: $9^{\frac{1}{2}}$ (read "9 to the $\frac{1}{2}$") is the square root of 9, which is 3.

We can set the exponents equal when the bases are equal because exponential functions are one-to-one. If the outputs of two exponential functions with the same base are the same, then the inputs, the exponents, are the same.

Now that we have a basic understanding of logarithms, we investigate some of the properties of logarithms. To do so, we will use logarithms with base 10 to do the computations. The **LOG** key on your calculator represents a logarithm with base 10. Base 10 logarithms—called ***common logarithms***—are often written without the base. The only other base that your calculator can use directly is base e, where e is the irrational number approximately equal to 2.7182818282846. A logarithm with base e— called a ***natural logarithm***— is represented by the **LN** key on your calculator.

INVESTIGATION

10. a. What is the value of $\log_{10}(1)$? Explain why.

b. What is the value of $\log_b(1)$ for any base b? Why?

11. **a.** What is the value of $\log_{10}(10)$? Explain why.

b. What is the value of $\log_b(b)$ for any base b? Why?

12. **a.** Complete Table 4 using your calculator.

TABLE 4 Logarithms of Exponentials

x	$\log_{10}(10^x)$
3	
7	
−5	
−9	
$\frac{1}{2}$	

b. Using the results of Table 4, what is the value of $\log_b(b^x)$? Why?

13. **a.** Complete Table 5 by selecting five different pairs of arguments a and b where $a \neq b$ and then calculating the corresponding base 10 logarithms.

TABLE 5 Properties of Logarithms I

a	b	$\log_{10}(a \cdot b)$	$\log_{10}(a) \log_{10}(b)$	$\log_{10}(a) + \log_{10}(b)$

b. Using the data in Table 5, state any relationships you see.

14. a. Complete Table 6 by selecting five different pairs of arguments a and b where $a \neq b$ and then calculating the corresponding base 10 logarithms.

TABLE 6 Properties of Logarithms 2

a	b	$\log_{10}\left(\frac{a}{b}\right)$	$\frac{\log_{10}(a)}{\log_{10}(b)}$	$\log_{10}(a) - \log_{10}(b)$

b. Using the data in Table 6, state any relationships you see.

15. a. Complete Table 7 by selecting five different pairs of arguments a and b, where $a \neq b$, and then calculating the corresponding base 10 logarithms.

TABLE 7 Properties of Logarithms 3

a	b	$\log_{10}(a^b)$	$(\log_{10}(a))^b$	$b(\log_{10}(a))$

b. Using the data in Table 7, state any relationships you see. Describe the relationships in words.

16. Consider the function $y = 10^{2x+1}$. Use logarithms to help you find x if $y = 12$.

In the previous series of investigations, you discovered some basic logarithmic properties.

- $\log_b(1) = 0$ since $b^0 = 1$. For example, $\log_7(1) = 0$ since $7^0 = 1$.
- $\log_b(b) = 1$ since $b^1 = b$. For example, $\log_7(7) = 1$ since $7^1 = 7$.
- $\log_b(b^x) = x$ since $b^x = b^x$. For example, $\log_7(7^3) = 3$ since $7^3 = 7^3$.
- $\log_b(x \cdot y) = \log_b(x) + \log_b(y)$ and $\log_b(x \cdot y) \neq \log_b(x) \cdot \log_b(y)$. For example, $\log_2(8 \cdot 16) = \log_2(8) + \log_2(16)$.
- $\log_b\left(\frac{x}{y}\right) = \log_b(x) - \log_b(y)$ and $\log_b\left(\frac{x}{y}\right) \neq \frac{\log_b(x)}{\log_b(y)}$. For example, $\log_2\left(\frac{16}{8}\right) = \log_2(16) - \log_2(8)$.
- $\log_b(x^y) = y \cdot \log_b(x)$ and $\log_b(x^y) \neq (\log_b(x))^y$ For example, $\log_2(4^3) = 3\log_2(4)$.

The last three can be fun to prove. Give it a try!

Finally, consider the exponential function $y = 10^{2x+1}$ in which the output is 12. The resulting equation is $12 = 10^{2x+1}$.

Rewrite in logarithmic form:	$\log_{10}(12) = 2x + 1$
Or omit the base of the log:	$\log(12) = 2x + 1$
Subtract 1 from both sides:	$\log(12) - 1 = 2x$
Divide both sides by 2:	$\frac{\log(12) - 1}{2} = x$

If we approximate log(12), we get $x \approx \dfrac{1.079 - 1}{2} = 0.0395$.

Finally, let's get to the core of what we are trying to accomplish—finding the inverse of the general exponential function.

17. Use logarithms to create the inverse function for $y = a(b)^x$. In other words, interchange x and y and solve for y.

We conclude by summarizing that the inverse of the basic exponential function $y = a(b)^x$ can be found as follows.

Interchange variables:	$x = a(b)^y$
Divide both sides by a:	$\dfrac{x}{a} = b^y$
Convert to logarithmic form:	$y = \log_b\left(\dfrac{x}{a}\right)$.

So for our Super Dog salary function $y = 23364.48598(1.07)^x$, the inverse function is

$$y = \log_{1.07}\left(\frac{x}{23364.48598}\right).$$

It is hard to do much with this function since its base is 1.07 and our calculators do not operate in such a base. We'll solve this problem in the next section.

STUDENT DISCOURSE

Pete: I had heard of logarithms before, but I had no idea what they were.

Sandy: Well, what do you think now?

Pete: The important idea to me is to think of a logarithm as an exponent.

Sandy: Say some more.

Pete: Well, consider something like $\log_5(75)$. I need to think about the power on 5 to get a result of 75.

Sandy: That's the way I see it. Continue.

Pete: Well, I know that 5^2 is 25 and, let's see, 5^3 is 125. Since 75 is between 25 and 125, then $\log_5(75)$ must be between 2 and 3. So I guess is about 2 point something.

Sandy: You seem to have the right idea.

Pete: But how would I find this on the calculator?

Sandy: Actually, you can't calculate this directly, but you can use something called the change-of-base formula, which appears in the next section.

Pete: I guess I can wait until then.

1. In the Glossary, define the words in this section that appear in ***italic boldface*** type.

2. Given $f(x) = e^x$ and $g(x) = \ln(x)$, create a table and graph for $f(x)$, $g(x)$, $f(g(x))$, and $g(f(x))$. Based on the results of $f(g(x))$ and $g(f(x))$, what is the relationship between $f(x)$ and $g(x)$?

3. Consider the expressions $\log_b(x \cdot y)$ and $(\log_b(x))(\log_b(y))$. Create a function machine for each that has two inputs x and y. Write the processes performed on x and y inside the machine so as to obtain each given expression as output. Based on your function machines, are these expressions equivalent?

4. Consider the expressions $\log_b\left(\frac{x}{y}\right)$ and $\frac{\log_b(x)}{\log_b(y)}$. Create a function machine for each that has two inputs x and y. Write the processes performed on x and y inside the machine so as to obtain each given expression as output. Based on your function machines, are these expressions equivalent?

5. Express in logarithmic form.

a. $81 = 3^4$

b. $25^{1/2} = 5$

6. Express in exponential form.

a. $2 = \log_3(9)$ **b.** $-3 = \log_5\left(\frac{1}{125}\right)$

7. Use algebraic reasoning to calculate the **exact** values of each of the following logarithms.

a. $\log_6(36)$ **b.** $\log_2\left(\frac{1}{8}\right)$

c. $\log_8(4)$

8. Given the logarithmic function $y = \log_2\left(\frac{x}{5}\right)$, create an exponential function that is the inverse of the given function. Describe what you did.

9. The function $A(t) = 5(0.6)^t$ models the information given in the **Drug in Bloodstream** problem. Use logarithms to create the inverse function algebraically.

10. The function $A(t) = 93{,}000(1.05)^{t/3}$ models the information given in the **Housing Value** problem. Use logarithms to create the inverse function algebraically.

11. Find the value of x in the following.

a. $3 = 10^x$ **b.** $17 = 10^{3x}$

c. $42 = 10^{4x-3}$ **d.** $5 = e^x$

e. $24 = e^{2x+7}$

12. Recently, the U.S. dollar was worth 1.73 German marks.

a. Write a function that defines the number of marks as a function of the number of dollars.

b. Create the inverse function for the function you wrote in part **a**.

c. For the inverse function, what does the input represent? What does the output represent?

13. An exponential function contains the points (0, 12) and (2, 108).

a. Write the equation for the function. Briefly explain what you did.

b. Create the inverse function. Briefly explain what you did.

14. Find the solution to the following equations/systems.

a. $2x^2 - 3x + 7 = 0$

b. $\frac{2}{3}x + 5 = 7 - 2x$

c. $x^2 + 5x = 1$

d. $2x + 5y = -3$
$3x - 10y = 1$

e. $\frac{z}{z+1} = 2$

f. $3 - \sqrt{2x+1} = 1$

REFLECTION

Write a paragraph describing everything you know about logarithms.

SECTION 8.4

Logarithmic Models

Purpose

- Develop and use a change-of-base formula for logarithms.
- Solve logarithmic and exponential equations algebraically.
- Investigate logarithmic models for data.
- Use the natural logarithmic function to linearize data.

Now that you know a little about logarithms, in this section we will investigate how they might be used. We will investigate how to approximate logarithms with various bases, how to solve equations using logarithms, how to model data with logarithmic functions, and how to use logarithms to determine if data can be modeled using an exponential function.

In the previous section we developed the logarithmic function

$$y = \log_{1.07}\left(\frac{x}{23364.48598}\right),$$

where the input represents the annual salary at Super Dog Burger Bits and the output represents the number of years on the job.

While this is a perfectly good function, we can't do much with it on our calculators since the only logarithmic functions available on most calculators are logarithm base 10 (the LOG key) and logarithm base e (the LN key). There's a good reason for this. Given a logarithm of any base, we can evaluate it using a fairly simple formula that uses a calculator logarithm base like 10 or e. Let's investigate.

1. Begin with the function $y = \log_{1.07}\left(\frac{x}{23364.48598}\right)$. Rewrite it in exponential form.

2. Apply the function logarithm base 10 (LOG) to both sides of the equation. This means that each side of the equation becomes an argument of logarithm base 10.

3. Use one of the major properties of logarithms to rewrite so that the variable y is not an exponent on the base 1.07.

4. Divide both sides of the equation by the logarithm that is multiplied by y, thus isolating y.

5. The output y should now be defined in terms of base 10 logarithms. Enter the result as a function on your calculator.

a. Display a table with salaries beginning at $25,000 in increments of $5000. What is the output when the input is $60,000? Interpret in terms of the problem situation.

b. Graph the function on the calculator. Sketch the general shape of the graph.

6. Repeat the preceding steps, applying a logarithm base e, rather than base 10, to both sides. Write the resulting function. Are the table and graph for this function similar to the table and graph you previously created?

7. Compare the original logarithm with base 1.07 to the base 10 logarithm representation. Without going through all the steps, describe to your confused friend Izzy how to convert a logarithm from one base to another.

8. Consider the number $\log_3(92)$.

a. Write two consecutive integers so that the answer to $\log_3(92)$ is between the integers. Explain your reasoning.

b. Use the change-of-base formula you just developed to calculate the value of $\log_3(92)$ on your calculator.

c. Raise the base 3 to a power equal to your answer in part **b**. How does the result of this exponentiation help you check your answer?

Let's look at changing the base for $y = \log_{1.07}\left(\frac{x}{23364.48598}\right)$.

Exponential form: $$1.07^y = \frac{x}{23364.48598}$$

Apply logarithm base 10 to both sides:

$$\log_{10}(1.07^y) = \log_{10}\left(\frac{x}{23364.48598}\right)$$

Logarithm property: $$y\,(\log_{10}(1.07)) = \log_{10}\left(\frac{x}{23364.48598}\right)$$

Isolate y: $$y = \frac{\log_{10}\left(\frac{x}{23364.48598}\right)}{\log_{10}(1.07)}$$

We must note that, when working with base 10 logarithms, the base of 10 is usually not written. Base 10 logarithms are called ***common logarithms***. So

$$y = \frac{\log\left(\frac{x}{23364.48598}\right)}{\log\,(1.07)}.$$

If we applied logarithm base e to both sides, the change would be from LOG to LN. LN is shorthand notation for logarithm base e. Base e logarithms are called ***natural logarithms***. The resulting function is

$$y = \frac{\ln\left(\frac{x}{23364.48598}\right)}{\ln\,(1.07)}.$$

The table and graph for both of these functions appear in Figure 1. Note that the input is annual salaries and the output is years.

Notice that the tables and graphs of both functions are the same. The two functions are equivalent.

Let's analyze the result of changing the base of a logarithm.

FIGURE I

```
Plot1 Plot2 Plot3
\Y1=log(X/23364.
48598)/log(1.07)

\Y2=ln(X/23364.4
8598)/ln(1.07)
\Y3=
\Y4=
```

X	Y1	Y2
35000	5.9731	5.9731
40000	7.9467	7.9467
45000	9.6875	9.6875
50000	11.245	11.245
55000	12.653	12.653
60000	13.939	13.939
65000	15.123	15.123

Y1=log(X/23364....

```
WINDOW
Xmin=20000
Xmax=100000
Xscl=10000
Ymin=-5
Ymax=25
Yscl=5
Xres=1
```

The original logarithm was $y = \log_{1.07}\left(\dfrac{x}{23364.48598}\right)$. The argument is $\dfrac{x}{23364.48598}$ and the base is 1.07.

Upon changing the base, the original argument becomes the argument of the logarithm in the numerator. The original base becomes the argument of the logarithm in the denominator. Our derivation suggests the change-of-base formula.

Change-of-base Formula: Convert from base b to base a

$$\log_b(x) = \frac{\log_a(x)}{\log_a(b)}$$

In most cases, the base a will be either 10 or e.

Consider $\log_3(92)$.

We are looking for an exponent on base 3 so that the exponentiation results in 92. We know that $3^4 = 81$ and $3^5 = 243$. So $\log_3(92)$ is between 4 and 5.

Using the change-of-base formula, we get

$$\log_3(92) = \frac{\log(92)}{\log(3)} \approx 4.1159093.$$

We would get the same result if we used base e logarithms rather than base 10 logarithms.

As a check, $3^{4.1159093} \approx 92$.

Let's use logarithms to solve some equations.

9. Drug in Bloodstream: The function $A(t) = 5(0.6)^t$ was used to define the amount A in milliliters of a drug in the bloodstream t hours after the drug was ingested. Determine algebraically the time it will take until there is only 1 ml of the drug left in the bloodstream.

a. Write the equation corresponding to the problem.

b. Solve for t by writing the equation in logarithmic form.

c. Use the change-of-base formula to express t in terms of logarithms of either base 10 or base e.

d. Calculate the answer. Check your result both tabularly and with a graph.

The problem asks us to find the input to $A(t) = 5(0.6)^t$ so that the output is 1. While it is fairly simple to solve this problem using a table or a graph, it is now possible to do the problem algebraically.

Equation: $1 = 5(0.6)^t$

Divide by 5 to isolate exponential expression. (You could substitute 0.2 for the fraction of each of these expressions.)

$$\frac{1}{5} = 0.6^t$$

Logarithmic form: $\log_{0.6}\left(\frac{1}{5}\right) = t$

Change-of-base formula: $\dfrac{\log\left(\frac{1}{5}\right)}{\log(0.6)} = t$

Compute: $3.15 \approx t$

The amount of drug in the bloodstream drops to 1 milliliter after approximately 3.15 hours. Figure 2 verifies this answer in a table and on a graph.

We return to using regression to model data by looking at how the logarithmic function can be used to determine if data are actually exponential.

FIGURE 2

X	Y_1
3	1.08
3.05	1.0528
3.1	1.0262
3.15	1.0003
3.2	.97511
3.25	.95052
3.3	.92655

X=3.15

WINDOW
Xmin=0
Xmax=10
Xscl=1
Ymin=-2
Ymax=6
Yscl=1
Xres=1

Intersection
X=3.1506601 Y=1

10. Consider the data displayed in Table 1.

TABLE 1 **Some Data**

x	y
0	3
1	6
2	12
3	24
4	48
5	96

a. Enter the data as two lists in your calculator. Construct a scatter plot of the data. Does the data appear to be exponential? Why or why not?

b. Create a third list by computing $\ln(y)$, where y represents the entries in the output column of Table 1. Construct a scatter plot using x as input and $\ln(y)$ as output. What do you notice about the scatter plot?

c. Find the best-fit exponential model for the original data. How "good" is the fit? How do you know?

The technique used to fit an exponential function to data requires the ***linearization of the data.*** This means that the data are transformed so the new data can be modeled by a linear function. To clarify, let's look at the last investigation.

Figure 3 displays the calculator's lists and the scatter plot for the original data. The scatter plot suggests that the data are exponential.

FIGURE 3

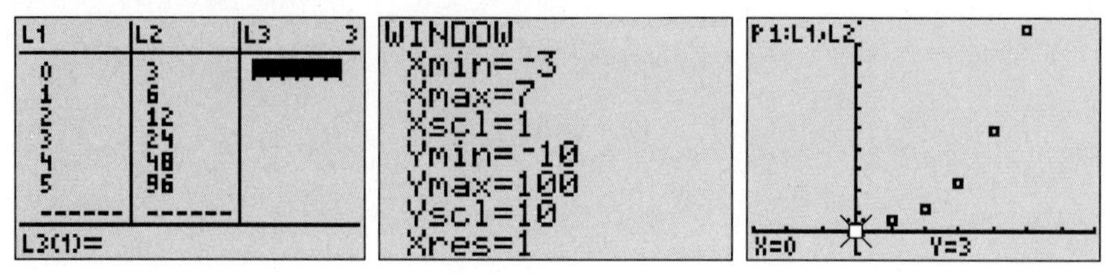

Figure 4 displays the calculator's lists in which L3 is computed as ln (L2). The scatter plot of L1 with L3 is displayed. Notice that the data lie in a straight line.

FIGURE 4

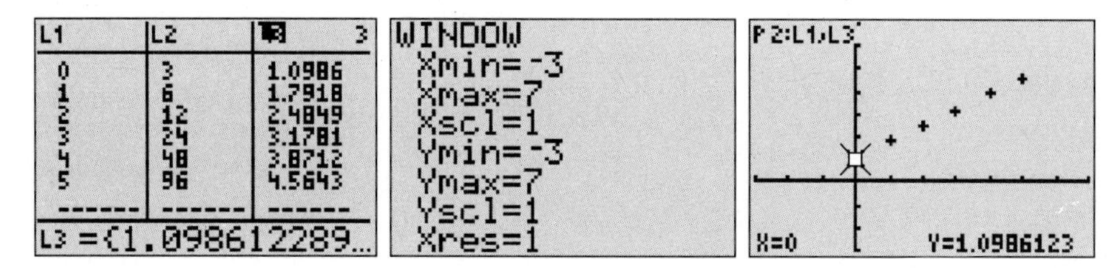

Figures 3 and 4 provide an example of the linearization of data. A set of data points (x, y) are modeled by an exponential function if the set of points $(x, \ln(y))$ are modeled by a linear function.

In the previous Investigation, we looked at data that was exactly exponential. Let's recall a problem we did in Chapter 4 with real-world data. We use linearization to determine how well an exponential function might model the data.

11. Outstanding Credit: The amount of installment credit in the United States continues to rise. Table 2 on page 655 displays the amount of money (in billions of dollars) in installment credit by year

a. Enter the data as two lists in your calculator. Construct a scatter plot of the data. Does the data appear to be exponential? Why or why not?

b. Create a third list by computing $\ln(y)$, where y represents the entries in the output column of Table 2. Construct a scatter plot

TABLE 2 Installment Credit Outstanding

Year	Installment Credit Outstanding ($ billions)
1986	572
1987	609
1988	663
1989	724
1990	735
1991	728
1992	731
1993	790
1994	903
1995	1025

Source: U.S. Bureau of the Census, *Statistical Abstract of the United States: 1996* (116th ed.). Washington D.C.: 1996, p. 516.

using x as input and $\ln(y)$ as output. Based on the scatter plot, would you conclude that the data may be modeled by exponential regression? Why or why not?

Figure 5 displays the calculator's lists and the scatter plot for the Outstanding Credit data. The scatter plot suggests that the data may be exponential.

FIGURE 5

L1	L2	L3
0	572	
1	609	
2	663	
3	724	
4	734	
5	728	
6	731	

L3(1)=

WINDOW
Xmin=-1
Xmax=12
Xscl=1
Ymin=500
Ymax=1200
Yscl=100
Xres=1

Installment credit

Years since 1986

Figure 6 displays the calculator's lists in which L3 is computed as $\ln(L2)$. The scatter plot of L1 with L3 is displayed.

FIGURE 6

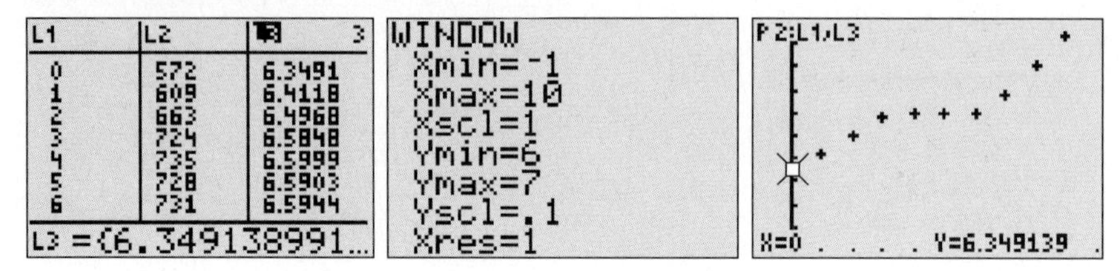

Notice that the data do not lie in a straight line, suggesting that an exponential model may not be a good fit for the data.

We complete the section by modeling some data using a logarithmic function.

12. Safety-belt Use: Table 3 displays data on the number of people (in thousands) by year whose lives were saved by wearing a safety belt.

TABLE 3 **Lives Saved by Safety-belt Use**

Year	Lives Saved (thousands)
1984	1
1985	1.8
1986	3.2
1987	3.9
1988	4.5
1989	4.5
1990	4.7
1991	4.8

Source: National Highway Traffic Safety Administration.

a. Choose a year *before* 1984 as an arbitrary zero. We suggest 1983. Enter the data in your calculator and construct a scatter plot. Does the graph seem similar to a logarithmic graph? Why?

b. Calculate the logarithmic regression model on your calculator. Write the regression model. (If you do not know how to do this, check the manual or ask your instructor or classmate who can help you.)

c. Use the regression model to predict the number of lives saved by safety belts in 1996. Explain what you did.

The data with scatter plot appear in Figure 7.

FIGURE 7

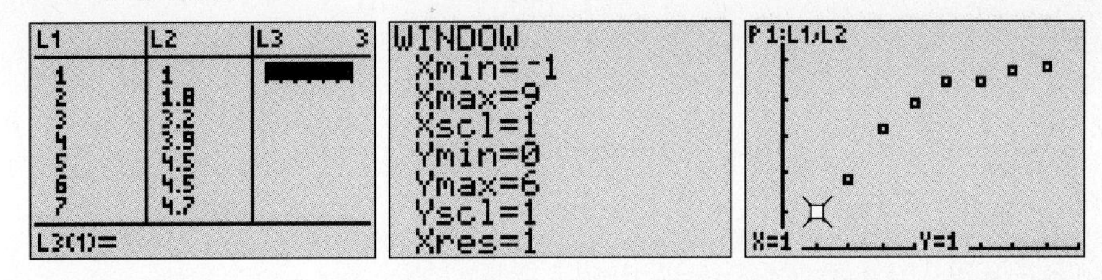

The graph closely resembles a logarithmic function. The model, found by using ***logarithmic regression***, and the graph appear in Figure 8.

FIGURE 8

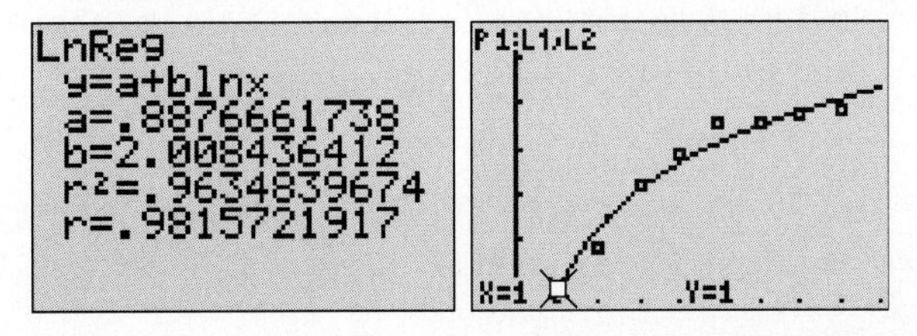

Let's round the parameters in the model to four decimal places. The model is

$$y = 0.8877 + 2.0084\ln(x).$$

To determine the number of lives saved in 1996, let $x = 13$.

$$y = 0.8877 + 2.0084\ln(13) \approx 6.0391.$$

Approximately 6040 lives were saved by safety-belt use in 1996, according to the model.

That concludes our introduction to logarithmic functions. We have now investigated linear functions, quadratic functions, exponential functions, logarithmic functions, rational functions, and radical functions. These classes of functions serve as an excellent core for continued study in mathematics.

STUDENT DISCOURSE

Pete: I want to go back to something at the beginning of this section.

Sandy: Go ahead.

Pete: My calculator has two keys for logarithms—the LOG key and the LN key.

Sandy: Mine too. LOG uses a base of 10 and LN uses a base of e.

Pete: I understand that, but why are these the only logarithms on the calculator?

Sandy: You really need only one logarithm key to find the logarithms of any base you like. Some early calculators only had the LN key since LOG could be found from LN.

Pete: Is that because of the change-of-base formula?

Sandy: Exactly. Using the change-of-base formula, a logarithm of any base can be converted to a logarithm of another base.

Pete: Okay, so why use bases of 10 and e?

Sandy: I would guess that base 10 is used because we work in a base 10 number system. My instructor this term mentioned that base 10 logarithms were used to do messy calculations prior to the widespread use of calculators.

Pete: And base e?

Sandy: I'm not sure, but I think it's due to the importance of e in modeling natural growth and decay processes.

Pete: I guess I can see that by looking at some of the applications in the Explorations. By the way, now that I have completed the course, thanks for all your help.

Sandy: It was fun. I learned a lot just by discussing the ideas with you. Let's continue to work together in other math courses.

Pete: Sounds good to me. Two heads are definitely better than one.

1. In the Glossary, define the words in this section that appear in ***italic boldface*** type.

2. **Town's Decreasing Population:** The population of a given town is 52,000 and is decreasing at a rate of 2.5% each year.

a. Write the population $P(t)$ as a function of time t. Use the general model $P(t) = P_0(1-r)^t$, were P_0 is the initial population and r is the annual rate of decay.

b. Use a graph or a table to determine approximately when the population will be 40,000. Write complete sentences stating your answer and explaining what you did.

c. Write an equation that represents the question given in part **b**. Solve the equation algebraically.

d. How long will it take for the population to be one-half of its original size? Write complete sentences stating your answer and explaining how you found the answer.

3. Construct the graphs of $y(x) = \log_b(x)$ for values of base b from 2 to 5 in increments of 1. Describe the graph of $y(x) = \log_b(x)$ for any bases $b > 1$. Be sure to discuss any intercepts, the domain, the range, and whether the graph is increasing or decreasing. In Chapter 7 we discussed asymptotes for rational functions. Do logarithmic functions have any asymptotes? If so, where are they?

4. Find the value(s) for x in the logarithmic equation

$$\log(5x-1) - \log(x-3) = 2.$$

a. Use a logarithmic property to rewrite the left side as one logarithm.

b. Use techniques discussed in this section to solve for the argument on the logarithm.

c. Solve for x.

d. Check answers numerically and graphically.

5. World Population: Table 4 displays the world's population, 1982–1993.

TABLE 4 **Worldwide Population**

Year	Population (billion)
1982	4.614
1983	4.695
1984	4.775
1985	4.856
1986	4.942
1987	5.029
1988	5.117
1989	5.206
1990	5.295
1991	5.381
1992	5.469
1993	5.557

Source: Lester Brown, Hal Kane, and David Malin Roodman, *Vital Signs 1994*, Worldwatch Institute (New York: W. W. Norton, 1994), p. 99.

Use the natural logarithm function to linearize the data. Given the graph of the linearized data, is an exponential function a good model for the data? Why or why not?

6. Walking Speed: Somewhat surprisingly, actual measurement has determined that the average walking speed s (feet per second) of pedestrians in a city is a function of the population p of the city where $s = 0.37 ln(p) + 0.05$. Determine algebraically the population of a city where, as an average, pedestrians walk no faster than 4 feet per second. Check your answer numerically and graphically. Show all algebraic steps and write a complete sentence summarizing your findings.

7. **Decay of a Language:** If a language has w basic words originally, of which N are still in use after t years measured in millennia (thousands of years), then the decay rate is about 19.5% per millennia.

a. Write the exponential decay function for this problem.

b. After how many years are one-half of the basic words still in use?

8. **AIDS Cases:** Table 5 displays global estimates of AIDS cases, 1982–1993.

TABLE 5 Worldwide AIDS Cases

Year	AIDS Cases (thousands)
1982	7
1983	21
1984	65
1985	143
1986	279
1987	493
1988	813
1989	1275
1990	1892
1991	2701
1992	3657
1993	4820

Source: Lester Brown, Hal Kane, and David Malin Roodman, *Vital Signs* 1994, Worldwatch Institute (New York: W. W. Norton, 1994), p. 99.

Use the natural logarithm function to linearize the data. Based on the graph of the linearized data, is an exponential function a good model for the data? Why or why not?

9. pH: Chemists use a number denoted by pH to describe quantitatively the acidity or alkalinity of solutions. By definition,

$$\text{pH} = -\log([H^+]) \text{for the hydrogen ion concentration } [H^+] \text{ in moles per liter.}$$

a. Approximate the pH of vinegar if $[H^+] \approx 6.3 \times 10^{-3}$.

b. Approximate the pH of carrots if $[H^+] \approx 1.0 \times 10^{-5}$.

c. Find the hydrogen ion concentration in apples if pH = 3.0.

d. Find the hydrogen ion concentration in beer if pH = 4.2.

10. Earthquakes and the Richter Scale: Earthquake intensity is reported using the Richter scale, which is computed $R = \log_{10}\left(\frac{a}{2}\right) + 4.25$, where output R is the magnitude (or intensity) and input a is the amplitude in microns of the vertical ground motion at the receiving station.

a. Find the magnitude on the Richter scale of an earthquake if the amplitude of the vertical ground motion is 250 microns. Write your answer as a complete sentence.

b. Use a table or graph to approximate the magnitude of the vertical ground motion when $R = 4$ and when $R = 6$. Write your answer as a complete sentence.

c. Construct the ratio of the vertical ground motion when $R = 6$ to the vertical ground motion when $R = 4$. Use this ratio to write a sentence describing how much more intense an earthquake measuring 6 on the Richter scale is than one measuring 4 on the Richter scale.

d. Create the inverse function so that the input is the intensity of the earthquake and the output is the amplitude of the vertical ground motion.

e. The Richter scale is limited to the range 1 to 10. Can you see any mathematical reason why the scale should be undefined below 1 and above 10? To answer this question, display a table for the inverse function created in part **d**. Looking at the table, guess why Richter and his successors restricted the earthquake scale. Explain your guess in three or four sentences.

11. Carbon Dating of Fossils: The technique of carbon-14 (^{14}C) dating is used to determine the age of archaeological and geological specimens. The function $T = -8310 \ln(x)$ is sometimes used to predict the age T (years) of a bone fossil where x is the percentage (expressed as a decimal) of ^{14}C still present in the fossil.

a. Estimate the age of a bone fossil where the percentage of ^{14}C still present in the fossil is 4%.

b. Approximate the percentage of ^{14}C present in a fossil that is 10,000 years old.

12. Safety-belt Laws: Table 6 on page 664 displays data on the number of people (in thousands) by year whose lives would *not* have been saved if the law did not require wearing a safety belt.

a. Choose a year *before* 1984 as an arbitrary zero. Enter the data in your calculator and construct a scatter plot. Does the graph seem to be similar to a logarithmic graph? Why?

b. Calculate a logarithmic regression model on your calculator. Write the regression model.

c. Use the regression model to predict the number of people whose lives would not have been saved in 1996 if the law did not require wearing a safety belt. Explain what you did.

TABLE 6 Safety-belt Laws

Year	Lives Not Saved (thousands)
1984	0.2
1985	1.2
1986	2.4
1987	3.0
1988	3.6
1989	3.7
1990	3.8
1991	3.8

Source: National Highway Traffic Safety Administration.

13. Consider the function $y(x) = \log(x)$.

a. Graph this function in a "friendly" window.

b. Overlay the graphs of $y(x) = \log(x - h)$, where h is replaced with one positive number and one negative number. What effect does the choice of h have on the graph of $y(x) = \log(x)$?

c. Graph $y(x) = \log(x)$. Overlay the graphs of $y(x) = \log(x) + k$, where k is replaced with one positive number and one negative number. What effect does the choice of k have on the graph of $y(x) = \log(x)$?

14. Refer to Exploration 13 to answer the following question.

a. What is the effect of the parameters h and k on the graph of $y(x) = \log(x - h) + k$ as compared to the graph of $y(x) = \log(x)$.

b. Does the shape of the graph change? Does the position of the graph change?

c. What is the domain and range of $y(x) = \log(x - h) + k$?

15. Recently, the Mexican peso was worth 0.125 U.S. dollar.

a. Write a function that defines the number of pesos as a function of the number of dollars.

b. Create the inverse function for the function you wrote in part **a**.

c. For the inverse function, what does the input represent? What does the output represent?

16. Express the following in logarithmic form:

a. $5^{-3} = \frac{1}{125}$

b. $8 = 4^{3/2}$

17. Express in exponential form:

a. $\log_2\left(\frac{1}{2}\right) = -1$

b. $4 = \log_5(625)$

18. Use algebraic reasoning to calculate the *exact* values of each of the following logarithms.

a. $\log_2(64)$

b. $\log_3\left(\frac{1}{9}\right)$

c. $\log_{11}(11)$

d. $\log_{15}(1)$

e. $\log_{16}(4)$

19. Find the approximate value of each of the following.

a. $\log_3(17)$

b. $\log_2(70)$

20. Find the solution to the following equations.

a. $\log_5(2x - 7) = 1$

b. $3(5)^{7x+1} = 12$

c. $\log_3(x + 1) + \log_3(x - 2) = 2$

d. $4x - \frac{2}{3} = 7 + \frac{10}{3}x$

e. $\log_2(2x - 5) - \log_2(x - 1) = 3$

f. $2\sqrt{x + 7} - 5 = 4$

g. $3x^2 - 2x - 6 = 0$

h. $\log_7(2x + 1)^3 = 2$

REFLECTION

Describe to your friend Izzy how you would go about solving an equation in which the variable is an argument of a logarithm. Compare the advantages and disadvantages of using a numerical, graphical, or algebraic approach.

SECTION 8.5

Transcendental Models: Review

Purpose

- Reflect on the key concepts of Chapter 8.

In this section you will reflect upon the mathematics in Chapter 8—what you've done and how you've done it.

1. State the most important ideas in this chapter. Why did you select each?

2. Identify all the mathematical concepts, processes, and skills you used to investigate the problems in Chapter 8.

There are a number of important ideas that you might have listed, including inverse functions, one-to-one functions, exponential growth, exponential decay, the number e, difference quotients, logarithms, common logarithms, natural logarithms, logarithmic properties, change-of-base formula, linearizing data, exponential equations, logarithmic regression, and logarithmic equations.

1. **Orange Juice Content:** Sixteen ounces of a liquid is 20% orange juice. Water is added to reduce the orange juice content. Let C be the concentration of orange juice and x be the number of ounces of water added.

   a. Write a function with input x and output C.

   b. Find the inverse function algebraically.

   c. What does the input of the inverse function represent? What does the output of the inverse function represent?

2. **Exercise Equipment Sale:** A fitness store marks down every item in the store 23%. Let s be the sale price and p be the original price.

   a. Write a function with input p and output s.

   b. Find the inverse function algebraically.

   c. What does the input of the inverse function represent? What does the output of the inverse function represent?

3. Consider the data in Table 1.

TABLE 1 Random Data 1

Input	−2	−1	0	1	2	3
Output	−10	0	4	2	−6	−20

a. Identify the data as linear, quadratic, or exponential. Create the specific algebraic function for the data.

b. Create a table for the inverse.

c. Find the inverse relation algebraically.

d. Graph both the function and the inverse relation.

e. Find the points of intersection of the function and its inverse.

4. Recently, the U.S. dollar was worth 126.77 Japanese yen.

a. Write a function that defines the number of yen as a function of the number of dollars.

b. Create the inverse function for the function you wrote in part **a**.

c. For the inverse function, what does the input represent? What does the output represent?

5. The graph of a function appears in Figure 1 along with the graph of $y = x$. Draw the graph of the inverse function.

FIGURE 1

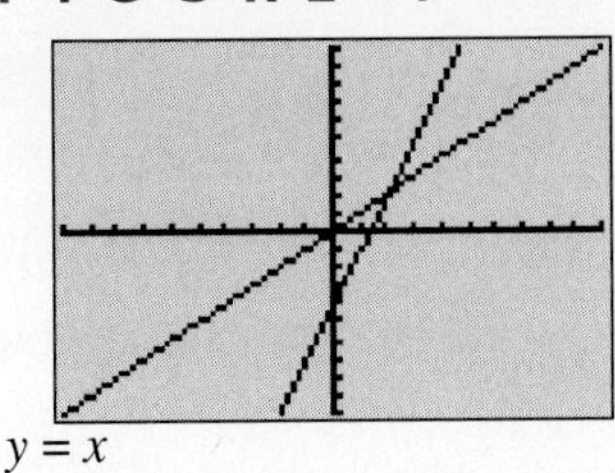

$y = x$

6. A partial table for a function appears below. Create a table for the inverse of this function.

TABLE 2 **A Random Function**

Input	Output
7	23
8	−9
47	89
−12	−1
0	14

7. Each of the following expresses a relationship between x and y. Let the input be x and the output be y. Find the inverse function/relation algebraically. Check your result numerically and graphically.

a. $y = 4 - 7x$

b. $7x - 2y = -9$

c. $y = 2x^2 - 9$

d. $y = \dfrac{x+2}{x-1}$

e. $y = \sqrt{9x - 1}$

8. Which of the inverse relations in Exploration 7 are functions? Why?

9. Housing Prices: In 1980, the average price of a house in a certain county was \$57,000. During the period 1980–1997, prices rose at an average rate of 2% per year.

a. Create a table with input the number of years since 1980 and output the average price of a house. Calculate the outputs for inputs 0, 1, 2, 3, and 4.

b. The data are modeled by what kind of function?

c. Create an algebraic equation for the function in which the input is the number of years since 1980 and the output is the average price (\$) of a house.

d. In what year did the average price of a house pass \$75,000? Describe how you found the answer.

10. Computer Depreciation: The value of a certain brand of computer depreciates at a rate of 13% per year. The original cost of the computer was \$1900. Let the input be the number of years since the computer was purchased and the output be the value (\$) of the computer.

a. Create a table for this function using inputs 0, 1, 2, 3, and 4.

b. What kind of function should be used to model the data?

c. Create an algebraic equation for the function.

d. How long will it take for the computer's value to fall below \$600? Describe how you found the answer.

11. Air Pressure: The air pressure p in pounds per square inch at an altitude of h feet above sea level is approximated by the function $p(h) = 14.7e^{-0.0000385h}$. Approximate the altitude at which the pressure is

a. 10 pounds per square inch.

b. one-half the pressure at sea level.

12. Express in logarithmic form.

a. $5^3 = 125$

b. $64^{1/2} = 8$

13. Express in exponential form.

a. $\log_4\left(\frac{1}{16}\right) = -2$

b. $3 = \log_7(343)$

14. Use algebraic reasoning to calculate the *exact* values of each of the following logarithms.

a. $\log_9(81)$

b. $\log_3\left(\frac{1}{81}\right)$

15. Find the approximate value of each of the following.

a. $\log_5(0.23)$

b. $\log_6(172)$

16. Given the points $(2, 7)$ and $(4, 5)$, create an exponential function that contains the points.

17. Figure 2 displays a table of an exponential function. Write the equation for this function. Explain what you did.

FIGURE 2

X	Y1
0	2
1	2.2
2	2.42
3	2.662
4	2.9282
5	3.221
6	3.5431

X=0

18. Figure 3 displays a graph of an exponential function. Write the equation for this function. Explain what you did.

FIGURE 3

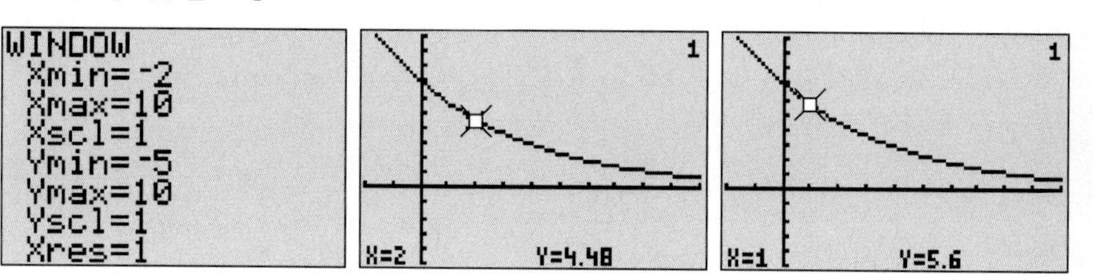

19. Find the solution to the following equations, inequalities, or systems of equations.

a. $3x^2 - 5x = 2$

b. $\dfrac{x+3}{x-2} - \dfrac{5}{x+2} = \dfrac{15}{x^2-4}$

c. $4 - 7x = 3x + 11$

d. $\sqrt{9-2x} + 5 = 17$

e. $2x^2 - x > -10$

f. $4x - 7y = 12$
$3x = 10y - 2$

g. $\dfrac{3}{4} - x = \dfrac{7}{5}x + \dfrac{1}{10}$

h. $4x^2 + 9x - 17 = 0$

i. $x - 3y = z + 2$
$3x + 2z = 1 - 4y$
$7x + 2 = y - 2z$

j. $3 - 11x > -7$

k. $10^{2x-5} = 1$

l. $x^2 < 8x - 7$

m. $-3 \leq 2 - (9 + 3x)$

n. $1 = 4e^{x-1}$

o. $\log_3(5x + 3) = 4$

p. $\log(x + 4) = 0$

q. $\log_5(2x + 1) + \log_5(x - 3) = 2$

r. $x^2 + 5x - 40 < 0$

s. $2(3)^x = 9$

t. $e^{x+7} = 3$

u. $4x - 3 \leq 7x + 2$

v. $4x - 1 = -2x^2$

20. Write the equation of the function satisfying the given conditions.

a. A linear function through $(7, -2)$ and $(4, 1)$.

b. A quadratic function with vertex $(-1, -5)$ and y-intercept $(0, 3)$.

c. An exponential function with base 1.4 and y-intercept $(0, 2)$.

d. A quadratic function with zeros 5 and -2 with y-intercept $(0, 6)$.

e. A linear function perpendicular to the function $5x - 2y = -1$ and through the point $(-3, -2)$.

f. A quadratic function through $(-2, -19)$ and $(3, -4)$ with vertex at $(8, 1)$.

g. A linear function with intercepts $(0, 5)$ and $(-2, 0)$.

h. An exponential function with base 0.23 and containing the point $(1, 21)$.

Construct a Concept Map centered on one of the following:

a. inverse function

b. exponential growth/decay

c. logarithm

d. logarithmic function

Select one of the important ideas you listed in Investigation 1. Write a paragraph to Izzy explaining your understanding of this idea. How has your thinking changed as a result of studying this idea?

Glossary

A *$a + bi$*

absolute value

a^{-1}

absolute maximum

absolute minimum

adding two matrices

additive inverse

algebraic functions

algebraic models

algebraic representation of the function

arbitrary zero

area

argument

arithmetic sequence

array

average velocity

B **base**

base of the exponent

best (least common) denominator

best-fit exponential model

bimodal

binary

bounded

C **chords**

circle graph

closure

coefficient

coefficient matrix

columns of table

common difference

common logarithms

common ratio

completing the square

complex conjugates

complex fraction

complex numbers

complex number system

complex plane

complex unit

composition of functions

concentration

confidence in an estimate

conjugate factors

constant

constant matrix

constant rate of change

continuous quantities

correlation coefficient

cube root

D Δ

Δ (Δ(output))

data

decreasing function

degree of the polynomial

demand equation

dependent variable

det

determinant

diagonal

difference quotient function

dimensions

discontinuous

discrete

discriminant

distributive property of roots over products

domain

domain of any sequence of numbers

domain of the relation or function

double zero

Δt

E

elements of a sequence

elimination method

end behavior model for the rational function

" = "

equation

equivalent equation

equivalent fraction

equivalent statements

evaluate

evaluating an expression

even

exponent

exponential decay function

exponential form

exponential function

exponential regression model

extraneous solutions

extrapolating

F

factored form

factoring out the common monomial factor

factorization tree

Fibonacci sequence

finite difference

formula

frequency plot

function

function machines

function notation

G

generalize

geometric sequence

global behavior of the function

goodness of fit

graphs

H **half-life**

histogram

horizontal asymptote

hypotenuse

I **identity**

identity matrix

imaginary axis

increasing function

independent variable

index

inequality

input

input (horizontal) axis

integers

interpolating

interquartile range

intersection

inverse function

inverse of matrix

irrational numbers

isolating a variable

L **length**

like radicals

line

line of symmetry

line plot

linear equation in standard form

linear function

linear transformation

linearization of the data

logarithmic form

logarithmic function with base *b*

logarithmic regression

M **magnitude**

mathematical expressions

mathematical model

mathematical statements

matrices

matrix

matrix multiplication

maximum

mean

measures of central tendency

median

midpoint of the line segment

mode

modular arithmetic

monomial

multiples

N **natural exponential function**

natural logarithm

natural numbers

negative rate of change

nested sets

net profit

non-colinear points

nonnegative

nonnegative real numbers

nonpolynomial algebraic functions

number system

O

oblique asymptote

odd number

one-to-one function

opposite reciprocals

order of a square matrix

order of operations

ordered pair

organization of data

output

output (vertical) axis

P

paired lists of data

parallel

parameters

parametric equations

Pascal's Triangle

perpendicular

piecewise function

points

polynomial function

positive rate of change

prime

prime factorization

prime number

principal *n*th root

principal square root

pure imaginary

Pythagorean Theorem

Q **quadratic formula**

quadratic models

R **®**

$\Re$

radical

radical equation

radical functions

radicand

range

range of the relation or function

rate of change

rational

rational functions

rational numbers

real axis

real numbers

real-world data

reciprocal

recursive definition of the sequence

regression line for the data

relation

removable discontinuity

residuals

revenue equation

reversibility of operations

rows of a table

S **scalar multiplication**

scatter plot

sequence

simplifying

singular matrix

slant asymptote

slope

slope-intercept form

solution to a system

solving an equation

sort in increasing order

square

square matrix

square trinomials

standard error of the estimate

statistic

statistical range

subset

substitute

substitution

successive outputs

sum of the squares of the residuals

superset

symmetry

system of equations

T **table**

tangent

third root

transcendental

triangular numbers

U **unary**

unbounded

unit

upper limit

V **variable matrix**

variables

vertex

vertical asymptote

W **whole numbers**

width

Z **zero of the function**

zero product property

zero rate of change

zeros of the denominator

Index

J

K

L

M

N

O

P

U

V

W

X

Y

Z